Reactive Oxygen Species in Plants

Swati Sachdev • Shamim Akhtar Ansari •
Mohammad Israil Ansari

Reactive Oxygen Species in Plants

The Right Balance

Springer

Swati Sachdev
School of Liberal Arts and Science
Faculty of Science, Era University
Lucknow, India

Shamim Akhtar Ansari
Institute of Forest Productivity
Ranchi, India

Mohammad Israil Ansari
Department of Botany
University of Lucknow
Lucknow, India

ISBN 978-981-19-9883-6 ISBN 978-981-19-9884-3 (eBook)
https://doi.org/10.1007/978-981-19-9884-3

This Springer imprint is published by the registered company Springer Nature Singapore Pte Ltd.
The registered company address is: 152 Beach Road, #21-01/04 Gateway East, Singapore 189721, Singapore

Preface

Plants often encounter multiple stresses such as extreme temperature, salinity, water scarcity, pathogenic attack, and many more either independently or simultaneously. Stress stimuli trigger production of different forms of reactive oxygen species (ROS) in vital compartments of the plant cell. The generated ROS react with other chemical entities leading to oxidative damage or modulating signaling pathway circumventing various stresses in plants. The molecular signatures associated with these damages and signaling pathways offer significant aid in elucidation of mechanisms of actions. To date, the emphasis has been laid only on ROS aspects damaging/disrupting cellular machinery and inflicting crop productivity loss. However, the recent researches have unambiguously established that the ROS at below threshold confers protection against both abiotic and biotic stress, augmenting crop productivity. This emphasizes for a proper understanding of ROS-based physio-molecular mechanisms and their upgradation in crops to adapt them to stressful conditions. As a result, the cultivation area of various economically important crops and their productivity and quality can be enhanced, arresting degradation of sites, improving environment quality, and mitigating ill impact of climate change.

The crafted book chapters deal with in-depth information underpinning molecular mechanisms involved in ROS actions including network of genes, proteins, and metabolites and serve as a gateway to mitigate adverse impacts of ROS on plant growth and productivity via redirecting signaling pathways for acclimatization to the stress stimuli. The information has been well illustrated and categorized in several chapters crafted lucidly, maintaining connectivity and synergy with each other.

The book highlights ROS in plant cells, focusing its dual nature under stressful conditions. The content demystifies the pathways triggering ROS generation, damaging reactions, signaling, cross-talk, and detoxification within the cell under single canopy. Overall, the book deals with dual role of ROS which is beneficial and harmful below and above threshold limits, respectively, and provides a complete scenario of ROS in cellular compartments under stressful environmental settings and encompasses the following thrust areas:

- Elucidation of mechanisms triggering ROS production in vital cellular compartments under multitude of stress
- Detoxification mechanism of ROS in plants

- Role of ROS in plants- damaging or stress acclimatization
- Effect of cross-talk between ROS and other compounds
- Recent technological trends in alleviation of ROS induced effects on plants and stress resilience development

Thus, the book comprises a compendium of up-to-date comprehensive scientific information on dual role of ROS, hitherto neglected, in crop abiotic and biotic stress management that would immensely benefit and educate graduate/post-graduate students, entrepreneurs, researchers, scientists, and faculty members alike.

Lucknow, Uttar Pradesh, India — Swati Sachdev
Ranchi, Jharkhand, India — Shamim Akhtar Ansari
Lucknow, Uttar Pradesh, India — Mohammad Israil Ansari

Contents

About the Authors

Swati Sachdev, presently working as Assistant Professor at the School of Liberal Arts and Science, Era University, Lucknow, India. She has completed her post-graduate and doctoral degree in Environmental Science from Babasaheb Bhimrao Ambedkar University, Lucknow, India. Her expertise and research area include sustainable agriculture, plant-microbe interaction, plant stress management, bioremediation, and monitoring of pesticides. She has four years of research and two years of teaching experience. She has published twelve research paper/review articles and seven book chapters. She has delivered oral and poster presentation in many national and international conferences. She was a gold medalist in post-graduate program (BBA University) and awarded with University Grant Commission–Junior Research Fellowship (UGC-NET/JRF).

Shamim Akhtar Ansari, Scientist G and Ex-Director, Institute of Forest Productivity, Ranchi, India is an erudite academician and researcher with an experience of >38 years. Dr. Ansari has specialized in Plant Physiology and Biochemistry during MS and doctoral work. Subsequently, he joined as Senior Scientist in the Indian Council of Forestry Research and Education under the Government of India Ministry of Environment, Forests and Climate Change. He has conducted extensive research, and developed expertise, in Clonal Forestry or Forest Biotechnology, including Tissue Culture and Adventitious Rhizogenesis and Population (molecular) Genetics and Association Mapping of Tropical Deciduous Trees. He has edited three books and published >100 research papers/review articles/popular science articles and received global citations. He has also guided

11 students for doctorate program. Dr. Ansari has received several academic accolades; notable among them are University Gold Medal (Aligarh Muslim University, Aligarh) and R.D. Asana Gold Medal (Indian Society for Plant Physiology, New Delhi). Dr. Ansari is an elected fellow of Indian Botanical Society, Indian Society for Plant Physiology, and the Linnean Society of London.

Mohammad Israil Ansari, is Professor of Botany at the Department of Botany, University of Lucknow, India. Before joining the University of Lucknow, he was Professor at Amity Institute of Biotechnology, Amity University, Lucknow Campus, Lucknow (2007–2016). Dr. Ansari graduated and post-graduated from the Department of Botany, Aligarh Muslim University, Aligarh, India and doctoral degree from Dr. Ram Manohar Lohia Avadh University, Faizabad, India. Dr. Ansari is an Academia Sinica Fellow from the Government of Taiwan. He has more than 27 years of research and 16 years of teaching experience with more than 125 research papers, several books, book chapters, review articles, Ph.D., post-graduate summer projects as well as funded research projects to his credit. Dr. Ansari has been associated with the Institute of Botany, Academia Sinica, Taiwan, Graduate Institute of Plant Biology, National Taiwan University, Taiwan and Institute of Plant and Microbial Biology, Academia Sinica, Taiwan (2000–2007). Dr. Ansari had made significant contributions in the field of Molecular Biology of Leaf Senescence, Dehydration Stress, Red Rot Disease of Sugarcane and Nanotechnology. Dr. Ansari is working on GABA-Transaminase gene that plays a very important role in leaf senescence process and role of nanotechnology in plant stress conditions. Dr. Ansari is editorial board member of several international journals as well as member of several Academic Board/Faculty Board/Board of Studies/Research and Development bodies. Dr. Ansari has delivered several invited lectures at national and international level. Dr. Ansari has completed several funded projects from the Department of Science and Technology, Govt. of India, New Delhi and the Department of Biotechnology, Govt. of India.

Reactive Oxygen Species (ROS): An Introduction

1

1.1 Introduction

Reactive oxygen species (ROS) are multifunctional and highly reactive metabolites of oxygen, ubiquitously produced in all aerobic organisms ranging from tiniest microbe to giant plants and animals (Hamilton et al. 2012; Ung et al. 2017). ROS is a collective term for reactive oxygen metabolites which includes free radicals like superoxide anion ($O_2^{\bullet-}$), hydroxyl ($^{\bullet}OH$), hydroperoxyl ($HO_2^{\bullet}$), alkyl ($RO^{\bullet}$), and peroxyl ($RO_2^{\bullet}$) radicals; and non-radicals such as hydrogen peroxide (H_2O_2), singlet oxygen ($^1O_2$), and excited carbonyl (RO^*) (Tripathy and Oelmuller 2012; Hasanuzzaman et al. 2020). Moreover, ROS also include certain acids such as hypochlorous acid (HOCl), hypoiodous acid (HOI) and hypobromous acids (HOBr) (Hasanuzzaman et al. 2020). Molecular oxygen (O_2) is the precursor of ROS genesis inside the cell and has been categorized as a free radical due to the presence of two unpaired electrons (Bayr 2005). These two unpaired electrons show same spin quantum number; therefore, their spin is forbidden (Bayr 2005; Gill and Tuteja 2010). To overcome this spin restriction, O_2 undergo reduction by accepting electron with opposite spin (Bayr 2005). This reduction occurs through acceptance of electron one by one in a sequential manner which ultimately results in the formation of diverse ROS (Bayr 2005; Gill and Tuteja 2010).

ROS are constantly generated in living cells as a result of aerobic metabolism under both normal and stressed conditions (Hamilton et al. 2012; Dumont and Rivoal 2019; Jalil et al. 2019; Kerchev et al. 2020; Shah et al. 2019). The reactive species are formed under oxidizing environment at many locations, in particular, chloroplast, nucleus, mitochondria, apoplast, plasma membrane, cytosol, and endoplasmic reticulum (Tripathy and Oelmuller 2012; Jalil and Ansari 2020a). ROS formation is mediated through numerous cellular pathways such as photosynthetic electron transport chain (ETC), mitochondrial electron transport chain (mtETC); activities of enzymes like NADPH oxidases (nicotinamide adenine dinucleotide phosphate oxidases) found in plasma membrane, xanthine oxidases (XOD),

S. Sachdev et al., *Reactive Oxygen Species in Plants*,
https://doi.org/10.1007/978-981-19-9884-3_1

glycolate oxidases (GOX), peroxidases (POX), etc., and photosensitizers like chlorophyll (Saed-Moucheshi et al. 2014; Podgorska et al. 2017). ROS formed within plant cell partake in a number of enzymatic reactions, mitochondrial electron transport, signaling transduction, plant growth, and programmed cell death (PCD) (Bayr 2005; Tripathy and Oelmuller 2012; Shah et al. 2019; Florez-Sarasa et al. 2020; Ansari et al. 2021). ROS are ambiguous molecules with ability to show dual performance within the cell, i.e., trigger both beneficial and detrimental effects; and therefore, are defined as double-edged sword. The dual activity of ROS depends upon their concentration, sites of production, and intrinsic defense system (Sachdev et al. 2021). Unencumbered production of ROS limit plant growth and productivity; whereas, their balanced accumulation mediates redox signaling and assists in plant growth and development even under stressful conditions (Florez-Sarasa et al. 2020). The controlled production of ROS mediates cell to cell communication by signals amplification via NADPH oxidase or respiratory burst oxidase homologue D (RBOHD) which results in the suppression of stress through modulation of protein structure and activation of defense-responsive genes (Shah et al. 2019).

Plants ordinarily transform 1–2% of the total oxygen intake into ROS (Roychowdhury et al. 2019; Shah et al. 2019). However, unfavorable conditions encompassing extremes of temperature, water scarcity, water-logging, nutrient deficiency, salinity, heavy metals, and pathogen attack impose stress on plants and exacerbate ROS formation, which may affect cellular homeostasis and culminate as oxidative stress (He et al. 2018; Hussain et al. 2018; Bulgari et al. 2019; Jalil and Ansari 2020a; Ansari et al. 2021; Sachdev et al. 2021). Environmental anomalies make plants highly vulnerable and trigger the over-production of ROS (Dikilitas et al. 2020; Jalil and Ansari 2020b; Maurya 2020), increasing their availability to biomolecules like lipids, proteins, nucleic acids, and carbohydrates. Due to reactive nature, ROS easily affect biomolecules and initiate oxidation events in plant cell leading to membrane damage, mutation, and altered cellular functioning which often culminates into plant death (Shah et al. 2019; Bhuyan et al. 2020; Jalil and Ansari 2020b). In requisite to attenuate ROS-induced damages, plants regulate the magnitude of ROS production through effective scavenging (Kumar et al. 2017; Jalil and Ansari 2020b; Ma et al. 2020). To quench excess ROS level, plants deploy intrinsic defensive system which includes enzymatic [superoxide dismutase (SOD), catalase (CAT), ascorbate peroxidase (APX), glutathione reductase (GR), POX] and non-enzymatic (phenolic compounds, flavonoids, vitamins, carotenoids) antioxidants, thereby alleviating oxidative stress (Kumar et al. 2017; Jalil and Ansari 2020b; Ma et al. 2020; Iqbal et al. 2021).

1.2 Genesis of Reactive Oxygen Metabolites in Plants

Reactive oxygen metabolites include a range of radical and non-radical oxygen species, most commonly, $O_2^{\bullet -}$, $^{\bullet}OH$, H_2O_2, and $^1O_2$. These ROS are generated sequentially on partial reduction or activation of O_2 (Bayr 2005; Mittler 2017; Maurya 2020; Sachdev et al. 2021). Molecular oxygen on acceptance of an electron

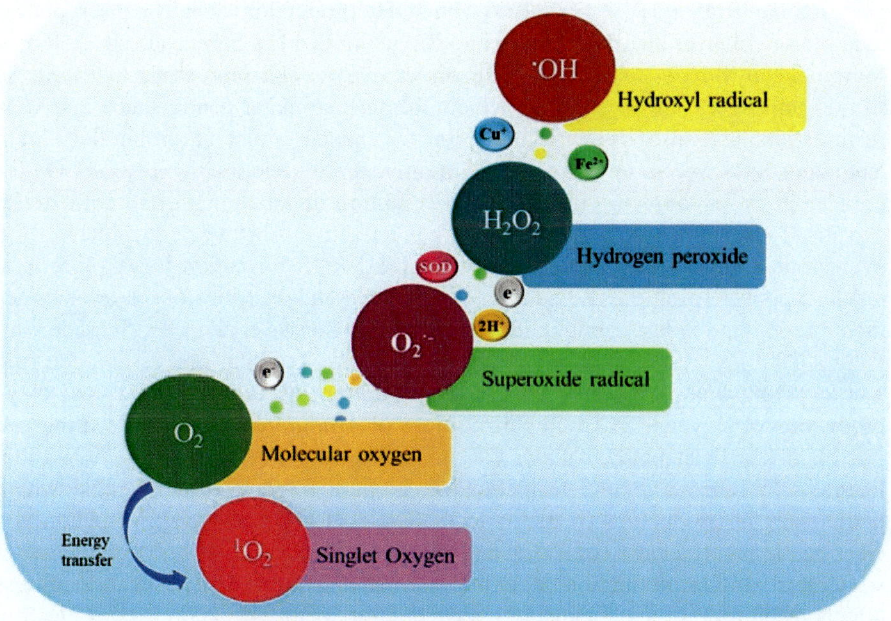

Fig. 1.1 Genesis of various forms of reactive oxygen species (ROS) from molecular oxygen through a series of partial reduction or activation. Molecular oxygen (O_2) either converts into singlet oxygen ($^1O_2$) on subsequent absorption of energy or results in formation of superoxide radical ($O_2^{\bullet -}$) on acceptance of an electron. The superoxide radical ($O_2^{\bullet -}$) formed is converted into hydrogen peroxide (H_2O_2) on dismutation either spontaneously or enzymatically in presence of enzyme superoxide dismutase (SOD). The hydrogen peroxide (H_2O_2) formed undergoes Fenton reaction in presence of transition metals (Cu^+/Fe^{2+}) and yields hydroxyl radical ($^{\bullet}OH$)

results in the production of $O_2^{\bullet -}$ radical, which contains single unpaired electron (Bayr 2005). The generated $O_2^{\bullet -}$ radical on dismutation, either enzymatically or spontaneously, give rise to the H_2O_2 (Hasanuzzaman et al. 2020). H_2O_2 further undergoes Fenton reaction in the presence of reduced copper or iron ions and results in the formation of highly toxic and reactive $^{\bullet}OH$ radical (Chapman et al. 2019). On subsequent absorption of energy from excited molecules in chloroplast, the O_2 molecules also induce production of $^1O_2$ (Chapman et al. 2019) (Fig. 1.1). Formation of various ROS in plant cell is highly localized and therefore takes place in various organelles engaging multiple enzymatic and non-enzymatic pathways (Podgorska et al. 2017; Janku et al. 2019). In normal circumstances, small fraction of oxygen available within plant cell is converted into ROS (Roychowdhury et al. 2019; Shah et al. 2019). Whereas, stress stimulus favors over-production and excess toxic accumulation of ROS (Gilroy et al. 2014; Huang et al. 2016; Saini et al. 2018).

Chloroplast is the prime location for ROS production in plant cell. Photosystem (PS) I and II reaction center are grossly involved in ROS over-production under stress due to readily available high energy-intermediates, reductants, and O_2 (Chen et al. 2012). Multiple stresses influence regular activity of the chloroplast, exacerbating ROS ($O_2^{\bullet-}$, $^1O_2$, H_2O_2 and $^{\bullet}OH$) production, which causes photo-oxidation (Miller et al. 2010; Foyer and Shigeoka 2011; Chen et al. 2012; Pinto-Marijuan and Munne-Bosch 2014; Hussain et al. 2018; Kamran et al. 2020). Stress stimuli impose closure of stomata which abridges stomatal conductance and CO_2 assimilation, and also triggers generation of excited triplet chlorophyll ($^3Chl^*$), mediating ROS over-production. Concurrently, the limited availability of CO_2 for the Calvin cycle coupled with high light condition under drought and cold stress, elevates O_2 accumulation, which acts as a key reductive substrate and results in ROS over-production (Mittler 2002). In peroxisomes, ROS formation takes place as a result of photorespiration, β-oxidation of fatty acid and activities of major enzymes including flavin oxidase, urate oxidase, and XOD (Corpas et al. 2001; Schrader and Fahimi 2006). Stress conditions accelerate the physiological processes occurring within peroxisomes triggering excessive generation of ROS. For instance, over-production of $O_2^{\bullet-}$ and H_2O_2 in peroxisomes of *Arabidopsis thaliana* seedling and pea plant has been noticed under Cd stress (Romero-Puertas et al. 1999, 2002; Corpas and Barroso 2014). Mitochondria is another plant cell organelle which participates in production of major ROS ($O_2^{\bullet-}$, H_2O_2, and $^{\bullet}OH$). Mitochondrial electron transport chain (mtETC) and photorespiration are two main pathways which lead to ROS production under both normal and stressed circumstances. Stress affects respiration process, causing over-reduction of mtETC and elevates electron leakage which reduces O_2 vis-à-vis elevation of ROS generation (Gill and Tuteja 2010; Miller et al. 2010). Plasma membrane and apoplast also act as potential sites for ROS production. NADPH oxidases found in plasma membrane mediate ROS generation through transfer of an electron from cytosolic NADPH to O_2, producing apoplastic $O_2^{\bullet-}$ (Choudhury et al. 2017; Podgorska et al. 2017; Janku et al. 2019), which dis-mutates into H_2O_2 (Das and Roychoudhury 2014; Choudhary et al. 2020). The plasma membrane localized NADPH oxidase under hypoxia condition stimulates H_2O_2 production in apoplastic regions (Wang et al. 2017). The production of various forms of ROS at different locations in plant cell through multiple mechanisms has been outlined in Fig. 1.2.

1.2.1 Superoxide Radical

Superoxide radical ($O_2^{\bullet-}$) is the primary ROS produced within the plant cell on monovalent reduction of O_2 (Das and Roychoudhury 2014; Hayyan et al. 2016) (Fig. 1.3). The $O_2^{\bullet-}$ is considered as both a radical and an anion due to presence of unpaired valance electron and negative (−1) oxidation state, respectively (Hayyan et al. 2016) and cellular organelles such as chloroplast, mitochondria, peroxisomes, apoplast and plasma membrane (Kohli et al. 2019). Chloroplast generates $O_2^{\bullet-}$ at PS I through partial reduction of O_2 by the photosynthetic ETC (Kohli et al. 2019), mitochondria by respiratory electron transport chain and plasma membrane by NADPH oxidase (Hayyan et al. 2016). Thus, processes like photosynthesis, respiration, and defense system actions are responsible for $O_2^{\bullet-}$ production (Hayyan et al. 2016).

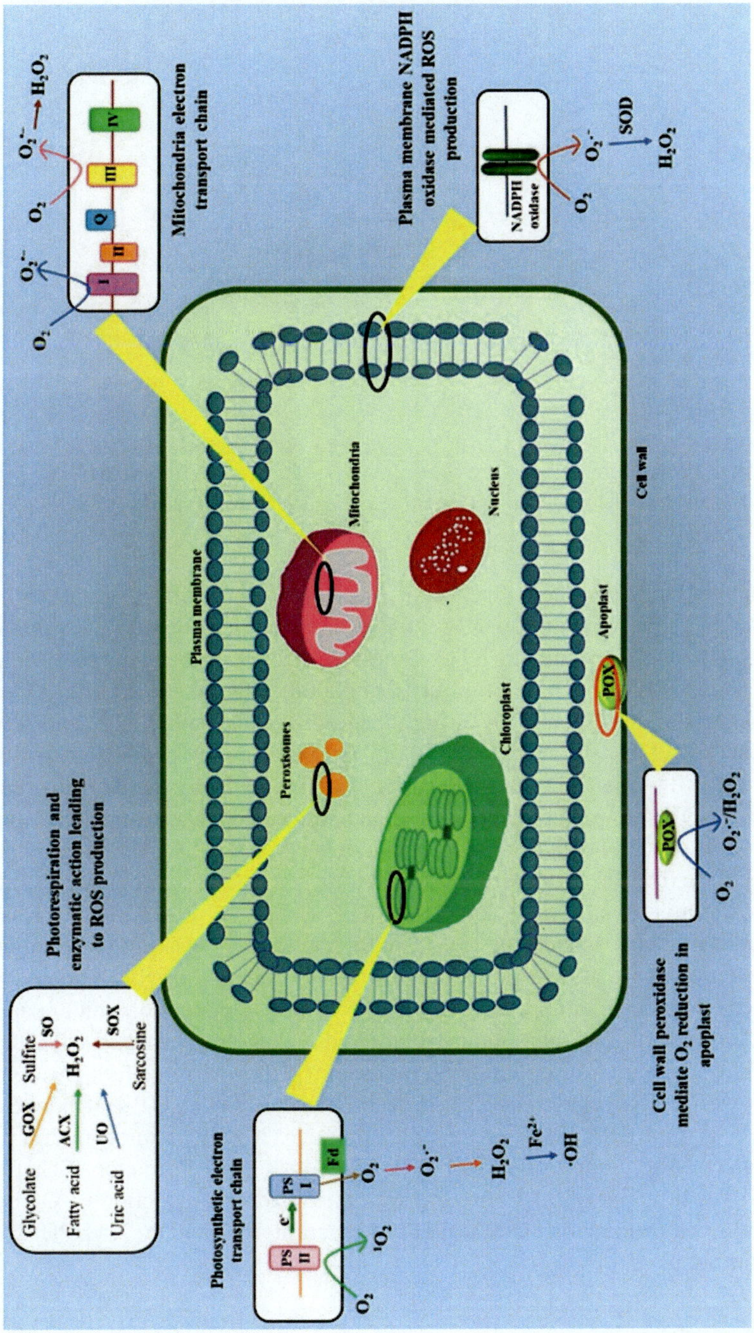

Fig. 1.2 Production of ROS at various cell organelles via different pathways and enzymes

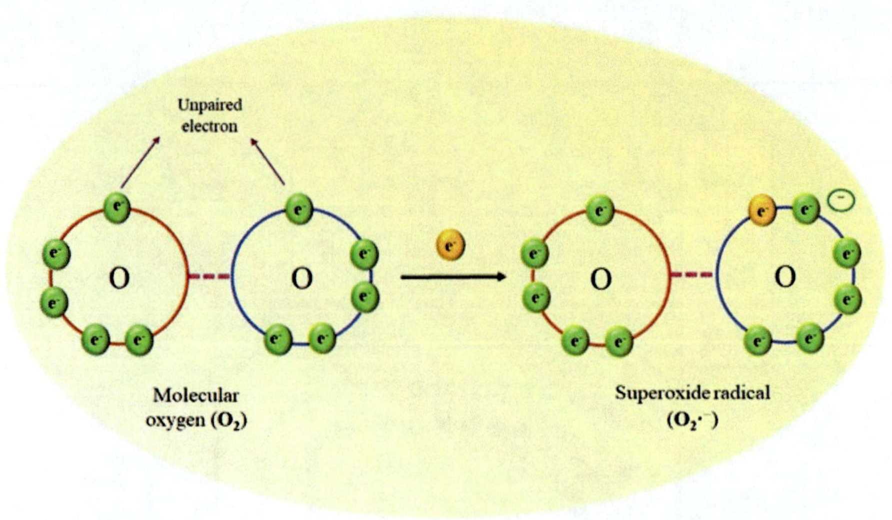

Fig. 1.3 Monovalent reduction of molecular oxygen by electron-yielding superoxide radical molecule

In chloroplast under the influence of unfavorable prevailing conditions, CO_2 concentration reduces. Consequently, $NADP^+$ availability abridges vis-à-vis electron leakage from photosynthetic ETC. Excessive electron leakage initiates Mehler's reaction and yields $O_2^{\bullet -}$ in chloroplast (Cruz 2008; Sharma et al. 2012; Xie et al. 2019). Moreover, excitation of O_2 by highly energized chlorophyll pigments leads to the production of $O_2^{\bullet -}$ (Maurya 2020). In peroxisomes, two different pathways generate $O_2^{\bullet -}$ (Ortega-Galisteo et al. 2012). The first pathway includes membrane-localized NADH (nicotinamide adenine dinucleotide) dependent small ETC which reduces O_2 to $O_2^{\bullet -}$ and then release into cytosol (Das and Roychoudhury 2014; Maurya 2020). The second pathway involves the enzymatic oxidation of xanthine and hypoxanthine to uric acid, generating $O_2^{\bullet -}$ in peroxisomes matrix (Das and Roychoudhury 2014; Maurya 2020). In mitochondria, $O_2^{\bullet -}$ is produced either on direct reduction of O_2 at flavoprotein region and Fe-S center of NADH dehydrogenase (Saed-Moucheshi et al. 2014) or by cytochrome c reductase, which generates electrons on reduction of ubiquinone. Increased leakage of electrons subsequently yields highly reducing ubisemiquinone radicals, which in turn reduces O_2 to generate $O_2^{\bullet -}$ (Das and Roychoudhury 2014; Saed-Moucheshi et al. 2014). Under excess light stress, glycine dehydrogenase complex produces NADP in mitochondria which donates electron to complex I of mtETC and leads to $O_2^{\bullet -}$ formation (Schertl and Braun 2014). In apoplast, $O_2^{\bullet -}$ production takes place on reduction of O_2 by the electron derived from cytosolic NADPH in a reaction mediated by NADPH oxidase (Choudhury et al. 2017).

1.2.2 Singlet Oxygen

Singlet oxygen ($^1O_2$) is a non-radical oxygen species, which is unstable and highly reactive (Dogra and Kim 2020) and participate in a variety of chemical and photochemical reactions (Lin et al. 2013). It is the first excited electronic state of O_2 and unlike other ROS, instead of electron transfer, it is formed by photo-activation of O_2 (Triantaphylidès and Havaux 2009; Dmitrieva et al. 2020). In plants, $^1O_2$ is mainly generated in chloroplast under excessive light conditions (Dmitrieva et al. 2020). However, under drought conditions, $^1O_2$ have been reported to be formed in root cells (Dmitrieva et al. 2020) and chloroplast (Mattos and Moretti 2015). Drought stress induces imbalance between ability of chloroplast to capture light and utilize it efficiently, which causes genesis of $^1O_2$ (Mattos and Moretti 2015). Moreover, certain phytotoxins such as phytoalexin, released during plant-microbe interactions mediate $^1O_2$ generation (Triantaphylidès and Havaux 2009). During photosynthesis, $^1O_2$ is derived through the energy transfer reaction from the ^{3}Chl to ground state O_2 molecule in the PS II reaction center and light-harvesting complex (LHC) of thylakoid appressed (?) membrane (Dogra and Kim 2020) (Fig. 1.4). Under illuminating conditions, LHC absorbs photons (energy) and generates a long-lived excited singlet chlorophyll ($^1Chl^*$) molecule which facilitates the conversion of excitation energy into electrochemical energy through charge separation. But

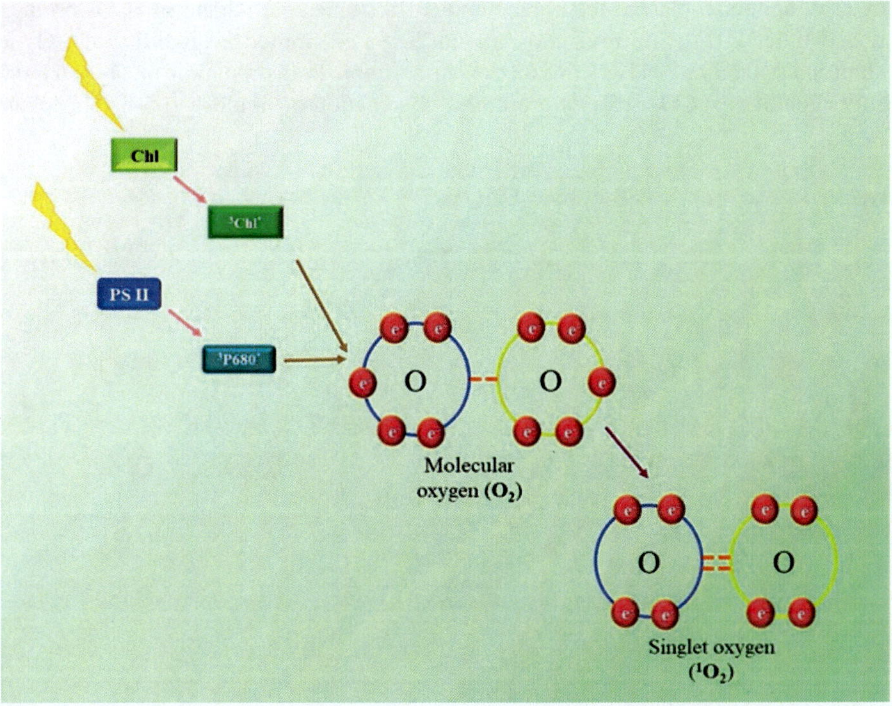

Fig. 1.4 Production of singlet oxygen in chloroplast under excessive illumination

under excess light, the energy absorbed by LHC at acceptor side of PS II surpasses its utilization threshold limit, resulting in $^3Chl^*$ generation (Pospisil 2016). The generated excited triplet chlorophyll readily reacts with O_2 leading to energy transfer, which causes shift in electron spin of O_2 in the molecular orbital and formation of $^1O_2$ (Dogra and Kim 2020). Stress circumstances limit CO_2 concentration in photosynthetic apparatus due to stomatal closure, which causes over-reduction of PS II (plastoquinone Q_A and Q_B) deterring the process of charge separation between P680 (chlorophyll molecules present at PS II) and pheophytin (Xie et al. 2019). This results in production of triplet chlorophyll (^{3}P680) at PS II reaction center, ultimately resulting in $^1O_2$ production (Sharma et al. 2012; Xie et al. 2019).

1.2.3 Hydrogen Peroxide

Hydrogen peroxide (H_2O_2) is a moderately reactive and stable ROS. Dismutation or disproportionation of $O_2^{\bullet -}$ is the major pathway for the production of H_2O_2 (Fig. 1.5) in various plant cell organelles (Foyer and Shigeoka 2011; Das and Roychoudhury 2014; Choudhary et al. 2020). This dismutation or disproportionation process takes place either spontaneously or enzymatically. The spontaneous dismutation of $O_2^{\bullet -}$ occurs under low pH condition and enzymatic dismutation, in presence of SOD enzyme. The key plant cell organelles, chloroplast, mitochondria, cytosol, apoplast, plasma membrane and peroxisomes participate in H_2O_2 production (Mittler 2017; Smirnoff and Arnaud 2019). A limited availability of CO_2 in chloroplast facilitates H_2O_2 production in addition to dismutation of $O_2^{\bullet -}$. Under stress conditions, CO_2 scarcity promotes oxygenation of ribulose 1, 5-bisphosphate

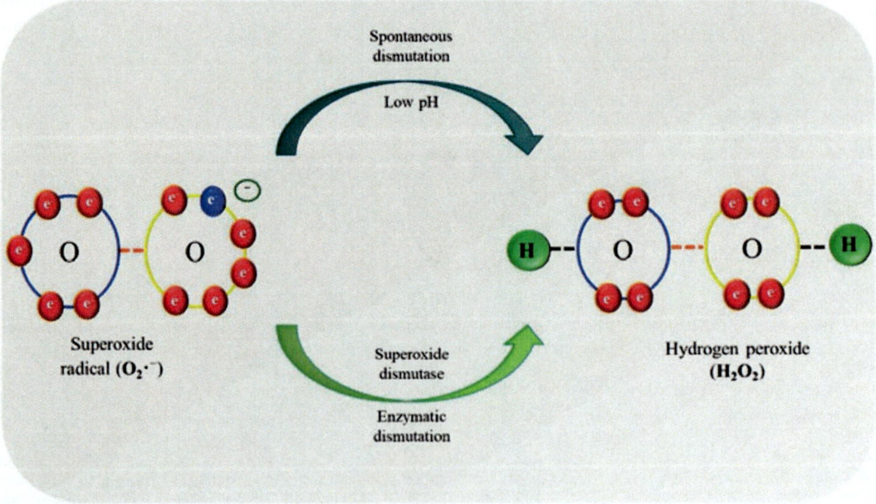

Fig. 1.5 Generation of hydrogen peroxide on dismutation of superoxide radical through both spontaneous and enzymatic pathways

(RuBP), triggering photorespiration and ultimately H_2O_2 formation (Voss et al. 2013; Suo et al. 2017). Peroxisome is another prime site for intracellular H_2O_2 production due to high oxidative metabolic activity (Das and Roychoudhury 2014). Photorespiration process occurring in chloroplast under unfavorable conditions produces glycolate, which is transported to peroxisomes, where it is oxidized by GOX, yielding H_2O_2 (Cruz 2008; Das and Roychoudhury 2014). Peroxisomes under the salinity are responsible for over-production of H_2O_2 arising from photorespiration process (Yamane et al. 2012). Similarly, >70% H_2O_2 in plants arises from photorespiration in plants exposed to drought stress condition (Voss et al. 2013). Peroxisomal enzymes like flavin oxidases, XOD, and acyl CoA oxidase (ACX) participating in fatty acid β-oxidation; and dismutation of $O_2^{\bullet -}$ are other pathways which trigger H_2O_2 generation in peroxisomes (Schrader and Fahimi 2006; Ortega-Galisteo et al. 2012; Del Río and Lopez-Huertas 2016). Stress conditions elevate activities of peroxisomal enzymes that overproduce H_2O_2. For instance, GOX activity elevation under heavy metal (Cd) stress enhances H_2O_2 level (Ortega-Galisteo et al. 2012). Analogously in herbicide (2,4-Dichlorophenoxy acetic acid) induced stress condition, activities of peroxisomal enzyme ACX and XOD and H_2O_2 production obtain simultaneous enhancement (Ortega-Galisteo et al. 2012). In mitochondria, production of H_2O_2 takes place due to disproportionation of $O_2^{\bullet -}$ by the action of enzyme Mn-SOD and APX (Sharma et al. 2012; Maurya 2020). Plasma membrane located NADPH oxidase triggers apoplastic extracellular H_2O_2 production under stress conditions (Wang et al. 2017). Further, the cell wall associated enzymes, including pH dependent extracellular POX, quinine reductase, lipoxygenases, amine oxidases, polyamine oxidases and germin-like oxalate oxidases also catalyze apoplast H_2O_2 production (Mika et al. 2004; Janku et al. 2019; Choudhary et al. 2020). Increase in apoplast H_2O_2 content in primary roots of maize subjected to water-deficient stress relates to the activity of the oxalate oxidase enzyme (Voothuluru and Sharp 2013). Martinez et al. (1998) have reported the production of H_2O_2 by cell wall related POX in *Arabidopsis* under potassium-deficient conditions.

1.2.4 Hydroxyl Radical

The $^{\bullet}OH$ radicals are the short-lived and most potent form of ROS. They have a single unpaired electron that enhances their reactivity with triplet ground state oxygen (Sharma et al. 2012). Haber–Weiss reaction and Fenton reaction are major pathways that mediate $^{\bullet}OH$ formation inside plant cells under both normal and stressed circumstances (Sharma et al. 2012) (Fig. 1.6). Production of $^{\bullet}OH$ radicals through Fenton reaction is favored by reduced forms of transition metals such as Cu^+ and Fe^{2+}. These transition metals catalyze the conversion of H_2O_2 into $^{\bullet}OH$ radical (Richards et al. 2015). In Haber–Weiss pathway, H_2O_2 and $O_2^{\bullet -}$ radical combine together to produce $^{\bullet}OH$ radicals along with hydroxyl ion and O_2 (Richards et al. 2015). Chloroplast, mitochondria and peroxisomes are the major cell organelles which are involved in $^{\bullet}OH$ generation in plant cell. Production of O_2 as a product of

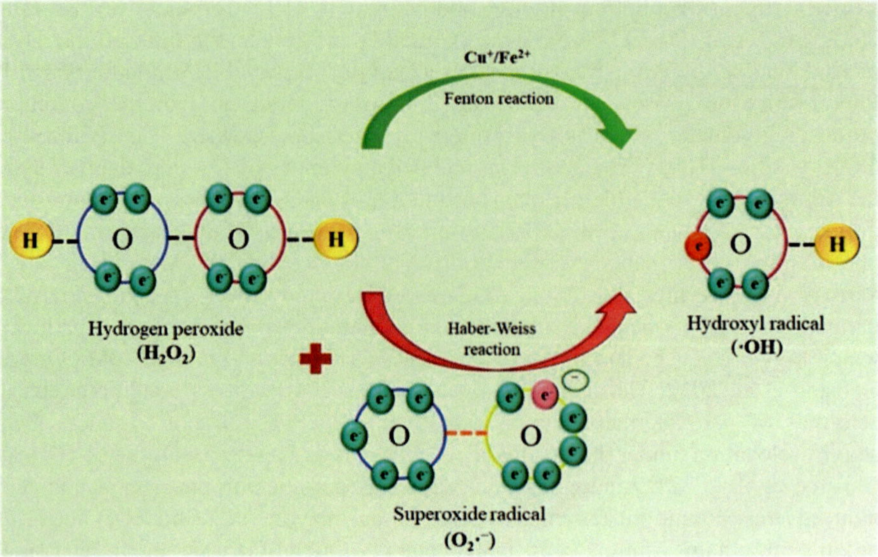

Fig. 1.6 Haber–Weiss and Fenton reaction pathways for the production of hydroxyl radical in plant cell

photosynthesis at PS II is responsible for formation of ${}^{\bullet}OH$ in chloroplast in presence of transition metals, which cause photoinhibition (Jakob and Heber 1996; Pospíšil 2009; Richards et al. 2015). Electron leakage from electron transport chain in mitochondria generates H_2O_2, which in presence of reduced Cu^+/Fe^{2+} produces ${}^{\bullet}OH$ (Keunen et al. 2011; Richards et al. 2015). Generation of ${}^{\bullet}OH$ radical in peroxisomes occurs due to fission of peroxynitrous arising from protonation of peroxynitrite ($ONOO^-$) (Chen et al. 2012; Richards et al. 2015). The extracellular generation of ${}^{\bullet}OH$ radical also takes place in plants which is mediated both enzymatically and non-enzymatically by the action of cell wall bounded enzymes including class III POX, SOD, and polyamine oxidases (PAO); and metals and ascorbate (AsA) (Richards et al. 2015). The non-enzymatic production of ${}^{\bullet}OH$ radical involves reduction of Cu^{2+}/Fe^{3+} into Cu^+/Fe^{2+} and O_2 to H_2O_2. The resultant H_2O_2 and Cu^+/Fe^{2+} undergoes Fenton reaction, producing ${}^{\bullet}OH$ radical (Richards et al. 2015). Enzymatic production of ${}^{\bullet}OH$ radical engages activity of POD, SOD, and PAO. The cell wall bounded POD enzymes in their compound III form convert $O_2^{\bullet-}$ and H_2O_2 into ${}^{\bullet}OH$ (Richards et al. 2015). The steady-state condition of POD enzymes and low pH in apoplast favors the over-production of ${}^{\bullet}OH$ radical (Richards et al. 2015). The excess production of ${}^{\bullet}OH$ radical could lead to the inactivation of enzymes (Richards et al. 2015). Other enzymes such as SOD and PAO also generate H_2O_2 which yield ${}^{\bullet}OH$ on subsequent Fenton rection (Richards et al. 2015). Another route for ${}^{\bullet}OH$ generation in plant cell is NADPH oxidase (NOX) which transports electron from cytosolic NADPH to outer side of the plasma membrane. The electron reduces O_2 into $O_2^{\bullet -}$ which finally produces $^{\bullet}OH$ in presence of reduced metal ions via H_2O_2 generation (Richards et al. 2015).

1.3 Chemistry and Activity of Reactive Oxygen Species

Each ROS display their own characteristic properties including reactivity, site of production and redox signaling (Mittler 2017). Some species are highly toxic and reactive, provoking extensive damages, whereas others are less toxic and display minor injuries. Although less toxic or less reactive species induces low vandalizing impact, yet undergoes conversion leading to the production of extremely reactive ROS that trigger oxidative stress by impinging stability of biomolecules causing disruption of membranes, and cellular activities (Das and Roychoudhury 2014). In addition to damaging act, ROS play role as signaling molecules under certain circumstances. Low or moderate concentration of ROS in cell has been reported to participate in signaling cascade which elicits resilience against stress (Sharma et al. 2012; Shah et al. 2019). Among various ROS, H_2O_2 performs perfectly as signal transducing molecules due to relatively long life-time and stability, which enable its diffusion through membranes via aquaporins (Sewelam et al. 2016; Kumar et al. 2017).

1.3.1 Superoxide Radical

The $O_2^{\bullet -}$ is a mildly reactive species with a half-life time of 2–4 μs (Phaniendra et al. 2015; Farooq et al. 2019). It has the potential to migrate to a distance of 30 nm and is often detected in the cytosol at a concentration of less than 0.5% (~0.3%) (De Grey 2002; Mittler 2017; Kohli et al. 2019). This anionic radical does not invoke considerable damage due to their moderate toxicity, but undergoes chemical or enzymatic dismutation and generates H_2O_2 which then converts into extremely toxic $^{\bullet}OH$ radicals causing oxidative damages (Das and Roychoudhury 2014). The enzymatic antioxidant SOD catalyzes the dismutation reaction, which happens at a rate of $2 \times 10^9\ M^{-1}\ s^{-1}$ or 10^4 times the rate constant for spontaneous dismutation (Kohli et al. 2019). Furthermore, $O_2^{\bullet -}$ undergoes protonation resulting in the formation of $HO_2^{\bullet}$ ($HOO^{\bullet}$), which is more harmful to plants and triggers oxidative stress (Kohli et al. 2019). As $O_2^{\bullet -}$ radicals have relatively low reactivity toward most of the biomolecules and are unable to cross lipid membrane, their level can be managed through effective compartmentalization.

1.3.2 Singlet Oxygen

The $^1O_2$ has a short half-life of 3 μs (Phaniendra et al. 2015), yet it can diffuse to the distance of 100 nm and damage biomolecules, in particular, proteins and lipids. The $^1O_2$ has potential to easily oxidize C–C double bond; hence predominantly targets the double bond of amino acid residues in proteins, PUFA (polyunsaturated fatty acids) in lipid membrane, guanine bases in nucleic bases and thiol groups of enzymes and proteins (Dmitrieva et al. 2020). As a result, hydroperoxides are generated, which set off a chain reaction of free radicals (Dmitrieva et al. 2020). The $^1O_2$ invokes severe damage to photosynthetic machinery, damaging PS I and PS II, which in turn result in loss of activity and under extreme condition, cell death (Das and Roychoudhury 2014). $^1O_2$ in chloroplast targets histidine residues of PS II proteins causing modification in PS II reaction center and its core component, D1 protein to proteolytic degradation (Dmitrieva et al. 2020). Moreover, $^1O_2$ trigger oxidation of cysteine (Cys) residue which suppresses protein biosynthesis in chloroplast, leading to photoinhibition. The oxidation of Cys residue, however, can be overcome through its reduction by antioxidant thioredoxins, thereby reversing restriction on protein biosynthesis (Dmitrieva et al. 2020). $^1O_2$ rapidly attacks on biomolecules like nucleic acid, proteins, pigments and lipid, causing oxidative damage (Dmitrieva et al. 2020). $^1O_2$ oxidizes C–C double bonds of biomolecules including amino acid residues, PUFA, guanine bases and thiol groups, forming hydroperoxides (Flors et al. 2006; Dmitrieva et al. 2020). These hydroperoxides trigger free radical chain reaction in presence of light or $^{\bullet}OH$ radicals, intensifying the damages (Dmitrieva et al. 2020). Oxidation of lipid by $^1O_2$ exacerbate peroxidation process which can directly invoke cell death (Triantaphylides et al. 2008). Under photooxidative stress, the *Arabidopsis* mutant with pronounced $^1O_2$ production capacity, exacerbated lipid peroxidation and induced cell death (Triantaphylides et al. 2008). On interaction with DNA, $^1O_2$ induce breakage in DNA strands, leading to chromosomal defects (Moan 1990). Reaction of $^1O_2$ with guanine bases result in the production of compounds such as 8-oxo-7,8-dihydroguanine, which are responsible for causing point mutation (Fischer et al. 2013). Aside from its negative consequences, $^1O_2$ has been linked to the activation of genes that protect against photo-oxidative stress (Das and Roychoudhury 2014).

1.3.3 Hydrogen Peroxide

H_2O_2 is a non-radical ROS and lacks unpaired electron, displaying moderate reactivity (Dar et al. 2017). It has substantially longer half-life duration of 1 ms or 1000 μs, allowing it to travel greater distance, even more than 1 m, and cross the cellular membranes via aquaporins (Das and Roychoudhury 2014; Farooq et al. 2019). The characteristics of H_2O_2 to easily traverse membranes, migrate to the long distance far from its production site, and comparative high stability allow H_2O_2 to display it actions at multiple locations (Farooq et al. 2019). H_2O_2 demonstrate both toxic and defensive response in the cell (Dar et al. 2017; Sachdev et al. 2021). At low concentration H_2O_2 mediate signaling pathways involved in processes like cell cycle, cell growth and development; stomatal movement; photosynthesis; senescence; photorespiration and acclimatization to stress (Das and Roychoudhury 2014; Dar et al. 2017). Under heat stress, H_2O_2 reportedly modulates heat shock genes expression in *Arabidopsis* cell culture, triggering production of APX2, HSP (heat shock protein) 17.6 and HSP 18.2 (Volkov et al. 2006). Analogously, excess accumulation of H_2O_2 in *Arabidopsis* cells induces tolerance toward light stress through increased production of HSPs, and APX1 which participate in H_2O_2 quenching (Pnueli et al. 2003). By contrast, at high concentration, H_2O_2 is known to cause severe damages in cell through oxidative modification of proteins and DNA (Farooq et al. 2019). H_2O_2 readily oxidized amino acid residues, in particular, Cys and methionine (Met), and thiol groups (Dar et al. 2017). Oxidation of thiol group of enzymes results in the inactivation of SOD involved in Calvin cycle, and may trigger plant cell death (Das and Roychoudhury 2014). As H_2O_2 is moderately reactive, hence most of the adverse impact associated with H_2O_2 results from its conversion into $^{\bullet}OH$ radical (Farooq et al. 2019).

1.3.4 Hydroxyl Radical

The $^{\bullet}OH$ radicals are short-lived ROS with a half-life time of 1ns or 10^{-9} s with strong positive redox potential and can migrate to a distance of 1 nm (Mittler 2017; Ding et al. 2020), thus often displaying reactivity near sites of production (Sharma et al. 2012). Due to a highly toxic and reactive nature, $^{\bullet}OH$ radicals interact with all biomolecules and subsequently damage cellular proteins, induce lipid peroxidation, and disrupt membranes (Sharma et al. 2012). Three types of reactions are displayed by $^{\bullet}OH$ radicals: (1) hydrogen abstraction, (2) addition, and (3) electron transfer (Treml and Smejkal 2016). All these reactions result in the production of new radicals and initiate chain reaction. For instance, $^{\bullet}OH$ radicals cause lipid damage via abstraction of hydrogen from methylene group of fatty acid, leading to the formation of carbon-centered lipid radical (Anjum et al. 2015). The lipid radical generated initiates chain reaction, ultimately causing peroxidation of PUFA as well as oxidation of protein and nucleic acids (Das and Roychoudhury 2014; Dar et al. 2017). The addition reaction displayed by the $^{\bullet}OH$ radicals results in the generation of hydroxy derivatives, wherein $^{\bullet}OH$ radicals are added to the double bond of aromatic compounds (Treml and Smejkal 2016). In particular, the addition of $^{\bullet}OH$ radicals to the double bonds of nucleic acid bases (purines and pyrimidines) results in DNA damage and production of various products including 8-hydroquinine (Halliwell 2006; Das and Roychoudhury 2014). The $^{\bullet}OH$ radicals also favor the cross-linkage between DNA and associated protein molecules, affecting structure and function of DNA (Das and Roychoudhury 2014). Stress conditions spark production of $^{\bullet}OH$ radicals in cell, which have been reported to affect plant growth and development. In pea seedlings, onset of dark condition results in the generation of $^{\bullet}OH$ radical possibly via Fenton reaction, which induces wilting of the epicotyl (Hideg et al. 2010). The $^{\bullet}OH$ also displays its reactivity through electron-transfer reaction (Treml and Smejkal 2016). Similar to H_2O_2, recently $^{\bullet}OH$ radicals have been viewed as a signaling molecule, which participate in plant resilience under environmental stress and also observed to play an important role in development processes, defense responses and microbes-assisted degradation of plant materials, alleviating agricultural waste (Richards et al. 2015).

1.4 Detoxification of Reactive Oxygen Species

Accumulation of ROS in cell organelles is regulated through the activities of antioxidants which quench excess ROS and aid in maintaining redox balance in cell (Hasanuzzaman et al. 2020). These antioxidants consist of several enzymatic and non-enzymatic compounds which function under both normal and stress conditions. The enzymatic antioxidants reduce ROS accumulation through conversion of ROS into water molecule, thus eliminating them from the cellular compartment (Nimse and Pal 2015). The key enzymatic antioxidants which operate in plant cell are SOD, CAT, APX, POX, MDHAR (monodehydroascorbate reductase), DHAR (dehydroascorbate reductase), GR, GST (glutathione S-transferase), and GPX (glutathione peroxidase) (Sachdev et al. 2021). By contrast, non-enzymatic antioxidants scavenge ROS by interrupting free-radical chain reaction (Nimse and Pal 2015). The non-enzymatic antioxidants include AsA, glutathione (GSH), compatible solutes, phenolic compounds, α-tocopherol, carotenoids, flavonoids, and stilbenes (Petrov et al. 2015; Martinez et al. 2016; Dumont and Rivoal 2019; Wongshaya et al. 2020; Iqbal et al. 2021).

1.4.1 Superoxide Radical

The effective scavenging of $O_2^{\bullet-}$ radical can be done through the sequential reduction of oxygen within the cell organelle by SOD, which possibly minimize the exposure of the cell to oxidative stress (Kehrer et al. 2010). SOD (EC: 1.15.1.1)? are metalloenzymes present in subcellular loci where free radicals and SOD substrates are generated (Luis et al. 2018). SODs are up-regulated during unfavorable conditions to prevent oxidative stress. Three forms of SOD or SOD isozymes are found in plants based on prosthetic metallic cofactors present at their active sites viz. CuZn-SOD (copper and zinc) localized in cytosol, chloroplast, peroxisomes, nuclei, mitochondria and apoplast; Fe-SOD (iron) in chloroplast, peroxisomes and mitochondria; and Mn-SOD (manganese) in peroxisomes, mitochondria and vascular tissues (Gill and Tuteja 2010; Ahmad et al. 2010; Zafra et al. 2018). SODs are well known to provide initial or first line of defense against toxic ROS (Gill and Tuteja 2010). They catalyze disproportionation of O_2^- free radicals by reducing one radical into H_2O_2 and oxidizing another into O_2 thereby eliminating the risk of production of more toxic free radical $OH^{\bullet}$ via Haber–Weiss reaction that take place in presence of metals such as Fe^{2+} or Cu^+ (Gill and Tuteja 2010).

1.4.2 Singlet Oxygen

The level of $^1O_2$ in cell can be reduced by efficient physical quenching through energy dissipation as heat or chemical quenching leading to generation of oxidized molecules (Dmitrieva et al. 2020). The non-enzymatic antioxidants including carotenoid, AsA, and tocopherols are major scavengers which protect plant cell from $^1O_2$

(Dmitrieva et al. 2020). Carotenoids are terpenoids that are present in photosynthetic organisms as light harvesting pigment (Hix et al. 2004; Uarrota et al. 2018). Carotenoids are photoprotective compounds that alleviate oxidative stress and protect plants from the damaging effects of high light illumination stress by quenching excessive energy and dissipating in form of heat (Hix et al. 2004; Uarrota et al. 2018). Carotenoid also avert over excitation of PS II in thylakoid membrane by efficiently scavenging $^1Chl^*$, $^3Chl^*$, and $^1O_2$ (Uarrota et al. 2018; Dmitrieva et al. 2020). Carotenoids not only protect cell from ROS but also participate in signaling during stress. During stress carotenoids scavenge over produced $^1O_2$ which results in breakdown products that act as stress-induced signals (Noctor et al. 2018).

1.5 Hydrogen Peroxide

Modulation in SOD activity under abiotic stress also triggers a cascade of expression of other antioxidant enzymes (Ahmad et al. 2010) that partake in conversion of H_2O_2 into H_2O. For instance, antioxidant enzyme APX, GPX, CAT catalyze conversion of H_2O_2 to H_2O (Zhang et al. 2016). CAT (EC: 1.11.1.6) are H_2O_2 scavenging tetrameric and heme-prosthetic group containing metalloenzyme that are located in peroxisomes and mitochondria (Anjum et al. 2016). The function of CAT enzymes is to catalyze decomposition of H_2O_2 into H_2O and O_2 (Chen et al. 2017; Su et al. 2018) and thus acts as signal preceptor (Hu et al. 2010; Liu et al. 2015; Anjum et al. 2016). The dismutation of H_2O_2 occurs in two steps in an energy-efficient manner due to unemployment of any reducing agent (Anjum et al. 2016). In first step heme-group of CAT is oxidized by H_2O_2 forming intermediate iron peroxide which followed by second step where intermediate formed is finally converted into H_2O and O_2 (Anjum et al. 2016). APX (EC: 1.11.1.11) are H_2O_2 scavenging heme-containing peroxidase enzyme (Anjum et al. 2016; Pandey et al. 2017) having protoporphyrin as prosthetic group (Pang and Wang 2010). APX are found in chloroplast, mitochondria, cytoplasmic matrix or cytosol, microbodies, and peroxisomes/glyoxysomes (Anjum et al. 2016). APX require reducing substrate AsA for proper functioning and maintaining its own stability (Pandey et al. 2017). Shigeoka et al. (2002) mention rapid inactivation of APX and loss of stability occurs when the concentration of AsA reaches below 20 μM. The catalytic reaction of APX takes place in few steps. Initially, the heme-group of APX is oxidized by H_2O_2 to oxyferryl which gets converted into water molecules by two successive one-electron reactions (Anjum et al. 2016). GR (EC: 1.6.4.2) is a flavo-protein oxidoreductase that participates as one of the important enzymes in H_2O_2 scavenging pathway, i.e., AsA-GSH cycle (Pang and Wang 2010; Yousuf et al. 2012). GR is generally localized in chloroplast, mitochondria, cytosol, peroxisomes and in non-photosynthetic tissues and organelles (Edwards et al. 1990), but predominantly in chloroplast accounting for 70–80% activity (Rao and Reddy 2008; Pang and Wang 2010). GR catalyzes conversion of oxidized glutathione or glutathione disulphide (GSSG) into reduced form known as glutathione (GSH) using NADPH as electron donor (Pang and Wang 2010; Yousuf et al. 2012) and maintains a balance between GSH/GSSG ratio necessary for the detoxification of H_2O_2 (Zandalinas et al. 2017).

1.6 Hydroxyl Radical

Excess production of $^{\bullet}OH$ radicals within cells can lead to oxidative damage (Sharma et al. 2012). Therefore, restricting their accumulation in cells below the damaging level is crucial for plant growth and development. Plants do not recruit any enzymatic mechanism to detoxify $^{\bullet}OH$ radicals (Das and Roychoudhury 2014), thus, effective quenching of H_2O_2 (substrate for $^{\bullet}OH$ radical generation) and limiting the availability of transitions metals for Fenton reaction can play central role in restricting formation of $^{\bullet}OH$ radical in plants (Richards et al. 2015). Further, plants employ non-enzymatic pathway for curbing concentration of $^{\bullet}OH$ radical within the cell. The non-enzymatic antioxidants including AsA, flavonoids, GSH, β-carotene, tocopherol, phenolic acids, and proline can effectively scavenge $^{\bullet}OH$ radical and protect cellular machinery from oxidative stress (Das and Roychoudhury 2014; Hasanuzzaman et al. 2020).

1.7 Conclusion

ROS are the common by-product of aerobic activity, inevitably found in every living cell in variable quantities. The clan of ROS includes a number of radical and non-radical forms of oxygen. Each form has its own characteristics and reactivity, leading to differential outputs in plant cell. These ROS may act as damaging entities or mediate signaling in plants under unfavorable conditions. Understanding the mechanism of production of different forms of ROS and their response under variable conditions is crucial to determine their ultimate fate. Elucidation of overall resultant impact of particular form of ROS on plant growth and productivity could facilitate identifying approaches to protect plants from damaging effects. Moreover, plants are endowed with antioxidant-based defense machinery which aids in protection of plant from the damaging effects of ROS. Array of enzymatic and non-enzymatic antioxidants are recruited by plant cell to quench excess accumulation of ROS, thereby maintaining redox homeostasis. However, adverse environmental circumstances disrupt this homeostasis by affecting functioning of antioxidants and trigger oxidative stress. Thus, to avoid ROS induced damage and strengthen plant defense system, a deep knowledge on properties of different types of antioxidants and factors influencing their activities is required. The overall knowledge and understanding on pathways involved in ROS production, mechanism of signaling and regulation of redox homeostasis under unfavorable circumstances can unlock opportunities to improve plant growth and productivity to reinforce global food security.

References

Ahmad P, Jaleel CA, Salem MA, Nabi G, Sharma S (2010) Roles of enzymatic and nonenzymatic antioxidants in plants during abiotic stress. Crit Rev Biotechnol 30(3):161–175

Anjum NA, Sofo A, Scopa A, Roychoudhury A, Gill SS, Iqbal M, Lukatkin AS, Pereira E, Duarte AC, Ahmad I (2015) Lipids and proteins—major targets of oxidative modifications in abiotic stressed plants. Environ Sci Pollut Res 22(6):4099–4121

Anjum NA, Sharma P, Gill SS, Hasanuzzaman M, Khan EA, Kachhap K, Mohamed AA, Thangavel P, Devi GD, Vasudhevan P, Sofo A, Khan NA, Misra AN, Lukatkin AS, Singh HP, Pereira E, Tuteja N (2016) Catalase and ascorbate peroxidase—representative H 2 O 2-detoxifying heme enzymes in plants. Environ Sci Pollut Res 23(19):19002–19029

Ansari MI, Jalil SU, Ansari SA, Hasanuzzaman M (2021) GABA shunt: a key-player in mitigation of ROS during stress. Plant Growth Regul 94:1–19

Bayr H (2005) Reactive oxygen species. Crit Care Med 33(12):S498–S501

Bhuyan MB, Hasanuzzaman M, Parvin K, Mohsin SM, Al Mahmud J, Nahar K, Fujita M (2020) Nitric oxide and hydrogen sulfide: two intimate collaborators regulating plant defense against abiotic stress. Plant Growth Regul 90(3):409–424. https://doi.org/10.1007/s10725-020-00594-4

Bulgari R, Franzoni G, Ferrante A (2019) Biostimulants application in horticultural crops under abiotic stress conditions. Agronomy 9(6):306. https://doi.org/10.3390/agronomy9060306

Chapman JM, Muhlemann JK, Gayomba SR, Muday GK (2019) RBOH-dependent ROS synthesis and ROS scavenging by plant specialized metabolites to modulate plant development and stress responses. Chem Res Toxicol 32(3):370–396

Chen S, Yin C, Strasser RJ, Yang C, Qiang S (2012) Reactive oxygen species from chloroplasts contribute to 3-acetyl-5-isopropyltetramic acid-induced leaf necrosis of *Arabidopsis thaliana*. Plant Physiol Biochem 52:38–51

Chen N, Teng XL, Xiao XG (2017) Subcellular localization of a plant catalase-phenol oxidase, AcCATPO, from *Amaranthus* and identification of a non-canonical peroxisome targeting signal. Front Plant Sci 8:1345. https://doi.org/10.3389/fpls.2017.01345

Choudhary A, Kumar A, Kaur N (2020) ROS and oxidative burst: roots in plant development. Plant Divers 42(1):33–43

Choudhury FK, Rivero RM, Blumwald E, Mittler R (2017) Reactive oxygen species, abiotic stress and stress combination. Plant J 90(5):856–867

Corpas FJ, Barroso JB (2014) Peroxynitrite (ONOO−) is endogenously produced in *Arabidopsis* peroxisomes and is overproduced under cadmium stress. Ann Bot 113(1):87–96

Corpas FJ, Barroso JB, del Río, L. A. (2001) Peroxisomes as a source of reactive oxygen species and nitric oxide signal molecules in plant cells. Trends Plant Sci 6(4):145–150

Cruz DCM (2008) Drought stress and reactive oxygen species: production, scavenging and signaling. Plant Signal Behav 3(3):156–165. https://doi.org/10.4161/psb.3.3.5536

Dar MI, Naikoo MI, Khan FA, Rehman F, Green ID, Naushin F, Ansari AA (2017) An introduction to reactive oxygen species metabolism under changing climate in plants. In: Reactive oxygen species and antioxidant systems in plants: role and regulation under abiotic stress. Springer, Singapore, pp 25–52

Das K, Roychoudhury A (2014) Reactive oxygen species (ROS) and response of antioxidants as ROS-scavengers during environmental stress in plants. Front Environ Sci 2:53. https://doi.org/10.3389/fenvs.2014.00053

De Grey AD (2002) $HO2^{\bullet}$: the forgotten radical. DNA Cell Biol 21(4):251–257

Del Río LA, Lopez-Huertas E (2016) ROS generation in peroxisomes and its role in cell signaling. Plant Cell Physiol 57(7):1364–1376

Dikilitas M, Simsek E, Roychoudhury A (2020) Modulation of abiotic stress tolerance through hydrogen peroxide. In: Roychoudhury A, Tripathi DK (eds) Protective chemical agents in the amelioration of plant abiotic stress: biochemical and molecular perspectives. Wiley, Newark, pp 147–173

Ding S, Li M, Gong H, Zhu Q, Shi G, Zhu A (2020) Sensitive and selective measurement of hydroxyl radicals at subcellular level with tungsten nanoelectrodes. Anal Chem 92(3): 2543–2549

Dmitrieva VA, Tyutereva EV, Voitsekhovskaja OV (2020) Singlet oxygen in plants: generation, detection, and signaling roles. Int J Mol Sci 21(9):3237

Dogra V, Kim C (2020) Singlet oxygen metabolism: from genesis to signaling. Front Plant Sci 10: 1640

Dumont S, Rivoal J (2019) Consequences of oxidative stress on plant glycolytic and respiratory metabolism. Front Plant Sci 10:166. https://doi.org/10.3389/fpls.2019.00166

Edwards EA, Rawsthorne S, Mullineaux PM (1990) Subcellular distribution of multiple forms of glutathione reductase in leaves of pea (*Pisum sativum* L.). Planta 180(2):278–284

Farooq MA, Niazi AK, Akhtar J, Farooq M, Souri Z, Karimi N, Rengel Z (2019) Acquiring control: the evolution of ROS-Induced oxidative stress and redox signaling pathways in plant stress responses. Plant Physiol Biochem 141:353–369

Fischer BB, Hideg E, Krieger-Liszkay A (2013) Production, detection, and signaling of singlet oxygen in photosynthetic organisms. Antioxid Redox Signal 18(16):2145–2162

Florez-Sarasa I, Fernie AR, Gupta KJ (2020) Does the alternative respiratory pathway offer protection against the adverse effects resulting from climate change? J Exp Bot 71(2):465–469

Flors C, Fryer MJ, Waring J, Reeder B, Bechtold U, Mullineaux PM, Nonell S, Wilson MT, Baker NR (2006) Imaging the production of singlet oxygen in vivo using a new fluorescent sensor, Singlet Oxygen Sensor Green®. J Exp Bot 57(8):1725–1734

Foyer CH, Shigeoka S (2011) Understanding oxidative stress and antioxidant functions to enhance photosynthesis. Plant Physiol 155(1):93–100

Gill SS, Tuteja N (2010) Reactive oxygen species and antioxidant machinery in abiotic stress tolerance in crop plants. Plant Physiol Biochem 48(12):909–930

Gilroy S, Suzuki N, Miller G, Choi WG, Toyota M, Devireddy AR, Mittler R (2014) A tidal wave of signals: calcium and ROS at the forefront of rapid systemic signaling. Trends Plant Sci 19(10):623–630

Halliwell B (2006) Reactive species and antioxidants. Redox biology is a fundamental theme of aerobic life. Plant Physiol 141(2):312–322. https://doi.org/10.1104/pp.106.077073

Hamilton CE, Gundel PE, Helander M, Saikkonen K (2012) Endophytic mediation of reactive oxygen species and antioxidant activity in plants: a review. Fungal Divers 54(1):1–10

Hasanuzzaman M, Bhuyan MHM, Parvin K, Bhuiyan TF, Anee TI, Nahar K, Hossen MS, Zulfiqar F, Alam MM, Fujita M (2020) Regulation of ROS metabolism in plants under environmental stress: a review of recent experimental evidence. Int J Mol Sci 21(22):8695

Hayyan M, Hashim MA, AlNashef IM (2016) Superoxide ion: generation and chemical implications. Chem Rev 116(5):3029–3085

He M, He CQ, Ding NZ (2018) Abiotic stresses: general defenses of land plants and chances for engineering multi stress tolerance. Front Plant Sci 9:1771. https://doi.org/10.3389/fpls.2018.01771

Hideg É, Vitányi B, Kósa A, Solymosi K, Bóka K, Won S, Inoue Y, Ridge RW, Böddi B (2010) Reactive oxygen species from type-I photosensitized reactions contribute to the light-induced wilting of dark-grown pea (*Pisum sativum*) epicotyls. Physiol Plant 138(4):485–492

Hix LM, Lockwood SF, Bertram JS (2004) Bioactive carotenoids: potent antioxidants and regulators of gene expression. Redox Rep 9(4):181–191

Hu YQ, Liu S, Yuan HM, Li J, Yan DW, Zhang JF, Lu YT (2010) Functional comparison of catalase genes in the elimination of photorespiratory H2O2 using promoter-and 3′-untranslated region exchange experiments in the Arabidopsis cat2 photorespiratory mutant. Plant Cell Environ 33(10):1656–1670

Huang S, Van Aken O, Schwarzlander M, Belt K, Millar AH (2016) The roles of mitochondrial reactive oxygen species in cellular signaling and stress response in plants. Plant Physiol 171(3): 1551–1559

Hussain HA, Hussain S, Khaliq A, Ashraf U, Anjum SA, Men S, Wang L (2018) Chilling and drought stresses in crop plants: implications, cross talk, and potential management opportunities. Front Plant Sci 9:393. https://doi.org/10.3389/fpls.2018.00393

Iqbal MS, Iqbal Z, Hashem A, Al-Arjani ABF, Abd-Allah EF, Jafri A, Ansari SA, Ansari MI (2021) Nigella sativa callus treated with sodium azide exhibit augmented antioxidant activity and DNA damage inhibition. Sci Rep 11(1):1–14

Jakob B, Heber U (1996) Photoproduction and detoxification of hydroxyl radicals in chloroplasts and leaves and relation to photoinactivation of photosystems I and II. Plant Cell Physiol 37(5): 629–635

Jalil SU, Ansari MI (2020a) Stress implications and crop productivity. In: Plant ecophysiology and adaptation under climate change: mechanisms and perspectives I. Springer, Singapore, pp 73–86

Jalil SU, Ansari MI (2020b) Isoprenoids in plant protection against abiotic stress. In: Roychoudhury A, Tripathi DK (eds) Protective chemical agents in the amelioration of plant abiotic stress: biochemical and molecular perspectives. Wiley, Newark, pp 424–436

Jalil SU, Zahera M, Khan MS, Ansari MI (2019) Biochemical synthesis of gold nanoparticles from leaf protein of *Nicotiana tabacum* L. cv. xanthi and their physiological, developmental, and ROS scavenging responses on tobacco plant under stress conditions. IET Nanobiotechnol 13(1): 23–29

Janku M, Luhova L, Petrivalsky M (2019) On the origin and fate of reactive oxygen species in plant cell compartments. Antioxidants 8(4):105. https://doi.org/10.3390/antiox8040105

Kamran M, Parveen A, Ahmar S, Malik Z, Hussain S, Chattha MS, Saleem MH, Adil M, Heidari P, Chen JT (2020) An overview of hazardous impacts of soil salinity in crops, tolerance mechanisms, and amelioration through selenium supplementation. Int J Mol Sci 21(1):148

Kehrer JP, Robertson JD, Smith CV (2010) Free radicals and reactive oxygen species. In: McQueen CA (ed) Comprehensive toxicology, 2nd edn. Elsevier, Oxford, pp 277–307

Kerchev P, van der Meer T, Sujeeth N, Verlee A, Stevens CV, Van Breusegem F, Gechev T (2020) Molecular priming as an approach to induce tolerance against abiotic and oxidative stresses in crop plants. Biotechnol Adv 40:107503. https://doi.org/10.1016/j.biotechadv.2019.107503

Keunen E, Remans T, Bohler S, Vangronsveld J, Cuypers A (2011) Metal-induced oxidative stress and plant mitochondria. Int J Mol Sci 12(10):6894–6918

Kohli SK, Khanna K, Bhardwaj R, Abd Allah EF, Ahmad P, Corpas FJ (2019) Assessment of subcellular ROS and NO metabolism in higher plants: multifunctional signaling molecules. Antioxidants 8(12):641

Kumar V, Khare T, Sharma M, Wani SH (2017) ROS-induced signaling and gene expression in crops under salinity stress. In: Khan MIR, Khan NA (eds) Reactive oxygen species and antioxidant systems in plants: role and regulation under abiotic stress. Springer, Singapore, pp 159–184

Lin H, Shen Y, Chen D, Lin L, Wilson BC, Li B, Xie S (2013) Feasibility study on quantitative measurements of singlet oxygen generation using singlet oxygen sensor green. J Fluoresc 23(1): 41–47

Liu Y, Yao Y, Hu X, Xing S, Xu L (2015) Cloning and allelic variation of two novel catalase genes (SoCAT-1 and SsCAT-1) in *Saccharum officinarum* L. and *Saccharum spontaneum* L. Biotechnol Biotechnol Equip 29(3):431–440. https://doi.org/10.1080/13102818.2015.1018839

Luis A, Corpas FJ, López-Huertas E, Palma JM (2018) Plant superoxide dismutases: function under abiotic stress conditions. In: In antioxidants and antioxidant enzymes in higher plants. Springer, Cham, pp 1–26

Ma J, Qiu D, Pang Y, Gao H, Wang X, Qin Y (2020) Diverse roles of tocopherols in response to abiotic and biotic stresses and strategies for genetic biofortification in plants. Mol Breed 40(2): 1–15

Martinez C, Montillet JL, Bresson E, Agnel JP, Dai GH, Daniel JF, Geiger JP, Nicole M (1998) Apoplastic peroxidase generates superoxide anions in cells of cotton cotyledons undergoing the hypersensitive reaction to *Xanthomonas campestris* pv. *malvacearum* race 18. Mol Plant-Microbe Interact 11(11):1038–1047

Martinez V, Mestre TC, Rubio F, Girones-Vilaplana A, Moreno DA, Mittler R, Rivero RM (2016) Accumulation of flavonols over hydroxycinnamic acids favors oxidative damage protection under abiotic stress. Front Plant Sci 7:838. https://doi.org/10.3389/fpls.2016.00838

Mattos LM, Moretti CL (2015) Oxidative stress in plants under drought conditions and the role of different enzymes. Enzym Eng 5:1–6

Maurya AK (2020) Oxidative stress in crop plants. In: Hasanuzzaman M (ed) Agronomic crops: stress responses and tolerance. Springer, Singapore, pp 349–380

Mika A, Minibayeva F, Beckett R, Luthje S (2004) Possible functions of extracellular peroxidases in stress-induced generation and detoxification of active oxygen species. Phytochem Rev 3(1-2): 173–193

Miller GAD, Suzuki N, Ciftci-Yilmaz SULTAN, Mittler RON (2010) Reactive oxygen species homeostasis and signalling during drought and salinity stresses. Plant Cell Environ 33(4): 453–467

Mittler R (2002) Oxidative stress, antioxidants and stress tolerance. Trends Plant Sci 7(9):405–410

Mittler R (2017) ROS are good. Trends Plant Sci 22(1):11–19

Moan J (1990) On the diffusion length of singlet oxygen in cells and tissues. J Photochem Photobiol B Biol 6(3):343–344

Nimse SB, Pal D (2015) Free radicals, natural antioxidants, and their reaction mechanisms. RSC Adv 5(35):27986–28006

Noctor G, Reichheld JP, Foyer CH (2018) ROS-related redox regulation and signaling in plants. Semin Cell Dev Biol 80:3–12

Ortega-Galisteo AP, Rodriguez-Serrano M, Pazmino DM, Gupta DK, Sandalio LM, Romero-Puertas MC (2012) S-Nitrosylated proteins in pea (*Pisum sativum* L.) leaf peroxisomes: changes under abiotic stress. J Exp Bot 63(5):2089–2103

Pandey S, Fartyal D, Agarwal A, Shukla T, James D, Kaul T, Negi YK, Arora S, Reddy MK (2017) Abiotic stress tolerance in plants: myriad roles of ascorbate peroxidase. Front Plant Sci 8:581. https://doi.org/10.3389/fpls.2017.00581

Pang CH, Wang BS (2010) Role of ascorbate peroxidase and glutathione reductase in ascorbate–glutathione cycle and stress tolerance in plants. In: Anjum NA, Chan MT, Umar S (eds) Ascorbate-glutathione pathway and stress tolerance in plants. Springer, Dordrecht, pp 91–113

Petrov V, Hille J, Mueller-Roeber B, Gechev TS (2015) ROS-mediated abiotic stress-induced programmed cell death in plants. Front Plant Sci 6:69. https://doi.org/10.3389/fpls.2015.00069

Phaniendra A, Jestadi DB, Periyasamy L (2015) Free radicals: properties, sources, targets, and their implication in various diseases. Indian J Clin Biochem 30(1):11–26

Pinto-Marijuan M, Munne-Bosch S (2014) Photo-oxidative stress markers as a measure of abiotic stress-induced leaf senescence: advantages and limitations. J Exp Bot 65(14):3845–3857

Pnueli L, Liang H, Rozenberg M, Mittler R (2003) Growth suppression, altered stomatal responses, and augmented induction of heat shock proteins in cytosolic ascorbate peroxidase (Apx1)-deficient *Arabidopsis* plants. Plant J 34(2):187–203

Podgorska A, Burian M, Szal B (2017) Extra-cellular but extra-ordinarily important for cells: apoplastic reactive oxygen species metabolism. Front Plant Sci 8:1353. https://doi.org/10.3389/fpls.2017.01353

Pospíšil P (2009) Production of reactive oxygen species by photosystem II. Biochim Biophys Acta Bioenerg 1787(10):1151–1160

Pospisil P (2016) Production of reactive oxygen species by photosystem II as a response to light and temperature stress. Front Plant Sci 7:1950. https://doi.org/10.3389/fpls.2016.01950

Rao AC, Reddy AR (2008) Glutathione reductase: a putative redox regulatory system in plant cells. In: Khan NA, Singh S, Umar S (eds) Sulfur assimilation and abiotic stress in plants. Springer, Berlin, pp 111–147

Richards SL, Wilkins KA, Swarbreck SM, Anderson AA, Habib N, Smith AG, McAinsh M, Davies JM (2015) The hydroxyl radical in plants: from seed to seed. J Exp Bot 66(1):37–46

Romero-Puertas MC, McCarthy I, Sandalio LM, Palma JM, Corpas FJ, Gomez M, Del Rio LA (1999) Cadmium toxicity and oxidative metabolism of pea leaf peroxisomes. Free Radic Res 31 (sup1):25–31

Romero-Puertas MC, Palma JM, Gómez M, Del Rio LA, Sandalio LM (2002) Cadmium causes the oxidative modification of proteins in pea plants. Plant Cell Environ 25(5):677–686

Roychowdhury R, Khan MH, Choudhury S (2019) Physiological and molecular responses for metalloid stress in rice—a comprehensive overview. In: Hasanuzzaman M, Fujita M, Nahar K, Biswas JK (eds) Advances in rice research for abiotic stress tolerance. Woodhead Publishing, Oxford, pp 341–369

Sachdev S, Ansari SA, Ansari MI, Fujita M, Hasanuzzaman M (2021) Abiotic stress and reactive oxygen species: generation, signaling, and defense mechanisms. Antioxidants 10(2):277

Saed-Moucheshi A, Shekoofa A, Pessarakli M (2014) Reactive oxygen species (ROS) generation and detoxifying in plants. J Plant Nutr 37(10):1573–1585

Saini P, Gani M, Kaur JJ, Godara LC, Singh C, Chauhan SS, Francies RM, Bhardwaj A, Kumar NB, Ghosh MK (2018) Reactive oxygen species (ROS): a way to stress survival in plants. In: Zargar SM, Zargar MY (eds) Abiotic stress-mediated sensing and signaling in plants: an omics perspective. Springer, Singapore, pp 127–153

Schertl P, Braun HP (2014) Respiratory electron transfer pathways in plant mitochondria. Front Plant Sci 5:163. https://doi.org/10.3389/fpls.2014.00163

Schrader M, Fahimi HD (2006) Peroxisomes and oxidative stress. Biochim Biophys Acta Mol Cell Res 1763(12):1755–1766

Sewelam N, Kazan K, Schenk PM (2016) Global plant stress signaling: reactive oxygen species at the cross-road. Front Plant Sci 7:187. https://doi.org/10.3389/fpls.2016.00187

Shah K, Chaturvedi V, Gupta S (2019) Climate change and abiotic stress-induced oxidative burst in rice. In: Hasanuzzaman M, Fujita M, Nahar K, Biswas JK (eds) Advances in rice research for abiotic stress tolerance. Woodhead Publishing, Oxford, pp 505–535

Sharma P, Jha AB, Dubey RS, Pessarakli M (2012) Reactive oxygen species, oxidative damage, and antioxidative defense mechanism in plants under stressful conditions. J Bot 2012:217037. https://doi.org/10.1155/2012/217037

Shigeoka S, Ishikawa T, Tamoi M, Miyagawa Y, Takeda T, Yabuta Y, Yoshimura K (2002) Regulation and function of ascorbate peroxidase isoenzymes. J Exp Bot 53(372):1305–1319

Smirnoff N, Arnaud D (2019) Hydrogen peroxide metabolism and functions in plants. New Phytol 221(3):1197–1214

Su T, Wang P, Li H, Zhao Y, Lu Y, Dai P, Ren T, Wang X, Li X, Shao Q, Zhao D, Zhao Y, Ma C (2018) The *Arabidopsis* catalase triple mutant reveals important roles of catalases and peroxisome-derived signaling in plant development. J Integr Plant Biol 60(7):591–607

Suo J, Zhao Q, David L, Chen S, Dai S (2017) Salinity response in chloroplasts: insights from gene characterization. Int J Mol Sci 18(5):1011. https://doi.org/10.3390/ijms18051011

Treml J, Smejkal K (2016) Flavonoids as potent scavengers of hydroxyl radicals. Compr Rev Food Sci Food Saf 15(4):720–738

Triantaphylidès C, Havaux M (2009) Singlet oxygen in plants: production, detoxification and signaling. Trends Plant Sci 14(4):219–228

Triantaphylides C, Krischke M, Hoeberichts FA, Ksas B, Gresser G, Havaux M, Breusegem MH, Mueller MJ (2008) Singlet oxygen is the major reactive oxygen species involved in photooxidative damage to plants. Plant Physiol 148(2):960–968

Tripathy BC, Oelmuller R (2012) Reactive oxygen species generation and signaling in plants. Plant Signal Behav 7(12):1621–1633

Uarrota VG, Stefen DLV, Leolato LS, Gindri DM, Nerling D (2018) Revisiting carotenoids and their role in plant stress responses: from biosynthesis to plant signaling mechanisms during stress. In: Gupta DK, Palma JM, Corpas FJ (eds) Antioxidants and antioxidant enzymes in higher plants. Springer, Cham, pp 207–232

Ung L, Pattamatta U, Carnt N, Wilkinson-Berka JL, Liew G, White AJ (2017) Oxidative stress and reactive oxygen species: a review of their role in ocular disease. Clin Sci 131(24):2865–2883

Volkov RA, Panchuk II, Mullineaux PM, Schoffl F (2006) Heat stress-induced H_2O_2 is required for effective expression of heat shock genes in *Arabidopsis*. Plant Mol Biol 61(4-5):733–746

Voothuluru P, Sharp RE (2013) Apoplastic hydrogen peroxide in the growth zone of the maize primary root under water stress. I. Increased levels are specific to the apical region of growth maintenance. J Exp Bot 64(5):1223–1233

Voss I, Sunil B, Scheibe R, Raghavendra AS (2013) Emerging concept for the role of photorespiration as an important part of abiotic stress response. Plant Biol 15(4):713–722

Wang F, Chen ZH, Liu X, Colmer TD, Shabala L, Salih A, Zhao M, Shabala S (2017) Revealing the roles of GORK channels and NADPH oxidase in acclimation to hypoxia in *Arabidopsis*. J Exp Bot 68(12):3191–3204

Wongshaya P, Chayjarung P, Tothong C, Pilaisangsuree V, Somboon T, Kongbangkerd A, Limmongkon A (2020) Effect of light and mechanical stress in combination with chemical elicitors on the production of stilbene compounds and defensive responses in peanut hairy root culture. Plant Physiol Biochem 157:93–104

Xie X, He Z, Chen N, Tang Z, Wang Q, Cai Y (2019) The roles of environmental factors in regulation of oxidative stress in plant. Biomed Res Int 2019:9732325. https://doi.org/10.1155/2019/9732325

Yamane K, Taniguchi M, Miyake H (2012) Salinity-induced subcellular accumulation of H_2O_2 in leaves of rice. Protoplasma 249(2):301–308

Yousuf PY, Hakeem KUR, Chandna R, Ahmad P (2012) Role of glutathione reductase in plant abiotic stress. In: Ahmad P, Prasad MNV (eds) Abiotic stress responses in plants. Springer, New York, NY, pp 149–158

Zafra A, Castro AJ, de Dios Alche J (2018) Identification of novel superoxide dismutase isoenzymes in the olive (*Olea europaea* L.) pollen. BMC Plant Biol 18(1):114. https://doi.org/10.1186/s12870-018-1328-z

Zandalinas SI, Balfagón D, Arbona V, Gomez-Cadenas A (2017) Modulation of antioxidant defense system is associated with combined drought and heat stress tolerance in citrus. Front Plant Sci 8:953. https://doi.org/10.3389/fpls.2017.00953

Zhang M, Smith JAC, Harberd NP, Jiang C (2016) The regulatory roles of ethylene and reactive oxygen species (ROS) in plant salt stress responses. Plant Mol Biol 91(6):651–659

2 ROS Generation in Plant Cells Orchestrated by Stress

2.1 Introduction

Plants are immobile organisms and lifetime remain intact to the place of germination. Due to static nature, plants constantly confront multifactorial abiotic and biotic elements. The abiotic elements comprehend physical and/or chemical units; while biotic elements include biological entities. Certain prevalent abiotic and biotic factors are antagonistic to plants and impose stress on them. The hostile environments alter plants' cellular metabolism, leading to reduced performance and low productivity. Unfavorable abiotic stimulus spanning water scarcity, waterlogging, nutrient deficiency, salinity, heat, chilling/freezing, and pollutants are ubiquitous stresses which affect most of the plant species growing on the globe, once or many times during their life-cycle (Sachdev et al. 2021). Moreover, biotic factors embracing pathogenic (bacterial, viral, fungal, nematodes) and herbivory attack; and competition with weeds also impose stress on plants, threatening their ability to survive (Rejeb et al. 2014; Pandey et al. 2017). Besides, biotic stress incurs more yield losses as it affects both pre-harvest and post-harvest products. The consequence of stresses on plants is often dependent on type of imposing stressful conditions. For instance, certain stresses such as extreme temperature events display their impact within few seconds to few hours after onset; stresses like water deficiency, pathogen attack takes few days to weeks to show their effects; and metal pollution or nutrient deficiency requires months to express their detrimental effects (Maurya 2020). Plants experience stresses that impact either individually or concurrently. Concurrent occurrence of stress often leads to severe damages, and even compels plants to struggle strenuously to survive due to trade-off between retaliations necessary to endure perilous conditions and increased susceptibility toward other stresses (Dresselhaus and Hückelhoven 2018; Sharma et al. 2019; Bulgari et al. 2019). The frequency of abiotic and biotic stresses and their intensities has been exacerbated recently and become a common phenomenon as a consequence of unsustainable human activities and climate change (Hasanuzzaman et al. 2020;

S. Sachdev et al., *Reactive Oxygen Species in Plants*,
https://doi.org/10.1007/978-981-19-9884-3_2

Ramegowda et al. 2020; Roeber et al. 2021). In particular, rise in global temperature has introduced plant diseases to new geographical areas, exposing plants to outlandish conditions for which they are not well-equipped (Dresselhaus and Hückelhoven 2018). Similarly, various anthropogenic activities such as industrial waste dumping, use of agrochemicals, mining, etc. have contributed to the accumulation of heavy metals and other xenobiotic pollutants in the soil. These heavy metals discharged into the soil ecosystem, are non-essential elements and show toxicity, resulting in declined growth and productivity (Rizvi and Khan 2018; Seneviratne et al. 2019; Sychta et al. 2021). Thus, there are more chances for abiotic and biotic stresses likely to occur in future, which could threaten the global agricultural productivity, intensifying the problem of food shortage.

Exposure of plants to abiotic and biotic stresses elicits various responses including over-production of ROS (Petrov et al. 2015; Maurya 2020). The common ROS generated within plant cell include $O_2^{\bullet-}$, $^{\bullet}OH$, H_2O_2, and $^1O_2$ (Iqbal et al. 2021). Inception of stress triggers ROS formation in plant cell via activation of multiple pathways in different cell compartments (Table 2.1). ROS usually generate through action of specific ROS-producing enzymes, in particular, nicotinamide adenine dinucleotide phosphate (NADPH) oxidase, germin-like oxalate oxidases, fatty acid oxidase, amino oxidase, and cell wall peroxidases; and/or due to over-reduction/impairment of photosynthetic and mitochondrial electron transport chain (ETC) (Pogány et al. 2006; Jalmi and Sinha 2015; Maurya 2020). For instance, under high light stress the excitement of chlorophyll molecules by excessive energy, generate $^1O_2$ in chloroplast (Pogány et al. 2006). Accumulation of the ROS in plants causes oxidative damage to biomolecules triggering oxidative stress to render cellular redox homeostasis (He et al. 2018). ROS show high affinity for lipid, protein, nucleic acid and carbohydrate molecules, denaturing their structure and functions (Salehi-Lisar and Bakhshayeshan-Agdam 2020). Due to structural damage and functional loss, plant growth and development is compromised, often resulting in reduced productivity and may threaten the survival of the plant as well (He et al. 2018). Recent studies have evidently demonstrated that the controlled production of ROS in cell proclaims positive impression on plants by mediating signaling pathways (Bose et al. 2014). Nevertheless, the detrimental footprints of uncontrolled ROS generation on plants are distressing and claim high agricultural productivity loss worldwide. Moreover, different stresses trigger different ROS signatures in subcellular compartments and ultimately determine their specific ill effects on plants (Choudhury et al. 2017). Thus, systemic studies on onset of multiple stresses could provide a clear picture on ROS signature in different compartments, which may help in developing improved strategies for ROS management below their damaging effects to render plants to thrive under future projected stress scenarios.

Table 2.1 Prime sites and major pathways leading to ROS production under various stress conditions

Natural and Anthropogenic factors	Stress	Prime site (s) of ROS generation	Major pathway (s) of ROS generation	References
Excessive greenhouse gases level Extreme temperature events (high) Fluctuation in light intensity Increase UV radiation exposure	Heat stress	Chloroplast, mitochondria, peroxisomes	Leakage of electrons from photosynthetic electron transport chain (ETC) and/or reduction of respiratory ETC; elevated photorespiration process generating H_2O_2 in peroxisomes	Hasanuzzaman et al. (2012), You and Chan (2015), Jalil et al. (2017), and Maurya (2020)
Extreme temperature events (low) Fluctuation in light intensity	Chilling/ freezing stress	Chloroplast, mitochondria, peroxisomes	Leakage of electrons from photosynthetic and mitochondrial ETC; peroxisomal enzymes activity	Suzuki and Mittler (2006), Coates et al. (2016), and He et al. (2021)
Irregular and reduced spells Freshwater scarcity	Water-deficit stress	Chloroplast, mitochondria, peroxisomes	Inhibition of photosynthetic and mitochondrial ETC; increased photorespiration; peroxisomal enzymes activity; peroxisomal ETC electron leakage	Sharma et al. (2012), Demidchik (2015), Jalil et al. (2017), and Maurya (2020)
Irregular and high intense spells Poor drainage	Waterlogging stress	Chloroplast, mitochondria	Inhibition of photosynthetic and mitochondrial ETC	Sasidharan et al. (2018) and Anee et al. (2019)
Irregular and reduced spells Sea water intrusion; Freshwater scarcity High temperature Excessive use of agrochemicals	Salinity stress	Chloroplast, mitochondria, peroxisome	Electron leakage from photosynthetic and mitochondrial ETC; increased photorespiration; peroxisomal enzymes activity	Bose et al. (2014), Parihar et al. (2015), and Jalil and Ansari (2020)
Excessive use of agrochemicals Nutrient leaching	Nutrient deficiency stress	Chloroplast, mitochondria, plasma membrane	Activity of nicotinamide adenine dinucleotide phosphate	Jung et al. (2009), Hajiboland (2012), and Gupta et al. (2017)

(continued)

Table 2.1 (continued)

Natural and Anthropogenic factors	Stress	Prime site (s) of ROS generation	Major pathway (s) of ROS generation	References
Salinity Water-deficit condition			(NADPH) oxidases; increased activation of ethylene signaling cascade; over-reduction of ETC	
Excessive use of agrochemicals Pollution	Heavy metals/ Pesticides stress	Chloroplast, mitochondria, peroxisomes, apoplast	Over-reduction of photosynthetic ETC causing photoinhibition and generation of ROS; increased availability of transition metals to catalyze reaction for $^{\bullet}OH$ radical generation; peroxisomal enzymes activity	Pandey et al. (2009), Steffens (2014), Trinh et al. (2014), Yuzbaşıoglu and Dalyan (2019), and Sharma et al. (2020)
Increase UV radiation exposure Fluctuation in light intensity	Light stress	Chloroplast, mitochondria, peroxisome	Photoinhibition and energy transfer from triplet chlorophyll to O_2, resulting in formation of $^1O_2$; over-reduction of mitochondrial ETC; peroxisomal ETC electron leakage	Demidchik (2015), Pospisil (2016), Yang et al. (2019), and Sharma et al. (2020)
Invasion and extensive growth of weeds Increase resistance in pests and pathogens	Biotic stress	Apoplast, Plasma membrane, peroxisome	Activity of enzymes including NADPH oxidases, amine oxidase, cell wall peroxidases, and oxalate oxidases; activity of xanthine oxidase and NADH/NADPH-dependent small ETC in peroxisomal membrane	Bolwell et al. (2002), Sharma et al. (2012), and Demidchik (2015)

2.2 Abiotic Stress and ROS

Abiotic factors such as temperature, water, salt, light, radiation, metal(loid)s, and nutrients act as stress when their intensity deviates from optimum, incurring pressure and debt on plant performance. Approximately world's 90% of arable lands have been engulfed with one or more kinds of stress which can affect major food crops and cause nearly 70% productivity loss at global scale (Waqas et al. 2019). Exacerbated occurrence of abiotic stress is common due to the gradual climate change, which in future may expand the strength and regularity of stress incidence (Kumar 2020). Plant response to environmental stress is very complex as well as dynamic, and may be reversible (elastic) or irreversible (plastic) (Cramer et al. 2011). The impression of abiotic stress on plant growth and productivity is determined by the stress involved, duration and intensity of stress, plants' genetic makeup, tissue or organ affected and the stage of plant's growth (Cramer et al. 2011; Razzaque et al. 2019). Normally, slight variation in abiotic factors temporarily restricts plant growth. However, extreme variation impinges plant's physiological performance and alter genomics status to the extent that abridges plant growth and development (Raza et al. 2019; Saijo and Loo 2020; Iqbal et al. 2021). The metabolic activities including photosynthesis and respiration lead to ROS production in sub-cellular organelles (chloroplast, mitochondria, and peroxisomes) at low concentration under normal condition (You and Chan 2015). But under stress stimuli production of ROS increases dramatically which hinders vital cellular processes, in particular, transpiration, respiration, and photosynthesis; and disrupts enzymatic activities, paralyzing plant productivity (Sachdev et al. 2021). The cellular accumulation of ROS exceeding the capacity of scavenging antioxidants under stress disturbs the cellular redox homeostasis consequently inducing oxidative stress owing to their ability to cause peroxidation of membrane lipids, proteins degradation, enzymes deactivation, and nucleic acid denaturation (You and Chan 2015; Chaki et al. 2020) (Fig. 2.1).

2.3 Extreme Temperature Events and Stress

Temperature is one of the crucial abiotic factors which influences plant growth and productivity. Plant growth and developmental process are very sensitive to temperature (Hasanuzzaman et al. 2013). Variation in temperature below or above optimum level induces cold or heat stress in plants, respectively (Maibam et al. 2020). An onset of temperature-related stresses triggers ROS production that affects the photosynthetic performance of plant, impairs plant architecture, reproduction, and fruit setting consequently reducing plant yield (Zaidi et al. 2014; Martinez 2016; Maurya 2020). ROS in chloroplast causes photoinhibition which diminishes the assimilation of carbon dioxide, leading subsequently to poor growth and productivity (Maurya 2020). Under both high and low temperature stress situation, plants demonstrate increased production of antioxidants and loss of chlorophyll content. However, an elevated temperature inflicts more dry weight loss than low temperature stress

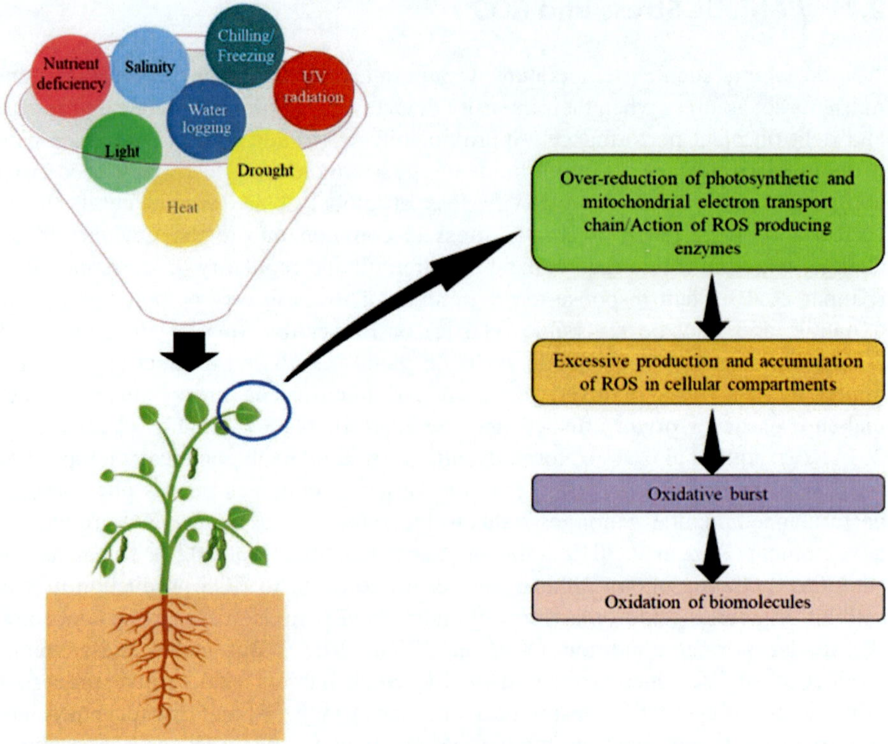

Fig. 2.1 The over-production of ROS under action of abiotic stresses and their effects on plants

possibly due to the inactivation of photosynthetic enzymes such as Rubisco as reported in *Brassica oleracea* (Soengas et al. 2018).

High Temperature/Heat Stress Heat stress is a leading environmental stress which limits plant growth and yields worldwide. Elevated levels of greenhouse gases in earth's atmosphere, high light intensities, and heat waves are the factors responsible for the occurrence of heat stress. The rise in global air temperature is anticipated to be 0.2 °C per decade which will result in 1.8–4 °C more increment in earth's temperature than present condition by the end of the twenty-first century (Hasanuzzaman et al. 2013). Plants encountered with heat stress show alteration in morphological, cellular, and metabolic performance, including petiole and hypocotyls elongation, reduced photosynthetic and respiratory activity, diminished enzymatic activities, up-regulated transcription and translation of heat shock proteins (HSP), elevated calcium influx, and ROS over-production (Bita and Gerats 2013). The stress-induced effects, sometimes, render plant growth severely, leading to low productivity and even plant death (Awasthi et al. 2015; Maibam et al. 2020). Nevertheless, the impact of heat stress depends upon the intensity of stress as well as on plant type (Hasanuzzaman et al. 2013; Awasthi et al. 2015). Heat stress also increases the susceptibility of plants toward biotic stress (Pandey et al. 2015; Suzuki and Katano 2018). High temperature escalates the production of ground-level ozone (O_3) which itself becomes an abiotic stress triggering over-production of ROS in plants (Coates et al. 2016).

Heat stress induces uncoupling of cellular enzymes and metabolic pathways that builds ROS concentration in the cell (Hasanuzzaman et al. 2013). High-temperature stimulus prompts closure of stomata which reduces carbon dioxide supply and fixation in photosynthetic apparatus and subsequently results in the production of H_2O_2 in peroxisome due to metabolic shift in the direction of photorespiration process. Along with this, over-production of $O_2^{\bullet -}$, H_2O_2, and $^1O_2$ attributable to over-reduced photosynthetic ETC also occurs (You and Chan 2015). ROS formed in thylakoid membrane disrupt the stability of cell membrane and enhances the production of malondialdehyde (MDA) (Maibam et al. 2020). High H_2O_2 concentration in *Phaseolus vulgaris* at elevated temperature (46–48 °C) has been observed to cause peroxidation of lipid membrane and increased accumulation of MDA (Nagesh Babu and Devaraj 2008). The injury-inducing impact of high-temperature stress on plants is either caused due to generation of ROS or reduced action of defensive antioxidant enzymes (Awasthi et al. 2015). In sorghum, high temperature (or heat stress) condition inactivates antioxidant enzymes, e.g., SOD by 22%, POX by 25%, and CAT by 15% compared to control plants (Djanaguiraman et al. 2018).

Low Temperature/Cold Stress Low temperature/cold stress is also another category of extreme temperature events. Low temperature includes both chilling and freezing stresses occurring at temperature below 20 °C and below 0 °C, respectively (Awasthi et al. 2015). Chilling/freezing stress are most damaging abiotic stresses affecting almost every aspect of physiology and biochemistry of plants, significantly reducing their growth, development, and productivity (Thakur et al. 2010; Awasthi et al. 2015; Megha et al. 2018). The low-temperature stress (chilling and freezing) hampers the photosynthetic as well as non-photosynthetic activities of the cell (Shewfelt 1992). It increases the solubility and accumulation of O_2, and promotes electron leakage from the photosynthetic and respiratory ETC, aggravating ROS production (Coates et al. 2016). Over-production of ROS under cold stress affects protein and lipid molecules that increase membrane fluidity and decrease enzyme activity (Jalil et al. 2017). Due to chilling stress in cucumber seedlings, increased electrolyte leakage along with reducing tissue water and chlorophyll content has been reported by Zhang et al. (2012). Similarly, under low-temperature, over-production of H_2O_2 and $O_2^{\bullet -}$ in tomato leaves has been reported with enhanced MDA level and *RBOH1* (*respiratory burst oxidase homolog 1*) expression. In maize, cold stress-induced tissue dehydration, provoked membrane disruption, and condensed growth and development due to increase in MDA level as an output of lipid peroxidation have been observed (Farooq et al. 2009). Low temperature severely damages photosynthetic apparatus (Liu et al. 2018); induces photoinhibition, reduces the activity of certain enzymes involved in Calvin cycle and uncouples thylakoids in plants sensitive to chilling conditions (Liu et al. 2018). Chilling stress perception triggers a series of events that disrupts ultrastructure and development of chloroplasts, impairing function such as physiological and biochemical activities (Liu et al. 2018). Exposure of *Vigna radiata* (mung bean) to low temperature damages PS II and diminishes photochemical efficiency due to photoinhibition and chloroplast degradation (Saleh 2007). Chilling stress in suspension-cultured mung bean cells perpetrates plastid enlargement and lipid drop buildup, eventually resulting in disruption of plastid (Ishikawa 1996).

2.4 Water-Deficit/Logging Stress

Water is the most limiting resource for plant productivity. Its availability in adequate amount is mandatory for the optimum growth and yield of agricultural crops. Precipitation is the major and natural source for the supply of fresh water to terrestrial plants. A gradual change in climate in the last decades has, however, enormously influenced the rainfall patterns causing erratic precipitation with altered magnitude and seasonal variations (Feng et al. 2013). The extreme changes have led to water deficit (drought) or waterlogging (flooding) situations at many locations on the globe. The water deficit condition is characterized by low water potential and turgor which may interfere with normal physiological and biochemical processes of the plant (Shao et al. 2008). The consequences of water deficit greatly vary with the severity and time length of the stress. It has been anticipated that the drought-stricken agricultural area is expected to increase twofold by 2050, reducing agricultural productivity significantly (Kumar et al. 2019). Water deficit situations accelerate the rate of stomatal closure and reduce CO_2 assimilation, promoting photoreduction of O_2 in the chloroplast. Consequently, photorespiration process accelerates that subsequently produces excess ROS, bringing plants under oxidative stress (Jalil et al. 2017). Sunflower cultivars grown under water deficit conditions produce low fresh and dry mass, reduced chlorophyll content with increased H_2O_2 and MDA content (Rasheed et al. 2020). Under water-deficit conditions, white clover leaves exhibit elevated MDA content and reduced dry mass together with over-production and accumulation of ROS (Lee et al. 2009).

Waterlogging is a state of soil saturation to such an extent that restricts air phase in soil switching to anaerobic condition. Waterlogging has become a major threat for agricultural crops worldwide due to alteration in rainfall pattern and magnitude. Along with erratic rainfall, poor drainage system, and low soil organic matter are some major factors that has amplified incidents of waterlogging. The consequences of waterlogging are determined not only on the basis of the severity of stress but on multiple factors including plant type, plant age, soil condition, prevailing temperature, humidity, and water level (Anee et al. 2019). Waterlogging causes morphological, physiological, and anatomical alterations in plants, all of which are detrimental to plant growth and development. Waterlogging, for instance, leads to wilting, chlorosis, necrosis, leaf senescence, stunted growth, delayed maturation, etc., and also makes susceptibility toward biotic stress (Anee et al. 2019). Waterlogging for a short duration creates hypoxia (deficiency of O_2) and for long duration, anoxia (absence of O_2). Hypoxia/anoxia suppresses the aerobic activity in the soil such as root respiration (Sasidharan et al. 2018), which in turn affect the energy status of the root cells, and induce root injury (Anee et al. 2019). Deficient O_2 conditions during waterlogging inhibit photosynthetic and mitochondrial ETC and consequently produce ROS to the level harmful to the plants (Chang et al. 2012; Sasidharan et al. 2018; Anee et al. 2019). Waterlogging also invokes leaching of essential nutrients from the soil, accumulation of salts, and elevated availability of heavy metals due to alteration in soil pH. These changes aggravate stress conditions and trigger nutrient deficiency, salinity, heavy metal, and oxidative stress in plants (Steffens 2014). Various plant species encountered with waterlogging scenarios have demonstrated altered physiological and growth processes, leading to impaired reproductive development and diminished yield (Ezin et al. 2010; Paltaa et al. 2010; Rasaei et al. 2012; Kumar et al. 2013). Anee et al. (2019) have demonstrated the onset of oxidative stress in sesame plants subjected to waterlogging that includes the accumulation of ROS and methylglyoxal content as well as increase in lipid peroxidation. Jerusalem artichoke exposed to waterlogging for 7 days expresses PS I and PS II photoinhibition which significantly reduces the photochemical efficiency and stability of photosynthetic apparatus (Yan et al. 2018).

2.5 Salinity Stress

Soil salinity is the utmost substantial environmental stress impinging 20–50% of crop yield as well as soil fertility worldwide (Yuan et al. 2015; He et al. 2018). Nearly 20% of the world's cultivable land has been affected by salinity (Etesami and Noori 2019). By the end of 2050, more than 50% of arable land would expectedly be salinized (Etesami and Noori 2019; Kumar et al. 2020). The annual global agricultural productivity loss associated with soil salinity is more than US$12 billion and is still rising (Bose et al. 2014). Climate change driven sea level rise, directed movement of salt water to the cultivable lands, increasing their salinity and exacerbating debt of salt stress on plant cultivation (Chima et al. 2015). Other environmental factors such as water scarcity (due to uneven precipitation and recessing water table) and high temperature promoting surplus water evaporation has also intensified the scenario of soil salinity (Vinke et al. 2017; Zhang et al. 2018). Salinity stress is caused by an overabundance of Na^+ and Cl^- ions, which create hypertonic conditions and cause osmotic stress (Ilangumaran and Smith 2017). Plant's homeostasis is hampered by induced osmotic stress, which impacts physiological processes such as germination, growth, photosynthesis, nutrient uptake and translocation, and yield (resulting in nutritional deprivation), and hormonal balance (Parihar et al. 2015; Ilangumaran and Smith 2017). Intake of excessive salt by plants may lead to accumulation of ions like Na^+ and Cl^- causing ion toxicity which hinders plant's ability to uptake essential nutrients, in particular, N, K^+, Ca^{2+}, and others inducing nutrient deficiency stress (Parihar et al. 2015). For instance, *Vicia faba* exposed to salinity stress increases accumulation of Na^+ altering Na^+/K^+ ratio (Gadallah 1999). Under the influence of salt stress, ability of plant to absorb water declines inducing drought-like circumstances (Jalil and Ansari 2020) which eventually shrink stomatal conductance and assimilation of CO_2, disrupting photosystem functioning, deactivating photosynthetic enzymes, and promoting superfluous production of ROS in plants (Hasanuzzaman et al. 2018; Riyazuddin et al. 2020). Moreover, a limited availability of CO_2 in plant triggers photorespiration which also participates in production of excess ROS (Bose et al. 2014). During salt stress, over-reduction of the ubiquinone pool in mitochondria induces electrons leakage from mitochondrial ETC complexes I and III to O_2, resulting in $O_2^{\bullet-}$ generation (Miller et al. 2010; Bose et al. 2014). Over-production of ROS under salinity alters structure and function of biomolecules and may even trigger programmed cell death (PCD) (Kumar et al. 2020). Salinity stress in wheat plant induces excess production of ROS resulting in lipid peroxidation and altered membrane stability that eventually led to electrolyte leakage (Kaur et al. 2017).

2.6 Nutrient Deficiency Stress

Optimum availability of essential nutrients is necessary for proper growth and development of the plant. However, due to increasing pressure on land to obtain more food products to feed growing population and input of unsustainable agricultural approaches has declined the nutrient reserve of the soil. Moreover, constantly changing climatic conditions have further aggravated the challenge of availability of adequate amount of plant nutrients in soil. Concurrent occurrence of abiotic stress like water deficiency, waterlogging, high temperature, toxic elements particularly Na^+ and excessive light triggers the adversity of nutrient deficiency in plants (Jung et al. 2009; Guo et al. 2016). Scarcity of nutrients in soil can attenuate the physiological activities of plant, in particular, ETC mechanism in chloroplast and mitochondria, water relation and gaseous exchange, triggering ROS production and oxidative stress as well (Hajiboland 2012). One of the major routes for ROS generation in plant cell is plasma membrane-bound NADPH oxidase enzyme activity that depends upon the availability of plant nutrients, particularly, Zn and K (Gupta et al. 2017). Hence, scarcity of these nutrients exacerbates the activity of NADPH oxidase enzyme resulting in production of $O_2^{\bullet-}$ (Hajiboland 2012) or H_2O_2 (Shin et al. 2005). Plants suffering with the dearth of nutrients also show increased production of ROS through ethylene signaling cascade. K deficient condition enhances biosynthesis of ethylene, eliciting over-production of ROS (Jung et al. 2009). Certain nutrients like N, Cu, Mg, Mn, Fe, and Zn are vital components of several enzymes such as Cu/Zn-SOD, Mn-SOD, etc., that perform multifarious function within cell, for instance, participation in energy metabolism, quenching of excessive ROS, and maintenance of cellular redox homeostasis (Shin et al. 2005; Tewari et al. 2006; Hajiboland 2012). Thus, nutrient deficiency impairs antioxidant machinery of plants to quench excess ROS, thereby indirectly accelerating the accumulation of ROS and oxidative stress (Tewari et al. 2006; Hajiboland 2012). For instance, abridged ability of enzymes to scavenge H_2O_2 and $O_2^{\bullet-}$ results in production of $^{\bullet}OH$ radical mediated through Heber–Weiss reaction (Hajiboland 2012). Deficiency of Mg, which is a major constituent of chlorophyll has also been found to be related to impaired activity of photosynthetic apparatus and ultimately encourages the formation of H_2O_2 (Tewari et al. 2006; Guo et al. 2016).

2.7 Metal/Metalloids Stress

Soil is the major reserve for the number of metals and metalloids encompassing copper (Cu), chromium (Cr), cobalt (Co), Cadmium (Cd), iron (Fe), nickel (Ni), manganese (Mn), magnesium (Mg) molybdenum (Mo), zinc (Zn), lead (Pb), mercury (Hg), vanadium (V), thallium (Tl), bismuth (Bi), tin (Sn), aluminum (Al), tungsten (W), antimony (Sb), silicon (Si), arsenic (As), and many more (Ali et al. 2019). Among these metal(loid)s, certain are essential for plants while others are non-essential and are considered toxic to living organisms, in particular, heavy metals (metals with atomic mass more than 50) like Cr, Cd, Ni, Pb, Al, Hg; and metalloids such as As (Ali et al. 2019). Even essential metals with known function may cause harmful effects on plants when present above the threshold limit (Angulo-Bejarano et al. 2021). Threshold limit denotes the maximum concentration of metal (loid)s in plant's environment, below which it has either non-toxic or beneficial role in plant growth and development (Ali et al. 2019; Angulo-Bejarano et al. 2021). The concentration exceeding the threshold limits may induce negative impact on plant health, even leading to plant death (Ali et al. 2019). Metals and metalloids are naturally present in soil; however, anthropogenic activities, in particular, industrial waste dumping, mining, etc., have raised their concentration several folds up (Xalxo et al. 2020; Angulo-Bejarano et al. 2021). Occurrence of metal(loid)s at high concentration, affect the growth, photosynthesis, biomass, and metabolic activity of plants due to over-production of ROS (Xalxo et al. 2020). Within plant cell chloroplast, mitochondria, apoplast, and peroxisome are the major sites of ROS formation under the influence of metal stress (Pandey et al. 2009; Steffens 2014; Trinh et al. 2014). The major-redox active metals like Fe, Cr, and Cu, which are considered as essential plant elements can induce metal toxicity and oxidative stress on excessive accumulation (Schutzendubel and Polle 2002; Steffens 2014). Fe in excessive amount can lead to the over-production of ROS via chain of events (Becana et al. 1998) leading damage to lipid membrane and chlorophyll (Hajiboland 2012). The reduced form of Fe within cellular environment, gets oxidized and generates H_2O_2 and $O_2^{\bullet-}$ via auto-oxidation (Schutzendubel and Polle 2002). The H_2O_2 oxidizes another reduced Fe molecule and generates highly toxic $^{\bullet}OH$ radical (Hajiboland 2012). Similar to Fe, Goodarzi et al. (2020) have reported phytotoxicity of Zn in safflower seedling that reduces biomass production and total chlorophyll content followed by oxidative damage due to over-production of H_2O_2, and increased MDA content and lipoxygenase activity. Cu, an essential metal, mediates the activity of enzymes such as superoxide dismutase, ascorbate oxidase and cytochrome oxidase. However, its excess disturbs redox equilibrium of the cell by generating ROS and triggering lipid peroxidation (Angulo-Bejarano et al. 2021). Moreover, ROS generated via Cu induced stress, react with thiol group through Fenton reaction, resulting in impairment of protein structure and function, damage to nucleic acid, lipids and other biomolecules, change in antioxidant defensive system, and altered metabolic pathways (Yadav et al. 2005; Hänsch and Mendel 2009; Viehweger 2014; Angulo-Bejarano et al. 2021). The non-essential metals induce toxicity in plants through ROS formation which disrupts the structure of cellular proteins and enzymes. For instance, Cd triggers over-production of ROS in plant cells by replacing Cu or Fe ions in antioxidant metalloenzymes which impair respiratory ETC, and disturb redox status of cell (Gupta et al. 2017). Wakeel et al. (2020) have reviewed the effect of Cr toxicity on plants and concluded that plants grown under Cr stress result in the formation of ROS which suppresses enzymatic antioxidant defense system, induces cytotoxicity and genotoxicity and causes alteration in ultrastructure and photosynthesis. As also induces oxidative stress in plants either on methylation or due to change in the oxidation state (Sharma 2012). Reduction of As in chloroplast and mitochondria changes oxidation state from arsenate to arsenite via action of cytochrome/cytochrome oxidase and triggers oxidative stress in plants (Sharma 2012).

2.8 Pesticides-Induced Stress

Like metals/metalloids, the residual chemical pesticides generate oxidative stress via over-production of ROS in plants (Sharma et al. 2020). The estimation indicates that out of the total pesticides applied, only 1% actually reaches to its target while the rest of the pesticides remain in the agricultural environment, contaminating soil, water, and the food chain (Sachdev and Singh 2018). Pesticides are well-known xenobiotic compounds which can retard plant growth and photosynthetic efficiency, resulting in molecular alterations, over-production of ROS, and abridge antioxidant machinery of plant (Yuzbaşıoglu and Dalyan 2019; Sharma et al. 2020). Many plants treated with high concentration of chemical pesticides display signs of toxicity and oxidative stress. For instance, tomato plant treated with chemical pesticide-Thiram increases H_2O_2 and MDA content, along with diminished chlorophyll content (Yuzbaşıoglu and Dalyan 2019). In another study, tomato seedling subjected to high concentrations of three different pesticides (imidacloprid, alpha-cypermethrin, and emamectin benzoate) accumulates significant ROS, elicits membrane damage due to lipid peroxidation and produces thiobarbituric acid reactive substances (TBARS), cellular injury and reduced root and shoot cell viability (Shakir et al. 2018). Sharma et al. (2019) have reported similar negative effects of pesticide imidacloprid in *Brassica juncea*. Pesticides can affect the photosynthetic apparatus of plant by causing degradation of photosynthetic (chlorophyll) pigments, which occur due to reducing stability of pigment-protein complex of photosynthetic apparatus (Dias et al. 2014). Pesticides also inhibit the transport of electron between primary quinone acceptor Q_A and secondary quinone acceptor Q_B of photosynthetic ETC, thus eliciting production of ROS (Devine and Shukla 2000). ROS subsequently reacts with lipid, nucleic acid, and protein molecules, ultimately resulting in oxidative stress. Triazine group of pesticides obstructs electron transport from PS II to PS I, limiting Q_A oxidation and, as a result, inhibiting photosynthesis by lowering maximum quantum yield (Kleczkowski 1994; Ralph 2000). Pesticides also affect the stability of D1 protein of PS II which cause photo-destruction of chlorophyll and trigger the formation of ROS (Sharma et al. 2020).

2.9 Light Stress

Light is a crucial abiotic factor that plays a key role in determining a plant's growth and development (Wang et al. 2017; Sharma et al. 2020). It is the main operating factor for the photosynthetic apparatus. Thus, any fluctuation in light intensity, below or above plant's requirement, induces low light or high light stress, respectively, impinging plant's physiological and biochemical functions. High light stress characterized by excessive light intensity occurs when the absorption of light is more than the photosynthetic requirement of the plant (Demmig-Adams and Adams 1992; Sharma et al. 2020), whereas low light stress is a condition of inadequate light intensity for driving activity of photosynthetic apparatus. Fluctuation in light intensities results in abridged photosynthetic performance along with poor growth and development (Yang et al. 2019). PS II and its reaction center are sensitive to fluctuation in light intensities and therefore are the prime site for photoinhibition (Yang et al. 2019). High light intensity causes over-loading of the light harvesting complex (LHC) system, which in turn generates surplus amount of ROS, due to generation of unused excitation energy. The excess ROS eventually leads to photo-oxidative damage and photoinhibition (Gururani et al. 2015; Pascual et al. 2017; Yang et al. 2019; Sharma et al. 2020). Under low light or insufficient light intensity, photosynthetic process gets conditioned, resulting in leaf etiolation, abridged photosynthetic efficiency, imbalance in carbon assimilation, reduced nitrogen metabolism, lowered stomatal conductance, water-use efficiency, and increased oxidative stress via over-production of ROS due to increased flow of electron to O_2 during photosynthetic and respiratory ETC process (Solymosi and Schoefs 2010; Li et al. 2017; Yang et al. 2019; Sharma et al. 2020). Prolonged exposure of *Brassica campestris* ssp. *Chinensis Makino* to low light stress results in enhanced accumulation of MDA content and significantly decreases the chlorophyll a, anthocyanin, and soluble protein content, stomatal conductance, net photosynthetic and transpiration rate, and enzymes activities (Zhu et al. 2017).

2.10 Biotic Stress and ROS

Occurrence of biotic stress is a ubiquitous phenomenon as plants often encounter several pests and pathogens during their life cycle. It has been estimated that on an average 26% of the plant yield is lost annually due to pest infestation and disease incidences (Piquerez et al. 2014). Moreover, it has been predicted that due to climate change, a poleward moment of pathogens could take place (Bebber et al. 2013), which may result in the development of new diseases that could have worst impact on global food security (Piquerez et al. 2014). ROS generated in plant cell as a response to unfavorable condition, display defensive role under biotic stress condition. Plants as a common response to pathogen attack trigger oxidative burst. ROS burst predominantly occurs at apoplast and at plasma membrane under biotic stress on action of NADPH oxidase enzyme (Bolwell et al. 2002), but may also originate in chloroplast and mitochondria (Sharma et al. 2012). The ROS burst under biotic stress is required for the development of resistance responses. When pathogens invade, plants trigger a resistance response that includes activation of local and systemic defenses as well as induction of a localized PCD at the infection site, known as hypersensitive response (HR). In plants when pathogen assaults, ROS scavenging enzymes are primarily downregulated, similar to what has been reported for alternative oxidases (AOX), which correspond to increasing ROS levels, triggering PCD to stop pathogen transmission.

ROS play positive role in plant protection against disease incidence. ROS act as antimicrobial agents and previous studies have demonstrated that accumulation of ROS in plants occurs as an early event during plant and pathogen interaction which facilitates control and inhibition of pathogen growth (Camejo et al. 2016). Production and accumulation of ROS on perception of pathogenesis activates PCD in region surrounding infected cells, thus quarantine biotrophic pathogens that thrives on living tissues, in specified zone and restrict their spread to non-infected areas (Andersen et al. 2018). ROS can kill invading bacterial phytopathogens directly (Paiva and Bozza 2014) and subsequently improves thickening of neighboring cell walls to inhibit pathogen spreading (Wang and Higgins 2005). To prevent fungal pathogen invasion, plants have evolved pattern-recognition receptor (PRRs) for chitin perception which triggers intracellular activation of mitogen-activated protein kinase (MAPK) cascades, resulting in rapid generation of ROS (Kawasaki et al. 2017). In *Arabidopsis*, PRR-Associated Kinase BIK1 directly phosphorylates NADPH oxidase and causes the pathogen-associated molecular pattern (PAMP)-induced ROS burst and fortify immunity against pathogen (Kadota et al. 2014).

Inhibition of biotrophic pathogens and stimulation of necrotrophic pathogens by the accumulation of H_2O_2 has been reported by Kumar et al. (2011). The elevated level of H_2O_2 in transgenic potato plants expressing gene coding for glucose oxidase (GO) managed bacterial soft root and fungal late blight diseases (Wu et al. 1995). During activation of defense system, accumulation of H_2O_2 in plant cell is linked with biosynthesis of phytoalexin and cross-linking of $^{\bullet}OH$-proline rich cell wall glycoprotein (Bradley et al. 1992; Davis et al. 1993; Lamb and Dixon 1997), that increase strength of cell wall and restrict entry of pathogens (Camejo et al. 2016; Andersen et al. 2018). Further, ROS act as signaling molecules triggering additional immune response in plants against pathogenic infection including modulation in gene expression and closure of stomatal opening (Gilroy et al. 2014; Camejo et al. 2016).

2.11 Conclusion

Reactive oxygen species (ROS) production during onset of variable stresses is one of the most common phenomena taking place within the plant cell. Generation of ROS under influence of abiotic and biotic stresses is mediated through different pathways determining diverse signature in the cell. These ROS signatures triggered by specific stress in particular compartment may have either positive or negative impact on the plant metabolic and physiological process, ultimately affecting overall plant's growth and development. The precise understanding regarding activation of specific pathway mediating ROS production at particular cellular site could facilitate to elucidate the fate and impact of stress-induced over-production of ROS and plant performance. Thus, deep knowledge on ROS signature within cellular component is imperative for effective management of plant stresses and safeguard food security.

References

Ali MA, Fahad S, Haider I, Ahmed N, Ahmad S, Hussain S, Arshad M (2019) Oxidative stress and antioxidant defense in plants exposed to metal/metalloid toxicity. In: Reactive oxygen, nitrogen and sulfur species in plants: production, metabolism, signaling and defense mechanisms. Wiley, Newark, pp 353–370

Andersen EJ, Ali S, Byamukama E, Yen Y, Nepal MP (2018) Disease resistance mechanisms in plants. Genes 9(7):339

Anee TI, Nahar K, Rahman A, Mahmud JA, Bhuiyan TF, Alam MU, Fujita M, Hasanuzzaman M (2019) Oxidative damage and antioxidant defense in *Sesamum indicum* after different waterlogging durations. Plants 8(7):196

Angulo-Bejarano PI, Puente-Rivera J, Cruz-Ortega R (2021) Metal and metalloid toxicity in plants: an overview on molecular aspects. Plants 10(4):635

Awasthi R, Bhandari K, Nayyar H (2015) Temperature stress and redox homeostasis in agricultural crops. Front Environ Sci 3:11

Bebber DP, Ramotowski MA, Gurr SJ (2013) Crop pests and pathogens move polewards in a warming world. Nat Clim Chang 3(11):985–988

Becana M, Moran JF, Iturbe-Ormaetxe I (1998) Iron-dependent oxygen free radical generation in plants subjected to environmental stress: toxicity and antioxidant protection. Plant Soil 201(1): 137–147

Bita C, Gerats T (2013) Plant tolerance to high temperature in a changing environment: scientific fundamentals and production of heat stress-tolerant crops. Front Plant Sci 4:273

Bolwell GP, Bindschedler LV, Blee KA, Butt VS, Davies DR, Gardner SL, Gerrish C, Minibayeva F (2002) The apoplastic oxidative burst in response to biotic stress in plants: a three-component system. J Exp Bot 53(372):1367–1376

Bose J, Rodrigo-Moreno A, Shabala S (2014) ROS homeostasis in halophytes in the context of salinity stress tolerance. J Exp Bot 65(5):1241–1257

Bradley DJ, Kjellbom P, Lamb CJ (1992) Elicitor-and wound-induced oxidative cross-linking of a proline-rich plant cell wall protein: a novel, rapid defense response. Cell 70(1):21–30

Bulgari R, Franzoni G, Ferrante A (2019) Biostimulants application in horticultural crops under abiotic stress conditions. Agronomy 9(6):306

Camejo D, Guzmán-Cedeño Á, Moreno A (2016) Reactive oxygen species, essential molecules, during plant–pathogen interactions. Plant Physiol Biochem 103:10–23

Chaki M, Begara-Morales JC, Barroso JB (2020) Oxidative stress in plants. Antioxidants 9(6):481

Chang R, Jang CJ, Branco-Price C, Nghiem P, Bailey-Serres J (2012) Transient MPK6 activation in response to oxygen deprivation and reoxygenation is mediated by mitochondria and aids seedling survival in *Arabidopsis*. Plant Mol Biol 78(1):109–122

Chima UD, Ofodile EAU, Naaluba M (2015) Evaluation of NaCl salinity tolerance of *Artocarpus altilis* (Parkinson) Fosberg and *Treculia africana* Decne for climate change adaptation in agroecosystems. Int J Sci Eng Res 6(5):1516–1523

Choudhury FK, Rivero RM, Blumwald E, Mittler R (2017) Reactive oxygen species, abiotic stress and stress combination. Plant J 90(5):856–867

Coates J, Mar KA, Ojha N, Butler TM (2016) The influence of temperature on ozone production under varying NO x conditions–a modelling study. Atmos Chem Phys 16(18):11601–11615

Cramer GR, Urano K, Delrot S, Pezzotti M, Shinozaki K (2011) Effects of abiotic stress on plants: a systems biology perspective. BMC Plant Biol 11(1):1–14

Davis D, Merida J, Legendre L, Low PS, Heinstein P (1993) Independent elicitation of the oxidative burst and phytoalexin formation in cultured plant cells. Phytochemistry 32(3):607–611

Demidchik V (2015) Mechanisms of oxidative stress in plants: from classical chemistry to cell biology. Environ Exp Bot 109:212–228

Demmig-Adams B, Adams WW III (1992) Photoprotection and other responses of plants to high light stress. Annu Rev Plant Biol 43(1):599–626

Devine MD, Shukla A (2000) Altered target sites as a mechanism of herbicide resistance. Crop Prot 19(8-10):881–889

Dias MC, Figueiredo P, Duarte IF, Gil AM, Santos C (2014) Different responses of young and expanded lettuce leaves to fungicide Mancozeb: chlorophyll fluorescence, lipid peroxidation, pigments and proline content. Photosynthetica 52(1):148–151

Djanaguiraman M, Boyle DL, Welti R, Jagadish SVK, Prasad PVV (2018) Decreased photosynthetic rate under high temperature in wheat is due to lipid desaturation, oxidation, acylation, and damage of organelles. BMC Plant Biol 18(1):1–17

Dresselhaus T, Hückelhoven R (2018) Biotic and abiotic stress responses in crop plants. Agronomy 8(11):267

Etesami H, Noori F (2019) Soil salinity as a challenge for sustainable agriculture and bacterial-mediated alleviation of salinity stress in crop plants. In: Kumar M, Etesami H, Kumar V (eds) Saline soil-based agriculture by halotolerant microorganisms. Springer, Singapore, pp 1–22

Ezin V, Pena RDL, Ahanchede A (2010) Flooding tolerance of tomato genotypes during vegetative and reproductive stages. Braz J Plant Physiol 22:131–142

Farooq M, Wahid A, Kobayashi N, Fujita D, Basra SMA (2009) Plant drought stress: effects, mechanisms and management. Agron Sustain Dev 29(1):185–212

Feng X, Porporato A, Rodriguez-Iturbe I (2013) Changes in rainfall seasonality in the tropics. Nat Clim Chang 3(9):811–815

Gadallah MAA (1999) Effects of proline and glycinebetaine on *Vicia faba* responses to salt stress. Biol Plant 42(2):249–257

Gilroy S, Suzuki N, Miller G, Choi WG, Toyota M, Devireddy AR, Mittler R (2014) A tidal wave of signals: calcium and ROS at the forefront of rapid systemic signaling. Trends Plant Sci 19(10):623–630

Goodarzi A, Namdjoyan S, Soorki AA (2020) Effects of exogenous melatonin and glutathione on zinc toxicity in safflower (*Carthamus tinctorius* L.) seedlings. Ecotoxicol Environ Saf 201: 110853

Guo W, Nazim H, Liang Z, Yang D (2016) Magnesium deficiency in plants: an urgent problem. Crop J 4(2):83–91

Gupta DK, Pena LB, Romero-Puertas MC, Hernández A, Inouhe M, Sandalio LM (2017) NADPH oxidases differentially regulate ROS metabolism and nutrient uptake under cadmium toxicity. Plant Cell Environ 40(4):509–526

Gururani MA, Venkatesh J, Tran LSP (2015) Regulation of photosynthesis during abiotic stress-induced photoinhibition. Mol Plant 8(9):1304–1320

Hajiboland R (2012) Effect of micronutrient deficiencies on plants stress responses. In: Ahmad P, Prasad MNV (eds) Abiotic stress responses in plants. Springer, New York, NY, pp 283–329

Hänsch R, Mendel RR (2009) Physiological functions of mineral micronutrients (cu, Zn, Mn, Fe, Ni, Mo, B, cl). Curr Opin Plant Biol 12(3):259–266

Hasanuzzaman M, Hossain MA, da Silva JAT, Fujita M (2012) Plant response and tolerance to abiotic oxidative stress: antioxidant defense is a key factor. In: Crop stress and its management: perspectives and strategies. Springer, Dordrecht, pp 261–315

Hasanuzzaman M, Nahar K, Alam M, Roychowdhury R, Fujita M (2013) Physiological, biochemical, and molecular mechanisms of heat stress tolerance in plants. Int J Mol Sci 14(5):9643–9684

Hasanuzzaman M, Oku H, Nahar K, Bhuyan MB, Al Mahmud J, Baluska F, Fujita M (2018) Nitric oxide-induced salt stress tolerance in plants: ROS metabolism, signaling, and molecular interactions. Plant Biotechnol Rep 12(2):77–92

Hasanuzzaman M, Bhuyan MHM, Zulfiqar F, Raza A, Mohsin SM, Mahmud JA, Fujita M, Fotopoulos V (2020) Reactive oxygen species and antioxidant defense in plants under abiotic stress: revisiting the crucial role of a universal defense regulator. Antioxidants 9(8):681

He M, He CQ, Ding NZ (2018) Abiotic stresses: general defenses of land plants and chances for engineering multistress tolerance. Front Plant Sci 9:1771

He A, Dean JM, Lodhi IJ (2021) Peroxisomes as cellular adaptors to metabolic and environmental stress. Trends Cell Biol 31(8):656–670

Ilangumaran G, Smith DL (2017) Plant growth promoting rhizobacteria in amelioration of salinity stress: a systems biology perspective. Front Plant Sci 8:1768

Iqbal Z, Iqbal MS, Hashem A, Abd Allah EF, Ansari MI (2021) Plant defense responses to biotic stress and its interplay with fluctuating dark/light conditions. Front Plant Sci 12:297

Ishikawa HA (1996) Ultrastructural features of chilling injury: injured cells and the early events during chilling of suspension-cultured mung bean cells. Am J Bot 83(7):825–835

Jalil SU, Ansari MI (2020) Physiological role of gamma-aminobutyric acid in salt stress tolerance. In: Hasanuzzaman M, Tanveer M (eds) Salt and drought stress tolerance in plants. Springer Nature, Cham, pp 337–350

Jalil SU, Ahmad I, Ansari MI (2017) Functional loss of GABA transaminase (GABA-T) expressed early leaf senescence under various stress conditions in *Arabidopsis thaliana*. Curr Plant Biol 9: 11–22

Jalmi SK, Sinha AK (2015) ROS mediated MAPK signaling in abiotic and biotic stress-striking similarities and differences. Front Plant Sci 6:769

Jung JY, Shin R, Schachtman DP (2009) Ethylene mediates response and tolerance to potassium deprivation in Arabidopsis. Plant Cell 21(2):607–621

Kadota Y, Sklenar J, Derbyshire P, Stransfeld L, Asai S, Ntoukakis V, Jones JDG, Shirasu K, Menke F, Jones A, Zipfel C (2014) Direct regulation of the NADPH oxidase RBOHD by the PRR-associated kinase BIK1 during plant immunity. Mol Cell 54(1):43–55

Kaur H, Bhardwaj RD, Grewal SK (2017) Mitigation of salinity-induced oxidative damage in wheat (*Triticum aestivum* L.) seedlings by exogenous application of phenolic acids. Acta Physiol Plant 39(10):1–15

Kawasaki T, Yamada K, Yoshimura S, Yamaguchi K (2017) Chitin receptor-mediated activation of MAP kinases and ROS production in rice and *Arabidopsis*. Plant Signal Behav 12(9):e1361076

Kleczkowski LA (1994) Inhibitors of photosynthetic enzymes/carriers and metabolism. Annu Rev Plant Biol 45(1):339–367

Kumar S (2020) Abiotic stresses and their effects on plant growth, yield and nutritional quality of agricultural produce. Int J Food Sci Agric 4:367–378

Kumar N, Prabhu PAJ, Pal AK, Remya S, Aklakur M, Rana RS, Gupta S, Raman RP, Jadhao SB (2011) Anti-oxidative and immuno-hematological status of Tilapia (*Oreochromis mossambicus*) during acute toxicity test of endosulfan. Pestic Biochem Physiol 99(1):45–52

Kumar P, Pal M, Joshi R, Sairam RK (2013) Yield, growth and physiological responses of mung bean [*Vigna radiata* (L.) Wilczek] genotypes to waterlogging at vegetative stage. Physiol Mol Biol Plants 19(2):209–220

Kumar A, Patel JS, Meena VS, Srivastava R (2019) Recent advances of PGPR based approaches for stress tolerance in plants for sustainable agriculture. Biocatal Agric Biotechnol 20:101271

Kumar A, Singh S, Gaurav AK, Srivastava S, Verma JP (2020) Plant growth-promoting bacteria: biological tools for the mitigation of salinity stress in plants. Front Microbiol 11:1216

Lamb C, Dixon RA (1997) The oxidative burst in plant disease resistance. Annu Rev Plant Biol 48(1):251–275

Lee BR, Li LS, Jung WJ, Jin YL, Avice JC, Ourry A, Kim TH (2009) Water deficit-induced oxidative stress and the activation of antioxidant enzymes in white clover leaves. Biol Plant 53(3):505–510

Li Y, Fan Y, Ma Y, Zhang Z, Yue H, Wang L, Li J, Jiao Y (2017) Effects of exogenous γ-aminobutyric acid (GABA) on photosynthesis and antioxidant system in pepper (*Capsicum annuum* L.) seedlings under low light stress. J Plant Growth Regul 36(2):436–449

Liu X, Zhou Y, Xiao J, Bao F (2018) Effects of chilling on the structure, function and development of chloroplasts. Front Plant Sci 9:1715

Maibam A, Nisar S, Zargar SM, Mahajan R (2020) High-temperature response and tolerance in agronomic crops. In: Agronomic crops, Stress responses and tolerance, vol 3. Springer, Cham, p 173

Martinez PA (2016) Genomics of temperature stress. In: Edwards D, Batley J (eds) Plant genomics and climate change. Springer, New York, NY, pp 137–147

Maurya AK (2020) Oxidative stress in crop plants. In: Hasanuzzaman M (ed) Agronomic crops. Springer, Singapore, pp 349–380

Megha S, Basu U, Kav NN (2018) Regulation of low temperature stress in plants by microRNAs. Plant Cell Environ 41(1):1–15

Miller GAD, Suzuki N, Ciftci-Yilmaz SULTAN, Mittler RON (2010) Reactive oxygen species homeostasis and signalling during drought and salinity stresses. Plant Cell Environ 33(4): 453–467

Nagesh Babu R, Devaraj VR (2008) High temperature and salt stress response in French bean (*Phaseolus vulgaris*). Aust J Crop Sci 2(2):40–48

Paiva CN, Bozza MT (2014) Are reactive oxygen species always detrimental to pathogens? Antioxid Redox Signal 20(6):1000–1037

Paltaa JA, Ganjealic A, Turnerb NC, Siddique KHM (2010) Effects of transient subsurface waterlogging on root growth, plant biomass and yield of chickpea. Agric Water Manag 97: 1469–1476

Pandey V, Dixit V, Shyam R (2009) Chromium effect on ROS generation and detoxification in pea (*Pisum sativum*) leaf chloroplasts. Protoplasma 236(1-4):85–95

Pandey P, Ramegowda V, Senthil-Kumar M (2015) Shared and unique responses of plants to multiple individual stresses and stress combinations: physiological and molecular mechanisms. Front Plant Sci 6:723

Pandey P, Irulappan V, Bagavathiannan MV, Senthil-Kumar M (2017) Impact of combined abiotic and biotic stresses on plant growth and avenues for crop improvement by exploiting physio-morphological traits. Front Plant Sci 8:537

Parihar P, Singh S, Singh R, Singh VP, Prasad SM (2015) Effect of salinity stress on plants and its tolerance strategies: a review. Environ Sci Pollut Res 22(6):4056–4075

Pascual J, Rahikainen M, Kangasjärvi S (2017) Plant light stress. In: eLS, pp 1–6

Petrov V, Hille J, Mueller-Roeber B, Gechev TS (2015) ROS-mediated abiotic stress-induced programmed cell death in plants. Front Plant Sci 6:69

Piquerez SJ, Harvey SE, Beynon JL, Ntoukakis V (2014) Improving crop disease resistance: lessons from research on *Arabidopsis* and tomato. Front Plant Sci 5:671

Pogány M, Harrach BD, Hafez YM, Barna B, Király Z, Páldi E (2006) Role of reactive oxygen species in abiotic and biotic stresses in plants. Acta Phytopathol Entomol Hung 41(1-2):23–35

Pospisil P (2016) Production of reactive oxygen species by photosystem II as a response to light and temperature stress. Front Plant Sci 7:1950

Ralph PJ (2000) Herbicide toxicity of *Halophila ovalis* assessed by chlorophyll a fluorescence. Aquat Bot 66(2):141–152

Ramegowda V, Da Costa MVJ, Harihar S, Karaba NN, Sreeman SM (2020) Abiotic and biotic stress interactions in plants: a cross-tolerance perspective. In: Priming-mediated stress and cross-stress tolerance in crop plants. Academic Press, London, pp 267–302

Rasaei A, Ghobadi ME, Jalali-Honarmand S, Ghobadi M, Saeidi M (2012) Waterlogging and its effects on nitrogen of soil and plant. Ann Biol Res 3:119–124

Rasheed R, Yasmeen H, Hussain I, Iqbal M, Ashraf MA, Parveen A (2020) Exogenously applied 5-aminolevulinic acid modulates growth, secondary metabolism and oxidative defense in sunflower under water deficit stress. Physiol Mol Biol Plants 26(3):489–499

Raza A, Razzaq A, Mehmood SS, Zou X, Zhang X, Lv Y, Xu J (2019) Impact of climate change on crops adaptation and strategies to tackle its outcome: a review. Plants 8(2):34

Razzaque S, Elias SM, Haque T, Biswas S, Jewel GNA, Rahman S et al (2019) Gene expression analysis associated with salt stress in a reciprocally crossed rice population. Sci Rep 9(1):1–17

Rejeb IB, Pastor V, Mauch-Mani B (2014) Plant responses to simultaneous biotic and abiotic stress: molecular mechanisms. Plants 3(4):458–475

Riyazuddin R, Verma R, Singh K, Nisha N, Keisham M, Bhati KK, Kim ST, Gupta R (2020) Ethylene: a master regulator of salinity stress tolerance in plants. Biomolecules 10(6):959

Rizvi A, Khan MS (2018) Heavy metal induced oxidative damage and root morphology alterations of maize (*Zea mays* L.) plants and stress mitigation by metal tolerant nitrogen fixing *Azotobacter chroococcum*. Ecotoxicol Environ Saf 157:9–20

Roeber VM, Bajaj I, Rohde M, Schmülling T, Cortleven A (2021) Light acts as a stressor and influences abiotic and biotic stress responses in plants. Plant Cell Environ 44(3):645–664

Sachdev S, Singh RP (2018) Isolation, characterisation and screening of native microbial isolates for biocontrol of fungal pathogens of tomato. Clim Change Environ Sustain 6(1):46–58

Sachdev S, Ansari SA, Ansari MI, Fujita M, Hasanuzzaman M (2021) Abiotic stress and reactive oxygen species: generation, signaling, and defense mechanisms. Antioxidants 10(2):277

Saijo Y, Loo EPI (2020) Plant immunity in signal integration between biotic and abiotic stress responses. New Phytol 225(1):87–104

Saleh AA (2007) Amelioration of chilling injuries in mung bean (*Vigna radiata* L.) seedlings by paclobutrazol, abscisic acid and hydrogen peroxide. Am J Plant Physiol 2(6):318–332

Salehi-Lisar SY, Bakhshayeshan-Agdam H (2020) Agronomic crop responses and tolerance to drought stress. In: Agronomic crops, vol 3. Springer, Cham, pp 63–91

Sasidharan R, Hartman S, Liu Z, Martopawiro S, Sajeev N, van Veen H, Yeung E, Voesenek LA (2018) Signal dynamics and interactions during flooding stress. Plant Physiol 176(2): 1106–1117

Schutzendubel A, Polle A (2002) Plant responses to abiotic stresses: heavy metal-induced oxidative stress and protection by mycorrhization. J Exp Bot 53(372):1351–1365

Seneviratne M, Rajakaruna N, Rizwan M, Madawala HMSP, Ok YS, Vithanage M (2019) Heavy metal-induced oxidative stress on seed germination and seedling development: a critical review. Environ Geochem Health 41(4):1813–1831

Shakir SK, Irfan S, Akhtar B, ur Rehman S, Daud MK, Taimur N, Azizullah A (2018) Pesticide-induced oxidative stress and antioxidant responses in tomato (*Solanum lycopersicum*) seedlings. Ecotoxicology 27(7):919–935

Shao HB, Chu LY, Jaleel CA, Zhao CX (2008) Water-deficit stress-induced anatomical changes in higher plants. C R Biol 331(3):215–225

Sharma I (2012) Arsenic induced oxidative stress in plants. Biologia 67(3):447–453

Sharma P, Jha AB, Dubey RS, Pessarakli M (2012) Reactive oxygen species, oxidative damage, and antioxidative defense mechanism in plants under stressful conditions. J Bot 2012:217037

Sharma A, Yuan H, Kumar V, Ramakrishnan M, Kohli SK, Kaur R, Thukral AK, Bhardwaj R, Zheng B (2019) Castasterone attenuates insecticide induced phytotoxicity in mustard. Ecotoxicol Environ Saf 179:50–61

Sharma A, Kumar V, Shahzad B, Ramakrishnan M, Singh Sidhu GP, Bali AS, Handa N, Kapoor D, Yadav P, Khanna K, Bakshi P, Rehman A, Kohli SK, Khan EA, Parihar RD, Yuan H, Thukral AK, Bhardwaj R, Zheng B (2020) Photosynthetic response of plants under different abiotic stresses: a review. J Plant Growth Regul 39(2):509–531

Shewfelt RL (1992) Response of plant membranes to chilling and freezing. In: Plant membranes. Springer, Dordrecht. https://doi.org/10.1007/978-94-017-2683-2_10

Shin R, Berg RH, Schachtman DP (2005) Reactive oxygen species and root hairs in *Arabidopsis* root response to nitrogen, phosphorus and potassium deficiency. Plant Cell Physiol 46(8): 1350–1357

Soengas P, Rodríguez VM, Velasco P, Cartea ME (2018) Effect of temperature stress on antioxidant defenses in *Brassica oleracea*. Acs Omega 3(5):5237–5243

Solymosi K, Schoefs B (2010) Etioplast and etio-chloroplast formation under natural conditions: the dark side of chlorophyll biosynthesis in angiosperms. Photosynth Res 105(2):143–166

Steffens B (2014) The role of ethylene and ROS in salinity, heavy metal, and flooding responses in rice. Front Plant Sci 5:685

Suzuki N, Katano K (2018) Coordination between ROS regulatory systems and other pathways under heat stress and pathogen attack. Front Plant Sci 9:490

Suzuki N, Mittler R (2006) Reactive oxygen species and temperature stresses: a delicate balance between signaling and destruction. Physiol Plant 126(1):45–51

Sychta K, Słomka A, Kuta E (2021) Insights into plant programmed cell death induced by heavy metals—discovering a terra incognita. Cells 10(1):65

Tewari RK, Kumar P, Sharma PN (2006) Magnesium deficiency induced oxidative stress and antioxidant responses in mulberry plants. Sci Hortic 108(1):7–14

Thakur P, Kumar S, Malik JA, Berger JD, Nayyar H (2010) Cold stress effects on reproductive development in grain crops: an overview. Environ Exp Bot 67(3):429–443

Trinh NN, Huang TL, Chi WC, Fu SF, Chen CC, Huang HJ (2014) Chromium stress response effect on signal transduction and expression of signaling genes in rice. Physiol Plant 150(2): 205–224

Viehweger K (2014) How plants cope with heavy metals. Bot Stud 55(1):1–12

Vinke K, Martin MA, Adams S, Baarsch F, Bondeau A, Coumou D, Donner RV, Menon A, Perrette M, Rehfeld K, Robinson A, Rocha M, Schaeffer M, Schwan S, Serdeczny O, Svirejeva-Hopkins A (2017) Climatic risks and impacts in South Asia: extremes of water scarcity and excess. Reg Environ Chang 17(6):1569–1583

Wakeel A, Xu M, Gan Y (2020) Chromium-induced reactive oxygen species accumulation by altering the enzymatic antioxidant system and associated cytotoxic, genotoxic, ultrastructural, and photosynthetic changes in plants. Int J Mol Sci 21(3):728

Wang J, Higgins VJ (2005) Nitric oxide modulates H2O2-mediated defenses in the *Colletotrichum coccodes*–tomato interaction. Physiol Mol Plant Pathol 67(3-5):131–137

Wang F, Chen ZH, Liu X, Colmer TD, Shabala L, Salih A, Zhao M, Shabala S (2017) Revealing the roles of GORK channels and NADPH oxidase in acclimation to hypoxia in *Arabidopsis*. J Exp Bot 68(12):3191–3204

Waqas MA, Kaya C, Riaz A, Farooq M, Nawaz I, Wilkes A, Li Y (2019) Potential mechanisms of abiotic stress tolerance in crop plants induced by thiourea. Front Plant Sci 10:1336

Wu G, Shortt BJ, Lawrence EB, Levine EB, Fitzsimmons KC, Shah DM (1995) Disease resistance conferred by expression of a gene encoding H2O2-generating glucose oxidase in transgenic potato plants. Plant Cell 7(9):1357–1368

Xalxo R, Chandrakar V, Kumar M, Keshavkant S (2020) Ecophysiological responses of plants under metal/metalloid toxicity. In: Plant ecophysiology and adaptation under climate change: mechanisms and perspectives I. Springer, Singapore, pp 393–428

Yadav SK, Singla-Pareek SL, Reddy MK, Sopory SK (2005) Methylglyoxal detoxification by glyoxalase system: a survival strategy during environmental stresses. Physiol Mol Biol Plants 11(1):1

Yan K, Zhao S, Cui M, Han G, Wen P (2018) Vulnerability of photosynthesis and photosystem I in *Jerusalem artichoke* (*Helianthus tuberosus* L.) exposed to waterlogging. Plant Physiol Biochem 125:239–246

Yang B, Tang J, Yu Z, Khare T, Srivastav A, Datir S, Kumar V (2019) Light stress responses and prospects for engineering light stress tolerance in crop plants. J Plant Growth Regul 38(4): 1489–1506

You J, Chan Z (2015) ROS regulation during abiotic stress responses in crop plants. Front Plant Sci 6:1092

Yuan F, Lyu MJA, Leng BY, Zheng GY, Feng ZT, Li PH, Zhu XG, Wang BS (2015) Comparative transcriptome analysis of developmental stages of the *Limonium bicolor* leaf generates insights into salt gland differentiation. Plant Cell Environ 38:1637–1657. https://doi.org/10.1111/pce.12514

Yuzbaşıoglu E, Dalyan E (2019) Salicylic acid alleviates thiram toxicity by modulating antioxidant enzyme capacity and pesticide detoxification systems in the tomato (*Solanum lycopersicum* Mill.). Plant Physiol Biochem 135:322–330

Zaidi NW, Dar MH, Singh S, Singh US (2014) *Trichoderma* species as abiotic stress relievers in plants. In: Biotechnology and biology of *Trichoderma*, pp 515–525. https://doi.org/10.1016/b978-0-444-59576-8.00038-2

Zhang Y, Jiang W, Yu H, Yang X (2012) Exogenous abscisic acid alleviates low temperature-induced oxidative damage in seedlings of *Cucumis sativus*. L. Trans Chin Soc Agric Eng 28(1): 221–228

Zhang M, Dong B, Qiao Y, Yang H, Wang Y, Liu M (2018) Effects of sub-soil plastic film mulch on soil water and salt content and water utilization by winter wheat under different soil salinities. Field Crop Res 225:130–140

Zhu H, Li X, Zhai W, Liu Y, Gao Q, Liu J, Ren L, Chen H, Zhu Y (2017) Effects of low light on photosynthetic properties, antioxidant enzyme activity, and anthocyanin accumulation in purple pak-choi (*Brassica campestris* ssp. Chinensis Makino). PLoS One 12(6):e0179305

Oxidative Stress Triggered Damage to Cellular Biomolecules

3

3.1 Introduction

Abiotic and biotic stresses are adverse situations that are ubiquitous and inescapable, influencing the growth and performance of the plant during ontogenesis. Plants during their journey of life, often exposed with multiple stresses, may be singly or simultaneously (incomplete sentence!!!!!). Perception of adverse stimuli triggers the production of reactive oxygen species (ROS) in various compartments of the cell, notably, chloroplast, mitochondria, peroxisomes, plasma membrane, cell wall, and apoplast through multiple pathways. The excessive magnitude of ROS generated within cell are neutralized by the antioxidant-based defense mechanism of plants. However, the onset of stress exacerbates production and accumulation of ROS causing oxidative burst. The most possible reasons for oxidative burst are: (a) over-production of ROS surpassing the capacity of antioxidants to detoxify them leads to alteration in cellular redox status and induces oxidative stress (Sweetlove and Møller 2009; Demidchik 2015; Jalmi and Sinha 2015); and (b) ROS production for signaling and fortification of immune response to induce tolerance against stress (Demidchik 2015).

The redox environment of a cell is a state where two complementary processes, i.e., oxidation and reduction, occur simultaneously (Paciolla et al. 2016). Redox reaction partakes in the maintenance of cellular and apoplastic metabolism, and any disturbance in cell leading to the imbalance in the redox state can affect cellular redox homeostasis (Paciolla et al. 2016). In plants, unfavorable environmental and biotic cues, disturbs redox status, altering physiological and biochemical processes, which eventually leads to poor growth and development; and may risk plant's survival under severe circumstances (Awasthi et al. 2015). The loss of redox equilibrium due to the oxidative burst often results in oxidative stress targeting key cellular biomolecules, particularly, proteins, enzymes, nucleic acids, carbohydrates, and lipids (Demidchik 2015). These biomolecules are central for plant activities and any undesirable alterations in their structure cause dysfunction,

S. Sachdev et al., *Reactive Oxygen Species in Plants*,
https://doi.org/10.1007/978-981-19-9884-3_3

withdrawing plant's ability to produce output and withstand negative circumstances. Oxidation of protein, enzymes, lipid membrane, carbohydrates, and nucleic acid, can affect cellular membrane integrity and permeability causing electrolyte leakage, degradation of chlorophyll, and inhibition of photosynthetic process that induce mutation and programmed cell death (PCD) (Demidchik 2015; Roychowdhury et al. 2019; Shah et al. 2019; Hasanuzzaman et al. 2020). The damages incurred by oxidative stress may be reversible or irreversible. Oxidative stress is the major notorious outcome of stressful events that intensifies the challenge of global food security. The present chapter evaluates the potential damages caused by the array of abiotic and biotic stress via oxidative stress on plant's cellular structure and function.

3.2 ROS Burst and Lipid Peroxidation

Lipids are the major constituents of both plant cell and organelle membrane. Their structure consists of both hydrophilic and hydrophobic ends. The hydrophilic end or the polar head of the lipid is linked with the glycerol backbone and the hydrophobic end contains two tails of fatty acid (may be saturated or unsaturated). The hydrophobic or non-polar head are the building blocks of the lipids and acts as a barrier between cell or organelles and the environment (Reszczyńska and Hanaka 2020). The plant lipid membranes are instrumental in maintaining cellular integrity and regulation of cell metabolism via signaling (Kim 2020). These membranes, however, are highly susceptible to ROS attack (Fig. 3.1), especially the polyunsaturated fatty acid (PUFA) tails, such as arachidonic acid, linolenic acid, linoleic acid, etc.,

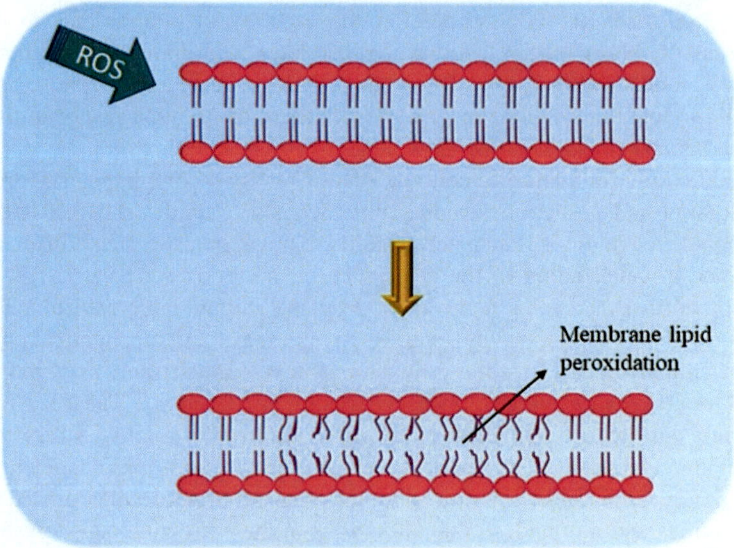

Fig. 3.1 Lipid membrane peroxidation induced by ROS

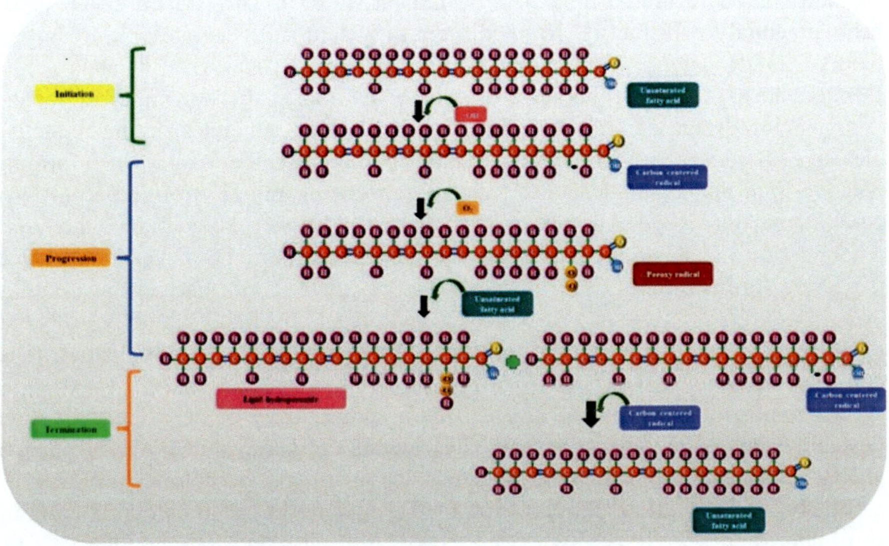

Fig. 3.2 The three stages of lipid peroxidation

containing C–C double bonds. Perception of adverse stimulus inducing oxidative burst causes oxidation of the lipid membrane. Lipid oxidation or lipid peroxidation is considered extremely harmful for the cell as it instigates the free radical's chain reaction. Being a source of extensive damage, lipid peroxidation is widely accepted as the "Hallmark" of oxidative stress (Farmer and Mueller 2013; Demidchik 2015). Lipid peroxidation reactions involve lipoxygenase activity, $^1O_2$ generation, and radical catalyzed mechanism, which differs quantitatively between underground and aboveground tissues depending on the nature of the ROS (de Dios Alche 2019).

Lipid peroxidation takes place when ROS, mainly, hydroxyl ($^{\bullet}OH$) radical attacks the methylene group of fatty acid and abstracts the hydrogen (H^+) atom forming carbon-centered lipid radical (Anjum et al. 2015), or ester linkage between glycerol and fatty acids, disintegrating membrane phospholipids (Sharma et al. 2012; Das and Roychoudhury 2014). The $^{\bullet}OH$ radical initiates the cyclic reaction resulting in the peroxidation of PUFA (Das and Roychoudhury 2014). The process of lipid peroxidation is mediated via three stages i.e., initiation, propagation, and termination (cleavage) (Anjum et al. 2015) (Fig. 3.2). Lipid peroxidation initiates on subsequent attack of ROS on target, which further triggers a cascade of reactions leading to the generation of other lipid radicals (lipid peroxyl radicals, hydroperoxides, etc.) and malondialdehyde (MDA). "Initiation," the first stage of lipid peroxidation kicks off with the abstraction of H^+ atom from lipid molecules and this abstraction is triggered by array of ROS including $^{\bullet}OH$, alkoxyl ($RO^{\bullet}$), and peroxyl radical ($ROO^{\bullet}$); and peroxynitrite (ONOOH), but excluding hydrogen peroxide (H_2O_2) and superoxide radical ($O_2^{\bullet -}$) (Halliwell and Gutteridge 2015). The initiation reaction results in the formation of lipid radicals (L^*), which in turn is activated by O_2, triggering the formation of O_2 centered lipid peroxyl radical (LOO^*). The formed LOO^* again bears potential to abstract H^+ from adjacent fatty acid and generate lipid hydroperoxide (LOOH) along with another (second) lipid radical (L*) (Catala 2006; Demidchik 2015). The generation of another L* marks the beginning of second "propagation" stage of lipid peroxidation. The LOOH generates during reductive cleavage in presence of transition metals, particularly, Fe^{2+} and Cu^+, forming reactive lipid alkoxyl radical (LO^*) capable of abstracting H^+ from adjacent fatty acid (Demidchik 2015). Another passage that causes lipid peroxidation is the direct reaction of C–C double bonds with $^1O_2$ generated from PS II reaction center, triggering formation of LOOH (Krieger-Liszkay et al. 2008; Farmer and Mueller 2013; Demidchik 2015). After a series of chain reactions, finally, lipid radicals end up forming lipid dimmers, indicating final termination stage of peroxidation (Das and Roychoudhury 2014).

Generation of lipid radical in a cell as a consequence of peroxidation causes destabilization of membrane, exacerbates membrane permeability and electrolyte leakage, deactivates membrane located enzymes and receptors, resulting in organelles disintegration and elevation of the oxidation as well as malfunctioning of proteins and nucleic acids (Sharma et al. 2012; Das and Roychoudhury 2014; Anjum et al. 2015; Demidchik 2015). These lipid radicals undergo enzymatic or non-enzymatic degradation, generating compounds like reactive carbonyl species (RCS) (Slade and Radman 2011), which selectively react with proteins through lipoxidation and cause protein dysfunction. Nevertheless, the process of lipoxidation can be prevented by reducing the degradation of lipid radicals through direct elimination from a cell (Sachdev et al. 2021). For instance, Gram-positive bacterium *Deinococcus radiodurans* have capacity to attenuate ROS induced damage to Fe–S proteins by transporting the cellular iron exterior to the cytosol (de Dios Alche 2019). Other lipid peroxidation end products considered as 'aggressive' substances, particularly, aldehydic secondary products (MDA, 4-hydroxy-2-hexenal, 4-hydroxy-2-nonenal, and acrolein) are the markers of oxidative stress (Sharma et al. 2012; Farmer and Mueller 2013; Das and Roychoudhury 2014; Sachdev et al. 2021). These markers can be measured through utilization of different approaches such as thiobarbituric acid assay for malondialdehyde (Hodges et al. 1999) or sensitive mass spectrometry-based techniques facilitating identification of individual lipid species and learning the chemical complexity of generated oxidative products (Shulaev and Oliver 2006; Farmer and Mueller 2013).

Oxidative stress-induced lipid peroxidation under a plethora of stress scenarios including salinity, temperature, metals and metalloids, drought, xenobiotic compounds, and UV radiation have been reported (Ali et al. 2005; Katsuhara et al. 2005; Singh et al. 2006, 2007; Hameed et al. 2011, 2013; Li et al. 2012; Majid et al. 2014; Kumari et al. 2015). For instance, ROS-induced lipid peroxidation and DNA damage occur in roots of the rice seedlings exposed to aluminum stress (Meriga et al. 2004). In another study, exposure of bean seedling to excess arsenic (As) triggers over-production of H_2O_2 causing lipid peroxidation and electrolyte leakage (Talukdar 2013). Moreover, the interaction of cytochrome oxidase with oxygen in plants treated with arsenic enhances the production of $O_2^{\bullet-}$ radicals and H_2O_2

accumulation in roots (Singh et al. 2007). In plants like *Sphagnum nemoreum* and *Pitters vittate*, arsenic-induced changes are noticed in the structure of chloroplast membrane, leading to thylakoid membrane breakdown and swelling (Li et al. 2006; Simola 1997). Over-accumulation of H_2O_2 in tomato plants subjected to salinity and heat stress individually or simultaneously promotes lipid peroxidation and MDA content (Martinez et al. 2016). *Solanum tuberosum* L. cv. Kufri chandramukhi cultivated under ambient CO_2 and increased O_3 records a considerable increase in lipid peroxidation, electrolyte leakage and a drop in chlorophyll content and photosynthetic rate as well (Kumari et al. 2015). Analogously on exposure to high temperatures, Ali et al. (2005) have recorded the loss of membrane integrity and K^+ leakage caused by membrane lipid peroxidation in *Phalaenopsis*.

3.3 Oxidation of Proteins

Proteins are functional and structural components of plant cell, which are crucial for plant growth and development and critically participate in conferring stress resilience in plants (Kosova et al. 2018). The amino acids which are building blocks of proteins are highly susceptible to ROS (Moller et al. 2007; Avery 2011). The redox oxidation of proteins can be described as covalent modification that brings changes in their structure, leading to aggregation, fragmentation, and/or conformational alterations (Zhang et al. 2013). These modifications impinge enzymatic and binding properties of protein, resulting in the loss of the major cellular functions, like specific metabolic activity; and structural, regulatory, or transport functions (Oracz et al. 2007; Anjum et al. 2015; Demidchik 2015). Oxidation of protein causes the accumulation of toxic protein aggregates, which can induce PCD under severe circumstances (Demidchik et al. 2010; Avery 2011).

Proteins are more prone to oxidation than other biomolecules because of abundance in cellular system and high reaction rate constant (de Dios Alche 2019). Protein are ubiquitous molecules and perform major role in plants, hence their oxidation has a widespread impact affecting important organelles and their function (Anjum et al. 2015). Therefore, the functional and structural deviations of cellular organelles are used as diagnostic markers for ROS-induced protein oxidation (Anjum et al. 2015). ROS including both radical and non-radical species induce protein oxidation. Certain ROS generate minor and selective damage to specific residues, while others trigger extensive and non-selective damages (de Dios Alche 2019). The non-radical oxidants target protein backbone inducing minor damage owing to slow oxidation rate. On the contrary, the widespread damage to the protein backbone is induced by radicals that instantly react and abstract H^+ atom from the α-carbon site leading to the genesis of stabilized carbon-centered radicals (de Dios Alche 2019). The oxidative damage to protein is mostly triggered by $^{\bullet}OH$ radical and $^1O_2$. These ROS generally oxidize protein molecules present in proximity to their site of production, probably owing to limited propagation ability (Sweetlove and Møller 2009). The oxidation of protein by ROS have also been found influenced by the presence of transition metals which catalyze Fenton reaction, resulting in the genesis of $^{\bullet}OH$ radical from H_2O_2 (Sweetlove and Møller 2009). Moller et al. (2007) have investigated the mechanisms of oxidative damage to protein complexes, in particular, photosystem I, D1 protein of photosystem II, RuBisCO, and superoxide dismutase (SOD). $^1O_2$ probably affects D1 protein only. On the other hand, $^{\bullet}OH$ radical generated by redox metals damages the remaining protein complexes. Furthermore, several proteins contain metals in their structure, which act as a key site for protein oxidation (Sweetlove and Møller 2009). Protein with Fe-S metal ion co-factors are highly susceptible to ROS-induced oxidation. For instance, the cubane $(4Fe–4S)^{2+}$ clusters of aconitase enzyme, participating in TCA (tricarboxylic acid) cycle, is essentially required for the activity of the enzyme. Cubane consists of four Fe, out of which three form ligand with the Cysteine (Cys) residue, and the fourth Fe remains exposed which can be readily targeted by the $O_2^{\bullet-}$ inducing one-electron oxidation, eventually inactivating the enzyme (Kennedy et al. 1983; Sweetlove and Møller 2009).

The oxidation of proteins can take place either in direct or indirect manner. Direct oxidation of protein involves oxidation of side chains of amino acid, in particular, those having sulfur and thiol groups attachments such as oxidation of Cys and methionine (Met) residue by $^1O_2$ and $^{\bullet}OH$; and degradation of peptide backbone causing carbonylation, disulfide bond formation, glutathionylation, and *S*-nitrosylation eventually leading to modification in protein activity (Das and Roychoudhury 2014). The indirect protein oxidation is triggered by the secondary products derived from lipid peroxidation, particularly, 4-hydroxynonenal and MDA (such as lysine or histidine) (Das and Roychoudhury 2014). Oxidation promotes prospect of proteins to undergo proteolytic digestion, making them prone to ubiquitination-mediated degradation by the proteasome (Moller et al. 2007; Das and Roychoudhury 2014). The oxidation of protein can either be reversible or irreversible (Das and Roychoudhury 2014). The reversible oxidation of protein leads to synthesis of *S*-sulfenation, *S*-nitrosylation, *S*-glutathionylation, disulfides, and methionine sulfoxide; and partakes in redox signaling and regulation of protein functions (Yan 2014). The irreversible protein oxidation generates carbonyls, nitrotyrosine, and sulfonic acid and causes aggregation and degradation, affecting biochemical functions of proteins (Yan 2014).

The irreversible oxidation of protein also includes carbonylation, tyrosine and tryptophan nitrosylation (nitration), and, protein-protein cross-linking which are responsible for causing functional loss of protein (Anjum et al. 2015). Protein carbonylation is the process that takes place due to oxidative cleavage of proteins backbone or direct oxidation of specific side chains of amino acids (Sweetlove and Møller 2009; Lounifi et al. 2013; Anjum et al. 2015; Ruiz-Ojeda et al. 2018). Oxidative cleavage of polypeptide backbone generates protein fragments with N-terminal α-ketoacyl amino acid residue; while the oxidation of side chains of amino acids, particularly, arginine (Arg), histidine (His), lysine (Lys), proline (Pro), tryptophan (Trp), or threonine (Thr) produces carbonyl (ketone or aldehyde) derivatives such as glutamic-semialdehyde, 4-OH-glutamate, and many others (Stadtman 1998; Sweetlove and Møller 2009; Lounifi et al. 2013; Anjum et al. 2015; Ruiz-Ojeda et al. 2018). Moreover, protein reaction with lipid peroxidation products like 4-hydroxy-2-nonenal, protein conjugation with reducing sugars or their oxidation products also result in protein carbonylation (Richter et al. 2005; Anjum et al. 2015). The primary protein carboxylation involving the oxidation of amino acids such as Arg, Lys, etc. generates 2,4-dinitrophenylhydrazine (DNPH) (Suzuki et al. 2010). Secondary protein carbonylation reaction taking place through addition of aldehydes generated during lipid peroxidation produces DNPH-derivatized protein (Suzuki et al. 2010).

Protein carbonylation via oxidative damage to protein polypeptide backbone is induced by $^{\bullet}OH$ radical which abstract H^+ atom from the α-carbon of polypeptide chains resulting in the formation of carbon-centered radicals. The carbon-centered radical reacts with O_2 forming peroxyl radical, which again reacts rapidly and produces protein peroxides. Further, the protein peroxide leads to formation of alkoxyl derivation via dismutation or due to the presence of Fe or $HO^{\bullet 2}$. The formed alkoxyl derivative undergoes protein fragmentation either through α-amidation or diamide (cleavage of peptide bond) pathway (Berlett and Stadtman 1997; Stadtman 2006; Höhn et al. 2013; Soladoye et al. 2015). In α-amidation pathway, alkoxyl derivative is converted to a hydroxyl derivative, finally resulting in formation an amide and *N*-alkyl carbonyl derivative (Höhn et al. 2013; Soladoye et al. 2015). In diamide pathway, alkoxyl derivative on reaction with O_2 cleaves peptide bond generating diamide derivative and isocyanate derivative (Höhn et al. 2013; Soladoye et al. 2015). Moreover, direct cleavage of peptide bond is also initiated by $^{\bullet}OH$ radical that oxidizes glutamyl and aspartyl residues in protein, forming *N*-pyruvyl derivative (Höhn et al. 2013; Soladoye et al. 2015). Oxidation of glucose $^{\bullet}OH$ radical is another route for protein backbone fragmentation through covalent binding of oxidized glucose with amino group followed by further reactions (Höhn et al. 2013). Protein carbonylation is the second most common form of protein oxidation after modification to Cys and Met (Lounifi et al. 2013) and has been reported in major cellular organelles like chloroplast, mitochondria, nucleus, cytosol, and peroxisomes (Anjum et al. 2015). Carbonylation of protein, in addition to irreversibility, also shows unrepairable damage and is considered as the best marker for determining oxidative damage (Moller et al. 2007). Carbonylation process requires more energy for oxidation of amino acids than the oxidation of sulfur-containing amino acids, i.e., Cys and Met, causing critical damage to both structure and function (Berlett and Stadtman 1997; Sweetlove and Møller 2009; Tanou et al. 2009; Lounifi et al. 2013). Carbonylation induces addition of a large and reactive group into the protein chain that displays multiple harmful effects on protein characteristics, such as covalent intermolecular cross-linking, cleavage, or alteration in protein degradation rate (Fig. 3.3). All these actions significantly modify or inhibit enzymatic and physiological activities of protein. Pronounced protein carbonylation in plants has been reported under multiple stresses including salinity, water-deficit, and metal(loid) toxicity (Romero-Puertas et al. 2002; Bartoli et al. 2004; Tanou et al. 2009; Pyngrope et al. 2013; Muszyńska et al. 2019). For instance, *B. juncea* leaves exposed to cobalt stress exhibit a significant elevation in the carbonylated protein content (Karuppanapandian and Kim 2013). Under water-deficit stress, carbonylated protein content increases several folds in mitochondria compared to chloroplast and

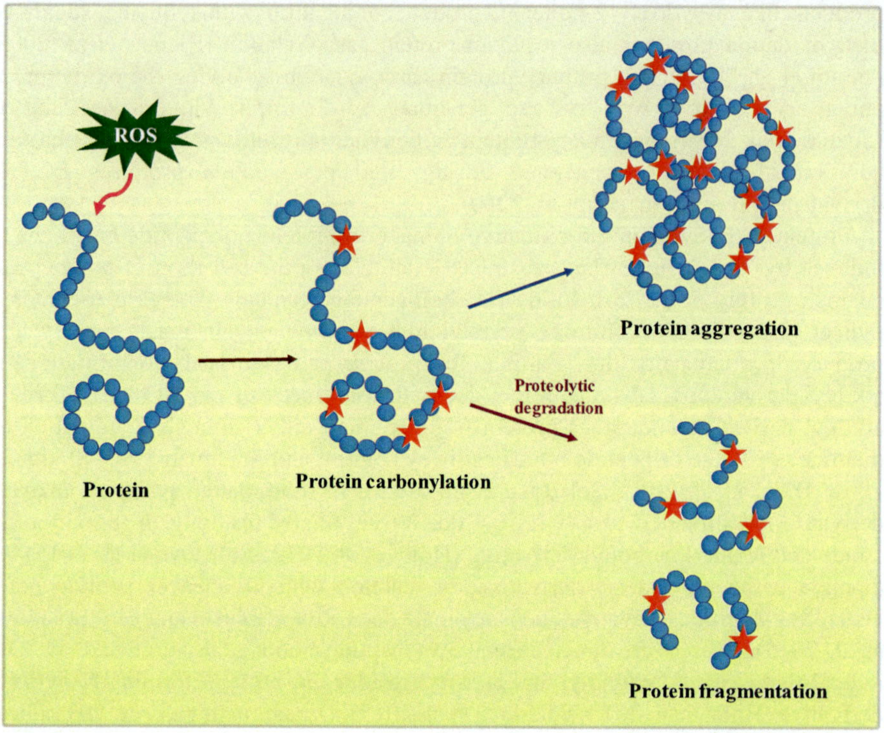

Fig. 3.3 Oxidation of protein inducing protein carbonylation, protein aggregation, and protein fragmentation

peroxisomes of wheat leaves (Bartoli et al. 2004). The exposure of cashew plants to salinity also enhances protein oxidation (Ferreira-Silva et al. 2012). Analogously, excessive CO_2 exposure induces protein carbonylation in *A. thaliana* and *Glycine max* (Qiu et al. 2008). Similarly, other stresses such as chilling, pesticides (paraquat) toxicity, and O_3 oxidative damage the proteins leading to functional (Prasad 1996; Kingston-Smith and Foyer 2000; Feng et al. 2008). Berlett and Stadtman (1997) consider protein carbonylation to be an irreversible process. Nevertheless, some transcription factors demonstrate enzymatic reversal of their carbonylation (Wong et al. 2008). Therefore, this reaction can potentially participate in ROS-mediated signaling pathways (Wong et al. 2008; Lounifi et al. 2013; Demidchik 2015). Another irreversible protein oxidation, i.e., nitration and protein-protein cross linkage diminishes protein functions and metabolism. For instance, nitration of tyrosine can severely affect key mechanisms of enzyme regulation and signal transduction (Stadtman 1998). Similarly, formation of protein-protein cross-linkage can result in the accumulation of protein polymers which are resistant to protease along with accumulation of protease inhibitors which participate in the degradation of oxidized form of oxidatively altered proteins (Stadtman 1998).

Although most of the protein oxidation damages within cell are irreversible, oxidation of proteins with sulfur containing Cys and Met amino acids is reversible (Bechtold et al. 2004; Moller et al. 2007; Hawkins et al. 2009; Onda 2013). The oxidation of Met and Cys amino acids produces various derivatives such as the formation of methionine sulfoxide on oxidation of Met by ROS and generation of disulfide, sulphonic acid, sulphinic acid, and sulphenic acid on Cys residue oxidation (Rinalducci et al. 2008; Höhn et al. 2013). Some of these derivatives show reversible nature, while others are irreversible (Rinalducci et al. 2008). For instance, $^{\bullet}OH$ radical and $^1O_2$ induce oxidation of Cys amino acid in a sequential manner, lead to formation of more oxidized derivatives, i.e., cysteine → cystine → cysteine sulfenic acid → cysteine sulfinic acid → cysteic acid (Sweetlove and Møller 2009; Demidchik 2015). All these oxidized derivatives show reversibility on action of glutaredoxin or thioredoxin enzymes, except cysteic derivative (Biteau et al. 2003; Tang et al. 2004; Buchanan and Balmer 2005; Moller et al. 2007; Onda 2013). The reversible oxidation of protein induces posttranslational modifications like glutathionylation and *S*-nitrosylation that can mediate redox regulation imparting tolerance against oxidative stress (Sweetlove and Møller 2009; Anjum et al. 2015; Musaogullari and Chai 2020).

3.4 ROS Induced Oxidation of Nucleic Acid

Nucleic acid oxidation in plants sustains extensive damages such as aging of seed stocks and even death of crop plants (Britt 1996). ROS stress is responsible for oxidation of the plant nucleic acid that even triggers mutation (Hollósy 2002). Because of their proximity to the ROS production site and lack of protective histones and associated proteins, chloroplastic and mitochondrial DNA are more sensitive to oxidation than nuclear DNA (Das and Roychoudhury 2014). Nucleic acid oxidation leads to three types of legions, mismatched bases, double-strand breaks, and chemically changed bases (Cooke et al. 2003; Yoshiyama et al. 2013) (Fig. 3.4). ROS inflicts sugar residue oxidation, nucleotide bases alteration (insertion or deletion), nucleotide abstraction, DNA strand break, and DNA and protein cross-linkage (Das and Roychoudhury 2014; Sachdev et al. 2021). For instance, in sensitive wheat genotype, drought-stressed leaves exhibit intersomal nDNA fragmentation that induces PCD (Hameed et al. 2013). Formation of deoxyribose radical on subsequent abstraction of H^+ from the C4 position of deoxyribose sugar backbone breaks DNA strand (Evans et al. 2004; Sachdev et al. 2021). Among various ROS, $^{\bullet}OH$ radical has been claimed to cause the most DNA damage because of its capacity to react with purines/pyrimidine bases, as well as deoxyribose sugar (Soares et al. 2019). The $^{\bullet}OH$ radicals are major cause of damage to polynucleic acids, reacting with them by adding to double bonds of nucleotide bases and abstraction of H^+ from each of the C–H bonds of 2′-deoxyribose and methyl group of thymine. The $^{\bullet}OH$ radical-induced DNA/RNA oxidation produces 8-oxo-7,8-dihydroguanine (8-oxoG) and 2,6-diamino-4-hydroxy5-formamidopyrimidine (FapyG) as major products (Cooke et al. 2003; Wang et al. 2010; Yoshiyama et al. 2013).

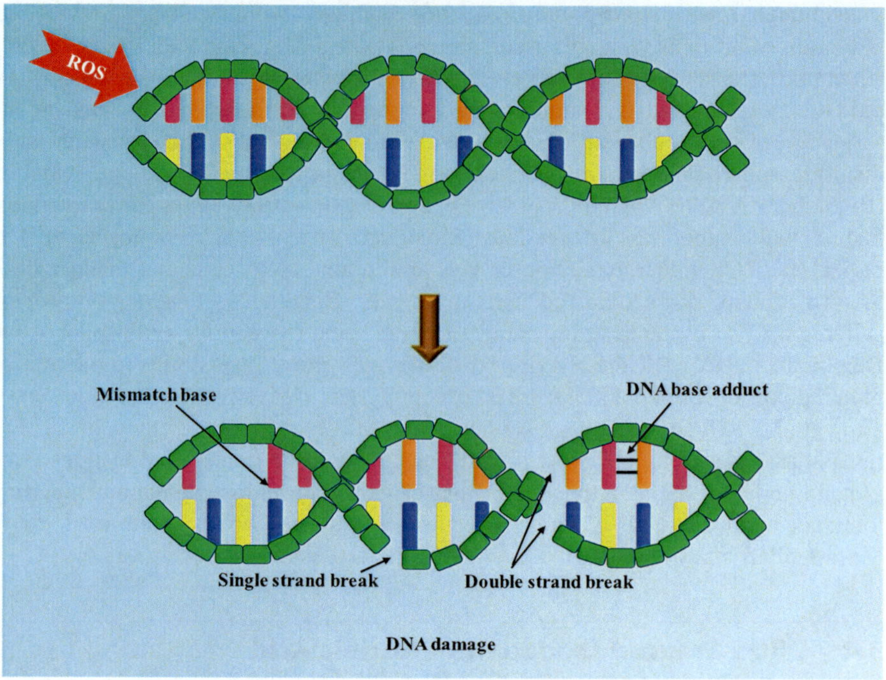

Fig. 3.4 DNA damage triggered by ROS resulting in DNA strand break, formation of base adducts and mismatching of base pairs

As for other ROS, $^1O_2$ exclusively reacts with guanine, but $O_2^{\bullet -}$ and H_2O_2 do not show any reactivity with purine or pyrimidine bases (Soares et al. 2019). The $^{\bullet}OH$ radical attacks the double bond of purines and pyrimidine bases (Halliwell 2006), causing DNA damage and generation of 8-hydroquinine along with other additional products such as hydroxyl methyl urea, thymine glycol, and others (Das and Roychoudhury 2014). The $^{\bullet}OH$ radical on reaction with DNA or related proteins causes DNA and protein cross-linking. This cross-linkage repairing is a problematic task; however, if not repaired before replication or transcription, can have lethal effect on plant cell (Das and Roychoudhury 2014). In addition to direct oxidation, the products formed during lipid peroxidation induce indirect DNA oxidation (Roldan-Arjona and Ariza 2009). MDA, a major lipid peroxidation product can react with the guanine residues of DNA, resulting in the genesis of M_1G i.e., pyrimidopurinone adduct (Fink et al. 1997). In addition to DNA, RNAs are also susceptible to ROS-induced oxidation (Mittler 2002). RNA oxidation causes the formation of 8-oxo-7,8-dihydroguanosine (8-OHG) which is used as a marker to detect the level of RNA oxidation. Oxidation of DNA/RNA subjected to stress has been reported in multiple plants. For instance, oxidation of RNA in soybean seedling and degradation of mRNA in *Lindernia subracemosa* subjected to Cd and water deficit stress (desiccation), respectively has been reported (Dinakar and Bartels 2012; Chmielowska-Bąk et al. 2018). Oxidation of DNA by ROS affect their replication and transcription ability resulting in unusual effect on protein synthesis, membrane stability, and signal transduction pathways, reducing metabolic efficiency and inducing genetic instability in a cell, along with abridging cellular homeostasis (Soares et al. 2019). The concentration of radicals created as a result of biomolecule oxidation has the potential to oxidize other biomolecules, thereby increasing oxidative damage in plant cells and leading to PCD in extreme cases. Furthermore, ROS-induced DNA damage activates particular transcription factors and stimulates signaling events, resulting in DNA repair (Yoshiyama et al. 2013).

3.5 Conclusion

Occurrence of oxidative stress in plants under stressful situations because of excessive production and accumulation of ROS is evident. The ROS target key biomolecules and cause dysfunction which increases debt on plants, potentially eclipsing global food production. To avoid extensive damage, it is a pre-requisite to eliminate free radicals generated through chain reaction and/or repair damaged biomolecules continuously. Thus, fortifying plants' intrinsic mechanisms to halt chain reactions and/or repair damaged biomolecules may facilitate the resistance toward stress and prevent productivity/economic losses as well.

References

Ali MB, Hahn EJ, Paek KY (2005) Effects of temperature on oxidative stress defense systems, lipid peroxidation and lipoxygenase activity in Phalaenopsis. Plant Physiol Biochem 43(3):213–223

Anjum NA, Sofo A, Scopa A, Roychoudhury A, Gill SS, Iqbal M, Lukatkin AS, Ahmad I (2015) Lipids and proteins—major targets of oxidative modifications in abiotic stressed plants. Environ Sci Pollut Res 22(6):4099–4121

Avery SV (2011) Molecular targets of oxidative stress. Biochem J 434(2):201–210

Awasthi R, Bhandari K, Nayyar H (2015) Temperature stress and redox homeostasis in agricultural crops. Front Environ Sci 3:11

Bartoli CG, Gómez F, Martínez DE, Guiamet JJ (2004) Mitochondria are the main target for oxidative damage in leaves of wheat (*Triticum aestivum* L.). J Exp Bot 55:1663–1669

Bechtold U, Murphy DJ, Mullineaux PM (2004) Arabidopsis peptide methionine sulfoxide reductase2 prevents cellular oxidative damage in long nights. Plant Cell 16(4):908–919

Berlett BS, Stadtman ER (1997) Protein oxidation in aging, disease, and oxidative stress. J Biol Chem 272(33):20313–20316

Biteau B, Labarre J, Toledano MB (2003) ATP-dependent reduction of cysteine–sulphinic acid by *S. cerevisiae* sulphiredoxin. Nature 425(6961):980–984

Britt AB (1996) DNA damage and repair in plants. Annu Rev Plant Biol 47(1):75–100

Buchanan BB, Balmer Y (2005) Redox regulation: a broadening horizon. Annu Rev Plant Biol 56: 187–220

Catala A (2006) An overview of lipid peroxidation with emphasis in outer segments of photoreceptors and the chemiluminescence assay. Int J Biochem Cell Biol 38(9):1482–1495

Chmielowska-Bąk J, Izbiańska K, Ekner-Grzyb A, Bayar M, Deckert J (2018) Cadmium stress leads to rapid increase in RNA oxidative modifications in soybean seedlings. Front Plant Sci 8: 2219

Cooke MS, Evans MD, Dizdaroglu M, Lunec J (2003) Oxidative DNA damage: mechanisms, mutation, and disease. FASEB J 17(10):1195–1214
Das K, Roychoudhury A (2014) Reactive oxygen species (ROS) and response of antioxidants as ROS-scavengers during environmental stress in plants. Front Environ Sci 2:53
de Dios Alche J (2019) A concise appraisal of lipid oxidation and lipoxidation in higher plants. Redox Biol 23:101136
Demidchik V (2015) Mechanisms of oxidative stress in plants: from classical chemistry to cell biology. Environ Exp Bot 109:212–228
Demidchik V, Cuin TA, Svistunenko D, Smith SJ, Miller AJ, Shabala S, Sokolik A, Yurin V (2010) Arabidopsis root K+-efflux conductance activated by hydroxyl radicals: single-channel properties, genetic basis and involvement in stress-induced cell death. J Cell Sci 123(9): 1468–1479
Dinakar C, Bartels D (2012) Light response, oxidative stress management and nucleic acid stability in closely related *Linderniaceae* species differing in desiccation tolerance. Planta 236(2): 541–555
Evans MD, Dizdaroglu M, Cooke MS (2004) Oxidative DNA damage and disease: induction, repair and significance. Mutat Res/Rev Mutat Res 567(1):1–61
Farmer EE, Mueller MJ (2013) ROS-mediated lipid peroxidation and RES-activated signaling. Annu Rev Plant Biol 64:429–450
Feng Y, Komatsu S, Furukawa T, Koshiba T, Kohno Y (2008) Proteome analysis of proteins responsive to ambient and elevated ozone in rice seedlings. Agric Ecosyst Environ 125:255–265
Ferreira-Silva SL, Voigt EL, Silva EN, Maia JM, Aragão TCR, Silveira JAG (2012) Partial oxidative protection by enzymatic and non-enzymatic components in cashew leaves under high salinity. Biol Plant 56:172–176
Fink SP, Reddy GR, Marnett LJ (1997) Mutagenicity in *Escherichia coli* of the major DNA adduct derived from the endogenous mutagen malondialdehyde. Proc Natl Acad Sci 94(16):8652–8657
Halliwell B (2006) Reactive species and antioxidants. Redox biology is a fundamental theme of aerobic life. Plant Physiol 141(2):312–322
Halliwell B, Gutteridge JM (2015) Free radicals in biology and medicine. Oxford University Press, New York
Hameed A, Bibi N, Akhter J, Iqbal N (2011) Differential changes in antioxidants, proteases, and lipid peroxidation in flag leaves of wheat genotypes under different levels of water deficit conditions. Plant Physiol Biochem 49(2):178–185
Hameed A, Goher M, Iqbal N (2013) Drought induced programmed cell death and associated changes in antioxidants, proteases, and lipid peroxidation in wheat leaves. Biol Plant 57(2): 370–374
Hasanuzzaman M, Bhuyan MHM, Zulfiqar F, Raza A, Mohsin SM, Mahmud JA, Fujita M, Fotopoulos V (2020) Reactive oxygen species and antioxidant defense in plants under abiotic stress: revisiting the crucial role of a universal defense regulator. Antioxidants 9(8):681
Hawkins CL, Morgan PE, Davies MJ (2009) Quantification of protein modification by oxidants. Free Radic Biol Med 46(8):965–988
Hodges DM, DeLong JM, Forney CF, Prange RK (1999) Improving the thiobarbituric acid-reactive-substances assay for estimating lipid peroxidation in plant tissues containing anthocyanin and other interfering compounds. Planta 207(4):604–611
Höhn A, König J, Grune T (2013) Protein oxidation in aging and the removal of oxidized proteins. J Proteome 92:132–159
Hollósy F (2002) Effects of ultraviolet radiation on plant cells. Micron 33(2):179–197
Jalmi SK, Sinha AK (2015) ROS mediated MAPK signaling in abiotic and biotic stress-striking similarities and differences. Front Plant Sci 6:769
Karuppanapandian T, Kim W (2013) Cobalt-induced oxidative stress causes growth inhibition associated with enhanced lipid peroxidation and activates antioxidant responses in Indian mustard (*Brassica juncea* L.) leaves. Acta Physiol Plant 35:2429–2443

Katsuhara M, Otsuka T, Ezaki B (2005) Salt stress-induced lipid peroxidation is reduced by glutathione S-transferase, but this reduction of lipid peroxides is not enough for a recovery of root growth in *Arabidopsis*. Plant Sci 169(2):369–373

Kennedy MC, Emptage MH, Dreyer JL, Beinert H (1983) The role of iron in the activation-inactivation of aconitase. J Biol Chem 258(18):11098–11105

Kim HU (2020) Lipid metabolism in plants. Plants (Basel) 9(7):871. https://doi.org/10.3390/plants9070871

Kingston-Smith AH, Foyer CH (2000) Bundle sheath proteins are more sensitive to oxidative damage than those of the mesophyll in maize leaves exposed to paraquat or low temperatures. J Exp Bot 51:123–130

Kosova K, Vítámvás P, Urban MO, Prášil IT, Renaut J (2018) Plant abiotic stress proteomics: the major factors determining alterations in cellular proteome. Front Plant Sci 9:122

Krieger-Liszkay A, Fufezan C, Trebst A (2008) Singlet oxygen production in photosystem II and related protection mechanism. Photosynth Res 98(1):551–564

Kumari S, Agrawal M, Singh A (2015) Effects of ambient and elevated CO2 and ozone on physiological characteristics, antioxidative defense system and metabolites of potato in relation to ozone flux. Environ Exp Bot 109:276–287

Li WX, Chen TB, Huang ZC, Lei M, Liao XY (2006) Effect of arsenic on chloroplast ultrastructure and calcium distribution in arsenic hyperaccumulator *Pteris vittata* L. Chemosphere 62:803–809

Li X, Zhang L, Li Y, Ma L, Bu N, Ma C (2012) Changes in photosynthesis, antioxidant enzymes and lipid peroxidation in soybean seedlings exposed to UV-B radiation and/or Cd. Plant Soil 352(1):377–387

Lounifi I, Arc E, Molassiotis A, Job D, Rajjou L, Tanou G (2013) Interplay between protein carbonylation and nitrosylation in plants. Proteomics 13(3-4):568–578

Majid U, Siddiqi TO, Iqbal M (2014) Antioxidant response of *Cassia angustifolia* Vahl. to oxidative stress caused by Mancozeb, a pyrethroid fungicide. Acta Physiol Plant 36(2):307–314

Martinez V, Mestre TC, Rubio F, Girones-Vilaplana A, Moreno DA, Mittler R, Rivero RM (2016) Accumulation of flavonols over hydroxycinnamic acids favors oxidative damage protection under abiotic stress. Front Plant Sci 7:838

Meriga B, Reddy BK, Rao KR, Reddy LA, Kishor PK (2004) Aluminium-induced production of oxygen radicals, lipid peroxidation and DNA damage in seedlings of rice (*Oryza sativa*). J Plant Physiol 161(1):63–68

Mittler R (2002) Oxidative stress, antioxidants and stress tolerance. Trends Plant Sci 7(9):405–410

Moller IM, Jensen PE, Hansson A (2007) Oxidative modifications to cellular components in plants. Annu Rev Plant Biol 58:459–481

Musaogullari A, Chai YC (2020) Redox regulation by protein S-Glutathionylation: from molecular mechanisms to implications in health and disease. Int J Mol Sci 21(21):8113

Muszyńska E, Labudda M, Hanus-Fajerska E (2019) Changes in proteolytic activity and protein carbonylation in shoots of *Alyssum montanum* ecotypes under multi-metal stress. J Plant Physiol 232:61–64

Onda Y (2013) Oxidative protein-folding systems in plant cells. Int J Cell Biol 2013:585431

Oracz K, Bouteau HEM, Farrant JM, Cooper K, Belghazi M, Job C, Job D, Corbineau F, Bailly C (2007) ROS production and protein oxidation as a novel mechanism for seed dormancy alleviation. Plant J 50(3):452–465

Paciolla C, Paradiso A, De Pinto MC (2016) Cellular redox homeostasis as central modulator in plant stress response. In: Redox state as a central regulator of plant-cell stress responses. Springer, Cham, pp 1–23

Prasad TK (1996) Mechanisms of chilling-induced oxidative stress injury and tolerance in developing maize seedlings: changes in antioxidant system, oxidation of proteins and lipids, and protease activities. Plant J 10:1017–1026

Pyngrope S, Bhoomika K, Dubey RS (2013) Oxidative stress, protein carbonylation, proteolysis and antioxidative defense system as a model for depicting water deficit tolerance in Indica rice seedlings. Plant Growth Regul 69(2):149–165

Qiu QS, Huber JL, Booker FL, Jain V, Leakey AD, Fiscus EL, Yau PM, Donald RO, Huber SC (2008) Increased protein carbonylation in leaves of *Arabidopsis* and soybean in response to elevated [CO2]. Photosynth Res 97:155

Reszczyńska E, Hanaka A (2020) Lipids composition in plant membranes. Cell Biochem Biophys 78:1–14

Richter T, Münch G, Lüth HJ, Arendt T, Kientsch-Engel R, Stahl P, Fengler D, Kuhla B (2005) Immunochemical crossreactivity of antibodies specific for "advanced glycation endproducts" with "advanced lipoxidation endproducts". Neurobiol Aging 26(4):465–474

Rinalducci S, Murgiano L, Zolla L (2008) Redox proteomics: basic principles and future perspectives for the detection of protein oxidation in plants. J Exp Bot 59(14):3781–3801

Roldan-Arjona T, Ariza RR (2009) Repair and tolerance of oxidative DNA damage in plants. Mutat Res/Rev Mutat Res 681(2-3):169–179

Romero-Puertas MC, Palma JM, Gómez M, Del Rio LA, Sandalio LM (2002) Cadmium causes the oxidative modification of proteins in pea plants. Plant Cell Environ 25(5):677–686

Roychowdhury R, Khan MH, Choudhury S (2019) Physiological and molecular responses for metalloid stress in rice—a comprehensive overview. In: Advances in rice research for abiotic stress tolerance. Elsevier, Amsterdam, pp 341–369

Ruiz-Ojeda FJ, Olza J, Gil Á, Aguilera CM (2018) Oxidative stress and inflammation in obesity and metabolic syndrome. In: Obesity. Academic Press, London, pp 1–15

Sachdev S, Ansari SA, Ansari MI, Fujita M, Hasanuzzaman M (2021) Abiotic stress and reactive oxygen species: generation, signaling, and defense mechanisms. Antioxidants 10(2):277

Shah K, Chaturvedi V, Gupta S (2019) Climate change and abiotic stress-induced oxidative burst in rice. In: Advances in rice research for abiotic stress tolerance. Woodhead Publishing, Witney, pp 505–535

Sharma P, Jha AB, Dubey RS, Pessarakli M (2012) Reactive oxygen species, oxidative damage, and antioxidative defense mechanism in plants under stressful conditions. J Bot 2012:217037

Shulaev V, Oliver DJ (2006) Metabolic and proteomic markers for oxidative stress. New tools for reactive oxygen species research. Plant Physiol 141(2):367–372

Simola LK (1997) The effect of lead, cadmium, arsenate and fluoride ions on the growth and fine structure of *Sphagnum nemoreum* in aseptic culture. Can J Bot 90:375–405

Singh S, Eapen S, D'souza SF (2006) Cadmium accumulation and its influence on lipid peroxidation and antioxidative system in an aquatic plant, *Bacopa monnieri* L. Chemosphere 62(2): 233–246

Singh HP, Batish DR, Kohli RK, Arora K (2007) Arsenic-induced root growth inhibition in mung bean (*Phaseolus aureus* Roxb.) is due to oxidative stress resulting from enhanced lipid peroxidation. Plant Growth Regul 53(1):65–73

Slade D, Radman M (2011) Oxidative stress resistance in *Deinococcus radiodurans*. Microbiol Mol Biol Rev 75(1):133–191

Soares C, Carvalho ME, Azevedo RA, Fidalgo F (2019) Plants facing oxidative challenges—a little help from the antioxidant networks. Environ Exp Bot 161:4–25

Soladoye OP, Juárez ML, Aalhus JL, Shand P, Estévez M (2015) Protein oxidation in processed meat: mechanisms and potential implications on human health. Compr Rev Food Sci Food Saf 14(2):106–122

Stadtman ER (1998) Free radical mediated oxidation of proteins. In: Free radicals, oxidative stress, and antioxidants. Springer, Boston, MA, pp 51–64

Stadtman ER (2006) Protein oxidation and aging. Free Radic Res 40(12):1250–1258

Suzuki YJ, Carini M, Butterfield DA (2010) Protein carbonylation. Antioxid Redox Signal 12(3): 323–325

Sweetlove LJ, Møller IM (2009) Oxidation of proteins in plants—mechanisms and consequences. Adv Bot Res 52:1–23

Talukdar D (2013) Arsenic-induced oxidative stress in the common bean legume, *Phaseolus vulgaris* L. seedlings and its amelioration by exogenous nitric oxide. Physiol Mol Biol Plants 19(1):69–79

Tang XD, Santarelli LC, Heinemann SH, Hoshi T (2004) Metabolic regulation of potassium channels. Annu Rev Physiol 66:131–159

Tanou G, Job C, Rajjou L, Arc E, Belghazi M, Diamantidis G, Molassiotis A, Job D (2009) Proteomics reveals the overlapping roles of hydrogen peroxide and nitric oxide in the acclimation of citrus plants to salinity. Plant J 60(5):795–804

Wang Z, Rhee DB, Lu J, Bohr CT, Zhou F, Vallabhaneni H, de Souza-Pinto NC, Liu Y (2010) Characterization of oxidative guanine damage and repair in mammalian telomeres. PLoS Genet 6(5):e1000951

Wong CM, Cheema AK, Zhang L, Suzuki YJ (2008) Protein carbonylation as a novel mechanism in redox signaling. Circ Res 102(3):310–318

Yan LJ (2014) Protein redox modification as a cellular defense mechanism against tissue ischemic injury. Oxidative Med Cell Longev 2014:343154

Yoshiyama KO, Sakaguchi K, Kimura S (2013) DNA damage response in plants: conserved and variable response compared to animals. Biology 2(4):1338–1356

Zhang W, Xiao S, Ahn DU (2013) Protein oxidation: basic principles and implications for meat quality. Crit Rev Food Sci Nutr 53(11):1191–1201

4 Senescence and Apoptosis: ROS Contribution to Stress Tolerance or Cellular Impairment

4.1 Introduction

Senescence and apoptosis are the developmental processes that occur during plant ontogenesis (Vanyushin et al. 2004; Gregersen et al. 2013). Senescence is an irreversible proliferative arrest (Raffetto et al. 2001; Childs et al. 2014), whereas apoptosis is a programmed cell death (PCD) (Elmore 2007). Senescence is the last stage of a developmental program that is regulated by both endogenous and exogenous factors. The endogenous factors like aging, reproductive development like seed and fruit development, and change in phytohormone level induce senescence (Ansari and Chen 2011; Jajic et al. 2015). The exogenous factors encompassing stressful events induced by heat, nutrient deficiency, drought, UV radiation, pathogen attack, etc., trigger senescence and often integrate with the developmental factors like aging (Lim et al. 2007; Ansari and Chen 2011; Jajic et al. 2015; Woo et al. 2018). Senescence is considered a strategy of plants that is instrumental in plant growth, differentiation, survival, and reproduction (Thomas 2013; Woo et al. 2018). Apoptosis is a genetically controlled cell suicide program triggered by internal and external signals. Internal signals induced by phytohormones, oxidative stress, aging, and DNA methylation (Vanyushin et al. 2004), which are perceived by a cell to eliminate itself during development, and the external stimulus indulge stressful events like pathogen attack, drought, high light, etc., encountered by plant cell. Senescence and Apoptosis are natural events that facilitate multiple beneficial functions, in particular, the translocation of plant nutrients to other cells and protecting nearby cells from stress-induced threats (Rogers 2015; Dickman et al. 2017). However, the occurrence of stressful conditions exacerbates the process of senescence and apoptosis, mediated by reactive oxygen species (ROS) induced signaling or oxidative stress that either causes cellular impairment (Gregersen et al. 2013; Jalil et al. 2017, 2019) or improve plant resilience toward stress.

S. Sachdev et al., *Reactive Oxygen Species in Plants*,
https://doi.org/10.1007/978-981-19-9884-3_4

Production of ROS within plant cells is one of the ubiquitous responses to the array of abiotic and biotic stresses (Sachdev et al. 2021). Under normal conditions, the production of ROS in cells is maintained at low levels by antioxidant enzymes. This balance can be disrupted by the depletion of antioxidants leading to the excess accumulation of ROS (Sachdev et al. 2021). Over-production and accumulation of ROS under stress induces oxidative stress and consequently result in cellular damage due to the oxidation of major biomolecules, viz. proteins, enzymes, lipid, carbohydrates, and nucleic acids (Sachdev et al. 2021). The oxidation of biomolecules alters their structure and function resulting in protein and enzyme dysfunction, lipid peroxidation, and DNA damage. However, ROS have been determined to play a dual role within the living system, by acting as both detrimental components and key regulators of various processes (Dat et al. 2000; Jajic et al. 2015). Extensive number of studies has determined the signaling role of ROS in plants under stressful conditions that serve as a mechanism of tolerance against biotic and abiotic assault (Kumar et al. 2011; Dias et al. 2011; Golldack et al. 2014). Change in the cellular concentration of ROS can facilitate signaling that modifies gene transcription leading to the acclimation of plants to stresses. A wide range of plant responses has been found to be triggered by hydrogen peroxide (H_2O_2) such as acclimation to drought, photooxidative stress, and induction of senescence (Jajic et al. 2015). Despite the vast research on ROS-mediated damage and signaling within plants, knowledge on the mechanism of their action is still not clear. In this chapter, an attempt has been made to summarize the functional role or contribution of ROS in context to senescence and apoptosis.

4.2 Senescence

Senescence is the dynamic, genetically instructed, irreversible plant development process that takes place in a highly coordinated manner on the expression of specific genes, which culminates into the end of functional life of the senescent plant, organ, tissue, or a cell (Tripathi and Tuteja 2007; Guo 2013; Thomas 2013; Ansari et al. 2014; Rogers 2015; Gan 2018; Bhattacharjee 2019). The main function of senescence is the remobilization or redistribution of nutrients from the senescent cell to other parts of the plant such as growing tissues and reproductive organs, through degradation of biomolecules (Ansari et al. 2014; Khan et al. 2021). The organs which are compromised or have already served their function and are no more useful for plants, often undergo senescence, generating nutrients for other plant parts, aiding in growth and development (Rogers 2015). For instance, the biomolecules like protein, lipids, and others present in leaves undergoing senescence are degraded and remobilized to the new developing structures such as seeds and fruits (Ansari et al. 2005, 2014). During the senescence process, the major events that took place within the cell include increase in the level of enzymes that can degrade biomolecules such as protease and nuclease, facilitating the breakdown of proteins and nucleic acids (Rubinstein 2000; Tripathi and Tuteja 2007; Del Duca et al. 2014). Additionally, the level of ROS is also elevated during senescence, which increases membrane permeability due to increased lipid peroxidation and reduced activity of enzymatic antioxidants (Rubinstein 2000; Del Duca et al. 2014; Rasool et al. 2019). Enhanced membrane permeability has been characterized as an early sign of plant senescence, which increases ions leakage, ultimately modifying the structural and functional attributes of the plant (Tripathi and Tuteja 2007). Overall senescence can be considered as a major nutrient recycling and even an energy saver process that is essentially needed for plant development (Tripathi and Tuteja 2007).

Senescence has been classified into two categories at the cellular level: (1) mitotic senescence and (2) postmitotic senescence (Gan 2003, 2018). Mitosis is a process where a cell undergoes several mitotic divisions to form a daughter cell. Cells' ability to undergo mitotic divisions is limited and after a certain number of divisions, it is arrested or ceases. The condition of the permanent arrest of mitotic division is known as mitotic senescence. In plants, arrest of proliferative activity of shoot apical meristem (SAM) is the perfect example of mitotic senescence, also known as proliferative senescence, where the SAM cells divide for a finite number of times and then cease (Gan 2018). The senescent SAM cells undergo differentiation to arise new organs like flowers or leaves. In plants, under certain conditions, in particular, the onset of unfavorable environmental cues, the ability of cells to divide is arrested temporarily, which is regained on termination of the stress. Such temporarily ceased conditions or process of the cell is known as cell quiescence (Gan 2018). Temporary cell proliferation arrest or cell quiescence is an evolutionary strategy of plants to sustain their life under unfavorable scenarios (Gan 2018). Further, the mitotically senescent cells may undergo degeneration, leading to death. This process is known as postmitotic senescence if it occurs as a factor of aging (Gan 2018). However, if degeneration of cells takes place rapidly, then the process of death is termed apoptosis (Gan 2018). Overall, the cells that undergo mitotic senescence are not dead, but the postmitotic senescence results in cell death (Gan 2018).

Senescence is a phase that occurs in all parts of the plant. It may take place either in the whole plant/organism; or may remain confined to a particular organ like flower, fruit, and leaf, or tissue such as xylem and phloem. The senescence process affecting the whole organism is known as organismal senescence or whole plant senescence, for example, monocarpic senescence that occurs in plants that show only one reproductive event in their life cycle (Gan 2018). On the other hand, the organ senescence is a senescence process occurring in a particular organ, for instance, leaf senescence, fruit senescence, flower senescence, root senescence, etc. (Tripathi and Tuteja 2007; Gan 2018). Flower senescence is a kind of organ senescence that occurs once the function of the flower has been served. Flowers are the organ that is responsible for continuing plant generations through reproduction and seeds production (Rubinstein 2000; Tripathi and Tuteja 2007). Therefore, flowers are only retained when needed, after which they undergo senescence, relocating the nutrients to other developing parts of the plant (Tripathi and Tuteja 2007). During flower senescence events similar to leaf senescence takes place such as an increase in the level of ROS, switch on of catabolic metabolism, etc., ultimately leading to the death of the flower (Tripathi and Tuteja 2007). Leaf senescence is the most studied form of organ senescence and determines the impact on agricultural produce, particularly,

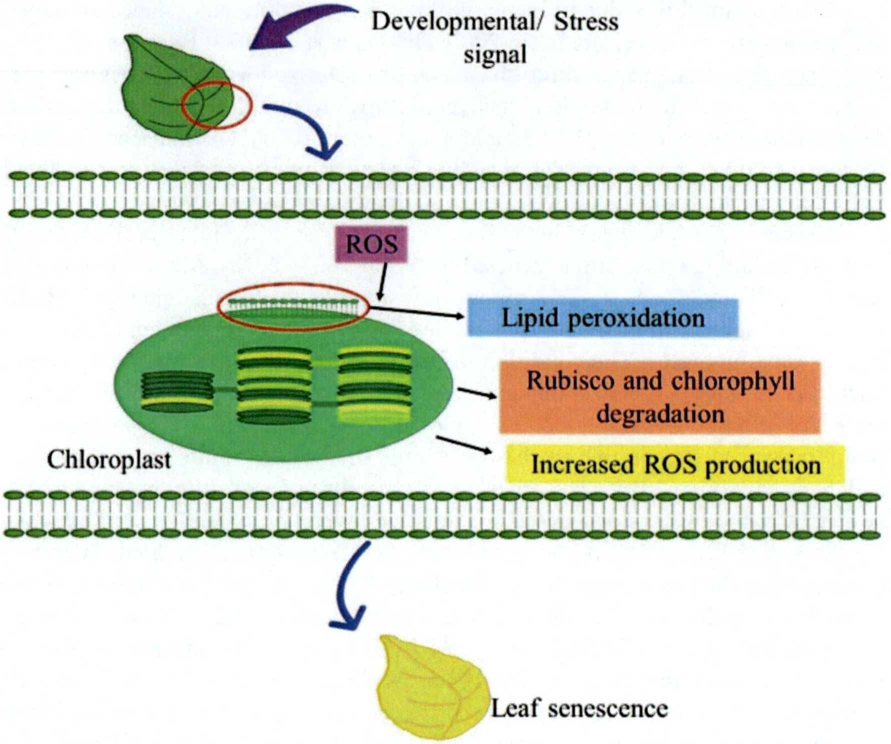

Fig. 4.1 Plants undergoing leaf senescence under developmental/stress signal show degradation of biomolecules in chloroplast and production of reactive oxygen species (ROS)

spoilage of leafy vegetables and limit plant yield by reducing the growth phase (Lim et al. 2007; Ansari and Chen 2011). The process of leaf senescence is mostly regulated by the plant age, but other internal and external signals can also contribute to the senescence (Lim et al. 2007). Under natural occurring leaf senescence, the process initiates from the tips or the margins of the leaf and progresses toward the base of the leaf. Nevertheless, stress target a particular area of leaf from where senescence begins (Lim et al. 2007). Chlorophyll degradation is the first noticeable feature of senescence. The process of senescence becomes apparent with the yellowing of leaf (Ougham et al. 2005; Ansari and Chen 2011). Other metabolic changes observed during leaf senescence include hydrolysis of lipids, proteins, and nucleic acids (Watanabe et al. 2013; Rasool et al. 2019) (Fig. 4.1). The most fascinating feature of leaf senescence is that the nucleus and mitochondria remain intact until the final stage reaches (Rasool et al. 2019). The occurrence of leaf senescence is obliged to well-coordinated action at cellular, tissue, organ, and organism levels regulated by the well-defined genetic program (Lim et al. 2007). During leaf senescence, cells undergo dramatic alterations at the physiological, biochemical, and molecular levels including switching metabolic process from anabolic to catabolic, resulting in degradation of chlorophyll and biomolecules; and nutrients redistribution to the parts elsewhere in the plant, particularly, new developing organs (Masclaux et al. 2000; Lim et al. 2007; Tripathi and Tuteja 2007; Woo et al. 2018, 2019). At the molecular level, it has been noted that during leaf senescence, the leaf transcriptome greatly differs from the onset and progression of senescence. In senescent leaves upregulation of various transcription factors (TFs) has been reported including families of NAC, C2H2 zinc finger (Cys2-His2), WRKY, AP2/EREBP, and MYB proteins (Guo and Gan 2005; Han et al. 2014; Li et al. 2021). For instance, the overexpression of *OsWRKY5* enhances leaf senescence and chlorophyll degradation in rice under normal as well as dark stress-induced senescence by upregulating associated genes, in particular, senescence-associated *NAC*, abscisic acid (ABA) biosynthesis, and chlorophyll degradation genes (Kim et al. 2019). Similarly, leaf senescence in rice is promoted by *OsWRKY42* that represses the expression of *OsMT1d*, related to scavenging of ROS and thereby elevating the accumulation of ROS (Han et al. 2014).

4.3 ROS and Senescence

Senescence has been found to be associated with plant growth and development, yet several reports suggest that the senescence occurring at a premature stage due to stress may abridge crop productivity (Gregersen et al. 2013; Ansari et al. 2014; Rasool et al. 2019; Jalil and Ansari 2020). Plant growing under stress conditions show tolerance by accelerating the process of senescence and abscission (Sade et al. 2018). Due to accelerated senescence, the plants or their next generation can survive under stress. For instance, under drought condition, leaf senescence triggers abscission of leaves, resulting in reduced loss of water through transpiration (Rasool et al. 2019). However, increased senescence affects agricultural yield manifesting significant losses, which impinge food security and elevate economic debt on farmers (Sade et al. 2018; Rasool et al. 2019). Production of ROS in plant cells is one of the earliest responses to unfavorable stimulus and senescence (Bhattacharjee 2005; Khanna-Chopra 2012; Ansari et al. 2021). Plant subjected to environmental cues, as a response, leads to over-production and accumulation of ROS, which mediate signaling and activate genetically programmed pathway of gene expression triggering cell suicide events in a regulated manner (Foyer and Noctor 2005; Khanna-Chopra 2012). Chloroplast is the major site for ROS production during stressful conditions (Partelli et al. 2011; Khan et al. 2021). Therefore, ROS primarily target chloroplast and induce degradation of major biomolecules and activation of signaling pathways associated with senescence, ultimately causing premature senescence and limiting crop productivity (Rasool et al. 2019) (Fig. 4.1). For instance, H_2O_2 which acts as a defensive molecule at a low concentration also enacts a detrimental impact on the plant at a higher concentration. The detrimental impact includes significantly degradation of certain crucial enzymes by oxidizing their thiol groups and triggering senescence (Quan et al. 2008). Apart from chloroplast, other cell organelles such as mitochondria, peroxisomes, cell wall and apoplast also generate

ROS (Khan et al. 2021). In plants, senescence-associated genes (SAGs), certain TFs, as well as kinase-coding genes mediate senescence (Rasool et al. 2019). The cells undergoing lipid peroxidation during senescence, further generate toxic ROS such as peroxyl radical ($RO_2^{\cdot}$), alkyl radical ($RO^{\cdot}$), and singlet oxygen ($^1O_2$) (Bhattacharjee 2005; Khanna-Chopra 2012). Moreover, the presence of Fe ions catalyzes the production of ROS within the cell (Bhattacharjee 2005). Under normal conditions, the level of Fe ions is low enough to catalyze the excess formation of ROS (Bhattacharjee 2005). However, during senescence, the availability of Fe ions increases owing to the increased proteolytic breakdown of metalloproteins, which enhance the level of ROS within a cell (Bhattacharjee 2005). The excess production of toxic ROS and their accumulation during senescence promote oxidative stress in the plant.

The rise in the level of ROS mediated through various pathways in the senescing cell can further accelerate the process of senescence (Bhattacharjee 2005). The over-production of ROS under stress can trigger degradation of Rubisco enzyme, other protein in the chloroplast (Khanna-Chopra 2012; Rasool et al. 2019). The oxidative damage to proteins, resulting in protein loss is an inherent characteristic of stress and senescence (Khanna-Chopra 2012). Cucumber seedlings subjected to a high temperature elevate the level of ROS, particularly, $^1O_2$ and H_2O_2 that induce premature senescence indicated by degradation of proteins, lipid peroxidation and reduction in chlorophyll content (Zhao et al. 2011). The sage plant influenced by drought displays symptoms of leaf senescence, in particular, increased peroxidation of lipids, loss of chlorophyll, abridged photosynthetic activity, and tremendous depletion of membrane-bounded chloroplastic antioxidants such as β-carotene and α-tocopherol, indicating the onset of oxidative stress in the plant (Munné-Bosch et al. 2001). Various investigations suggest that the accumulation of H_2O_2 and over-production of $^1O_2$ as putative ROS lead to oxidative stress in chloroplast, which mediates leaf senescence in field-grown sage under Mediterranean climate. The H_2O_2 molecule participates in leaf senescence via two different modes, i.e., by acting as signaling molecules that promote senescence and by participating in the breakdown of biomolecule during later stages of senescence (Jajic et al. 2015). In plants when bolting and flowering initiates, at that point H_2O_2 level increases along with the reduction in the activity of *ascorbate peroxidase 1* (*APX1*), which subsequently induces activation of SAGs and expression of TFs (Zimmermann et al. 2006). For instance, the leaves of *Pelargonium* showing senescence under the influence of dark stress demonstrates chlorophyll degradation and over-production of ROS, which subsequently follows the activation of two SAGs transcripts viz., cysteine protease homolog PeSAG12-1 and senescence-related TF PeWRKY6-1 (Rosenvasser et al. 2006).

4.4 Apoptosis

Programmed cell death (PCD) is a genetically controlled process of cell suicide where cells eliminate themselves on induction of developmental signals or as a response to unfavorable abiotic and biotic cues (Dickman et al. 2017). PCD is an important physiological process, necessary to maintain cellular homeostasis in both prokaryotic and eukaryotic organisms (Emanuele et al. 2018). The mechanism of PCD facilitates developmental morphogenesis and the elimination of cells that are damaged, mutated, or infected (Jan and Khurshid 2013). Apoptosis is a type of PCD that takes place on activation of an intracellular proteolytic cascade in response to the developmental process, environmental insult, or incompatible plant-microbes interaction (Hilario et al. 2010; Emanuele et al. 2018). The process of apoptosis is executed by the proteases and is characterized by cell (cytoplasm and nucleus) shrinkage, cell membrane blebbing, chromatin condensation and nuclear fragmentation, plasma membrane blebbing, the release of cytochrome (Cyt) c from mitochondria to cytosol, production of ROS, formation of apoptotic bodies, and finally cell degradation (Dickman et al. 2017; Emanuele et al. 2018; Valandro et al. 2020) (Fig. 4.2). The mechanism of apoptosis in an animal cell is well-recognized and studied, however, the insight to apoptosis mechanism in plants is still at an infant stage. Nevertheless, recent works have attempted to elucidate the pathways and function of arrays of enzymes as well as genes participating in apoptosis in plants. Through these studies, it has been concluded that the pathway of apoptosis in animals and plants share a certain degree of resemblance. In contrast to the animal cell, where caspases have been reported to mediate the process of apoptosis, in plants, the family of cysteine proteases called metacaspases, vacuolar processing enzymes (VPEs), etc., analogous to caspases partake in apoptosis (Emanuele et al. 2018; Valandro et al. 2020). The action of these enzymes in plants is still not fully elucidated. However, it has been deciphered that the metacaspases attack and cleave arginine (Arg) and lysine (Lys) residues instead of aspartic acid (Asp) (Salvesen et al. 2016; Yanık et al. 2020). These enzymes are generally found in the cytosol but during PCD they can re-locate to other organelles (Woltering 2004; Yanık et al. 2020). The metacaspases precursor is activated on elevation in Ca^{2+} level (Tsiatsiani et al. 2011; Yanık et al. 2020).

In plants, Apoptosis-like PCD is carried out via a series of events that takes place in a sequential manner (Reape and McCabe 2010) (Fig. 4.3). The first stage of induction of apoptosis is initiated on the perception of developmental or stress signals (Reape and McCabe 2010). This raises the influx of cellular calcium, increases the accumulation of ROS, and modifies phosphate and/or ATP level (Reape and McCabe 2010). These cellular alterations trigger the opening of permeability transition pore (PTP), a polyprotein complex at contact sites between inner mitochondrial membrane (IMM) and outer mitochondrial membrane (OMM) (Reape and McCabe 2010). The opening of PTP affects the mitochondrial transmembrane potential (MTP) ($\Delta\psi_m$), resulting in mitochondrial swelling and mitochondrial membrane permeabilization (MMP) (Reape and McCabe 2010). MMP causes the release of Cyt c and Mg^{2+} dependent nuclease into the cytosol, which in turn

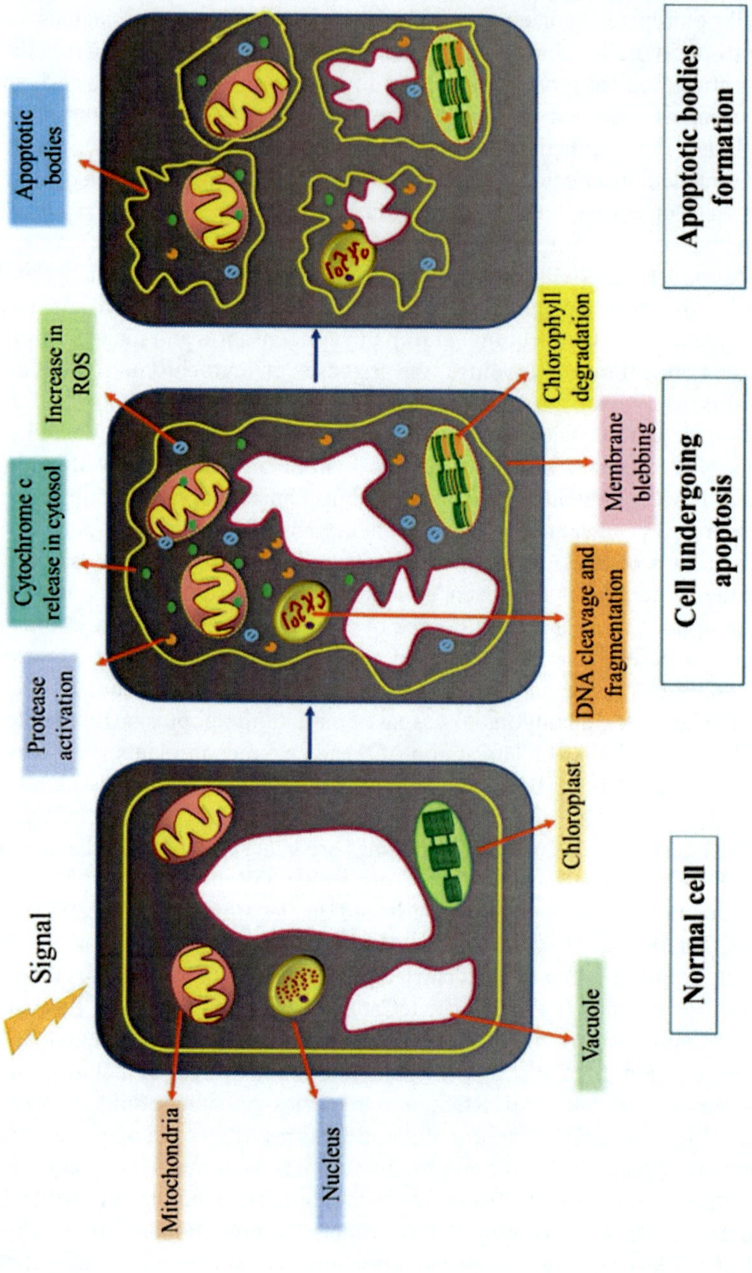

Fig. 4.2 Cell undergoing apoptosis showing hallmark characteristics

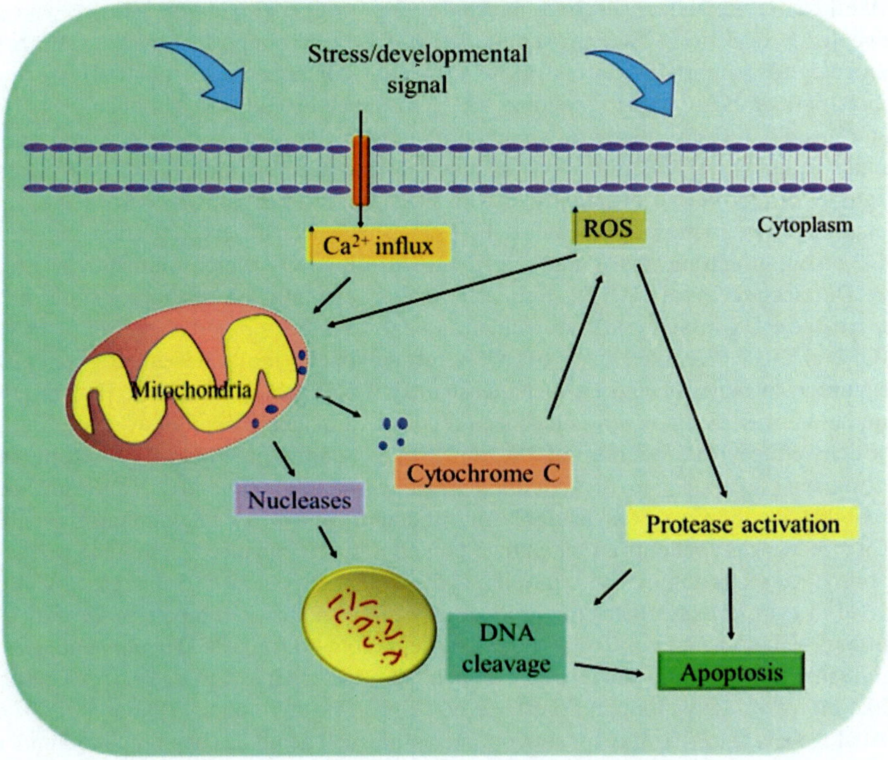

Fig. 4.3 The events involved in plant apoptosis on perception of developmental/stress signal

increases production and exacerbated accumulation of ROS in cytosol due to disruption of electron transport chain (ETC) and cleavage of high molecular weight (HMW) DNA, respectively (Reape and McCabe 2010). Amplification of ROS levels within cells activates proteases that trigger cellular destruction and eventually lead to cell death (Reape and McCabe 2010; Balakireva and Zamyatnin 2019; Yanık et al. 2020). For example, in *Nicotiana tabacum* increased activity of caspase-like proteases due to formation of lipid-derived reactive carbonyl species, downstream products of ROS trigger PCD (Biswas and Mano 2016). The ROS generated via Cyt c in cytosol initiate a feedback loop that again triggers PTP opening, amplifying the mechanism of loss of MT, MMP, and PCD (Reape and McCabe 2010).

4.5 ROS and Apoptosis Under Stress

Plants in their journey of life encounter a wide array of abiotic and biotic stresses that establish increased production of ROS within cellular compartments. ROS are ambiguous molecules that show dual character within the cell. They can mediate tolerance against unfavorable stimuli at low concentration; however, concentration significantly exceeding beyond threshold limit triggers oxidative stress, which eventually leads to PCD (Petrov et al. 2015). PCD is an important mechanism that operates to support the normal growth and development of the plant. However, induction of PCD under directives of environmental cues or pathogenic attack significantly reduces plant growth and productivity (Petrov et al. 2015). The occurrence of unfavorable stimuli, over-production, and accumulation of ROS in plants have been reported to eventually lead to PCD via various pathways. For instance, plants exposed to prolonged water-deficit stress have been observed to show reduced CO_2 fixation and increased electron leakage to triplet O_2, which ultimately trigger PCD (Gechev et al. 2012). Similarly, under extreme temperature conditions, increased ROS production and cellular damage have been reported to induce PCD (Qi et al. 2011). Elevated levels of ROS affect cellular biomolecules and trigger PCD by upregulating the activities of protease enzymes (Yanık et al. 2020). High accumulation of ROS under stress has been reported to induce oxidation of cardiolipins which reduces the stability of the inner mitochondrial membrane, resulting in diffusion of Cyt c into the intramembrane region. Subsequently, ROS enhance mobilization of Ca^{2+} from endoplasmic reticulum lumen to mitochondria, where Ca^{2+} reduces mitochondrial membrane potential which enhances membrane permeability and extrusion of Cyt c into the cytosol (Yanık et al. 2020). Increased Cyt c level in cytosol exacerbates the production of ROS which participate in the execution of PCD (Yanık et al. 2020). For instance, in tomato fruit, PCD occurs under the influence of heat stress and displays sequential events such as Cyt c release, caspase-like proteases activation, and presence of TUNEL-positive nuclei (Qu et al. 2009). Analogously, in *Pleurotus* species, apoptosis-like cell death has been reported under high-temperature stress (Song et al. 2014). In peanut root tip cells, exposure to aluminum stress triggers ROS burst, upregulated expression of *Respiratory burst oxidase homolog* (Rboh) and *Cyclooxygenase* (COX) genes, increased opening of PTP, reduced IMM potential, released Cyt c into the cytosol, and activated caspase 3-like proteases activity (Huang et al. 2014). *Arabidopsis thaliana* exposed to excess ultraviolet-C radiation also displays ROS burst, which remains restricted to chloroplast and adjacent mitochondria and induces apoptosis-like PCD characterized by loss of MTP, the activity of caspases-like proteases, DNA laddering, and nucleus fragmentation (Gao et al. 2008). The plant demonstrates another form of PCD on the perception of biotic stress, known as the hypersensitive response (HR). HR is a localized cell death that takes place as a resistance strategy and remains restricted to the areas surrounding pathogen infection sites (Kim et al. 2008; Balint-Kurti 2019) and propagates systemic acquired resistance (Mukhtar et al. 2016). HR is a hybrid type of cell death that shows characteristics of both apoptosis-like death and necrosis (Mukhtar et al. 2016). Interaction of plants with *Sclerotinia sclerotiorum* that synthesize oxalic acid has been reported to induce over-production and accumulation of ROS along with apoptotic-like cell death (Kim et al. 2008).

Increased ROS concentrations act as an upstream messenger leading to apoptotic-like PCD (Yanık et al. 2020). Thus, supervising the over-production, accumulation, and propagation of ROS is pivotal to prevent unnecessary cell death. Downregulating the pathways leading to ROS over-production can be one of the strategies to alleviate ROS production in plants (Farrant et al. 2007). Moreover, quenching excess ROS by the antioxidants could be effective in regulating ROS-induced PCD (Yanık et al. 2020). *Cochliobus victoriae*-derived toxic victorin, which induces nuclear apoptosis in oat seedlings due to H_2O_2 burst has been reported to be alleviated on treatment with catalase (CAT) or *N*-acetyl-cysteine, owing ability to scavenge excess ROS (Yao et al. 2002). The importance of antioxidants in abridging ROS-induced PCD has been supported by the observations of Nguyen et al. (2009). In the study it has been reported that the exposure of rice anther to short-term drought increases the accumulation of H_2O_2 and downregulated the transcripts of enzymatic antioxidants, reducing the concentration of scavenging antioxidants that in turn aggravate DNA fragmentation in the tissues eventually leading to PCD.

4.6 Conclusion

Senescence and apoptosis are natural developmental events that are necessary to maintain plant growth and development. Increased incidences of such events in the plant at early stages, however, can shrink agricultural productivity and disable global food security. Plants encountered with stressful stimuli often display the production of ROS. Accumulation of ROS in higher concentrations accelerates the process of senescence and apoptosis-like PCD via different pathways such as enhanced oxidation of cellular biomolecules, increased mitochondrial membrane potential loss, activation of protease enzymes, etc. Other pathways involved in ROS-induced senescence and/or apoptosis in plants are yet to be identified. Therefore, more studies at the metabolic and molecular levels are needed to be undertaken to understand the plant growth-arresting role of ROS under stress.

References

Ansari MI, Chen SC (2011) Leaf senescence-an overview. Int J Recent Trends Sci Technol 1(3): 110–114

Ansari MI, Lee RH, Chen SCG (2005) A novel senescence-associated gene encoding γ-aminobutyric acid (GABA): pyruvate transaminase is upregulated during rice leaf senescence. Physiol Plant 123(1):1–8

Ansari MI, Hasan S, Jalil SU (2014) Leaf senescence and GABA shunt. Bioinformation 10(12):734

Ansari MI, Jalil SU, Ansari SA, Hasanuzzaman M (2021) GABA shunt: a key-player in mitigation of ROS during stress. Plant Growth Regul 94:1–19

Balakireva AV, Zamyatnin AA Jr (2019) Cutting out the gaps between proteases and programmed cell death. Front Plant Sci 10:704

Balint-Kurti P (2019) The plant hypersensitive response: concepts, control and consequences. Mol Plant Pathol 20(8):1163–1178

Bhattacharjee S (2005) Reactive oxygen species and oxidative burst: roles in stress, senescence and signal transducation in plants. Curr Sci 89:1113–1121

Bhattacharjee S (2019) ROS in aging and senescence. In: Reactive oxygen species in plant biology. Springer, New Delhi, pp 65–79

Biswas MS, Mano JI (2016) Reactive carbonyl species activate caspase-3-like protease to initiate programmed cell death in plants. Plant Cell Physiol 57(7):1432–1442

Childs BG, Baker DJ, Kirkland JL, Campisi J, Van Deursen JM (2014) Senescence and apoptosis: dueling or complementary cell fates? EMBO Rep 15(11):1139–1153

Dat J, Vandenabeele S, Vranova EVMM, Van Montagu M, Inzé D, Van Breusegem F (2000) Dual action of the active oxygen species during plant stress responses. Cell Mol Life Sci 57(5): 779–795

Del Duca S, Serafini-Fracassini D, Cai G (2014) Senescence and programmed cell death in plants: polyamine action mediated by transglutaminase. Front Plant Sci 5:120

Dias CV, Mendes JS, Dos Santos AC, Pirovani CP, da Silva Gesteira A, Micheli F, Gramacho KP, Hammerstone J, Mazzafera P, de Mattos Cascardo JC (2011) Hydrogen peroxide formation in cacao tissues infected by the hemibiotrophic fungus *Moniliophthora perniciosa*. Plant Physiol Biochem 49(8):917–922

Dickman M, Williams B, Li Y, de Figueiredo P, Wolpert T (2017) Reassessing apoptosis in plants. Nat Plants 3(10):773–779

Elmore S (2007) Apoptosis: a review of programmed cell death. Toxicol Pathol 35(4):495–516

Emanuele S, Oddo E, D'Anneo A, Notaro A, Calvaruso G, Lauricella M, Giuliano M (2018) Routes to cell death in animal and plant kingdoms: from classic apoptosis to alternative ways to die—a review. Rend Lincei Sci Fis Nat 29(2):397–409

Farrant J, Brandt WF, Lidsey GG (2007) An overview of the mechanisms of dessication tolerance in selected angiosperm resurrection plants. Plant Stress 1:72–84

Foyer CH, Noctor G (2005) Redox homeostasis and antioxidant signalling: a metabolic interface between stress perception and physiological responses. Plant Cell 17:1866–1875

Gan S (2003) Mitotic and postmitotic senescence in plants. Sci Aging Knowledge Environ 2003(38):re7

Gan S (2018) Concepts and types of senescence in plants. Methods Mol Biol 1744:3–8

Gao C, Xing D, Li L, Zhang L (2008) Implication of reactive oxygen species and mitochondrial dysfunction in the early stages of plant programmed cell death induced by ultraviolet-C overexposure. Planta 227(4):755–767

Gechev TS, Dinakar C, Benina M, Toneva V, Bartels D (2012) Molecular mechanisms of desiccation tolerance in resurrection plants. Cell Mol Life Sci 69(19):3175–3186

Golldack D, Li C, Mohan H, Probst N (2014) Tolerance to drought and salt stress in plants: unraveling the signaling networks. Front Plant Sci 5:151

Gregersen PL, Culetic A, Boschian L, Krupinska K (2013) Plant senescence and crop productivity. Plant Mol Biol 82(6):603–622

Guo Y (2013) Towards systems biological understanding of leaf senescence. Plant Mol Biol 82(6): 519–528

Guo Y, Gan S (2005) Leaf senescence: signals, execution, and regulation. Curr Top Dev Biol 71: 83–112

Han M, Kim CY, Lee J, Lee SK, Jeon JS (2014) OsWRKY42 represses OsMT1d and induces reactive oxygen species and leaf senescence in rice. Mol Cells 37(7):532

Hilario E, Cañavate ML, Lacalle J, Alonso-Alconada D, Lara-Celador I, Alvarez-Granda L, Alvarez A (2010) Cell death. A comprehensive approximation. Delayed cell death. In: Microscopy: science, technology, applications and education. Formatex Research Centre, Badajoz, pp 1025–1032

Huang W, Yang X, Yao S, LwinOo T, He H, Wang A, Li C, He L (2014) Reactive oxygen species burst induced by aluminum stress triggers mitochondria-dependent programmed cell death in peanut root tip cells. Plant Physiol Biochem 82:76–84

Jajic I, Sarna T, Strzalka K (2015) Senescence, stress, and reactive oxygen species. Plants 4(3): 393–411

Jalil SU, Ansari MI (2020) Stress implications and crop productivity. In: Plant ecophysiology and adaptation under climate change: mechanisms and perspectives I. Springer, Singapore, pp 73–86

Jalil SU, Ahmad I, Ansari MI (2017) Functional loss of GABA transaminase (GABA-T) expressed early leaf senescence under various stress conditions in *Arabidopsis thaliana*. Curr Plant Biol 9: 11–22

Jalil SU, Khan MIR, Ansari MI (2019) Role of GABA transaminase in the regulation of development and senescence in *Arabidopsis thaliana*. Curr Plant Biol 19:100119

Jan N, Khurshid I (2013) Programmed cell death or apoptosis: do animals and plants share anything in common. Biotechnol Mol Biol Rev 3(5):111–126

Khan MIR, Jalil SU, Chopra P, Chhillar H, Ferrante A, Khan NA, Ansari MI (2021) Role of GABA in plant growth, development and senescence. Plant Gene 26:100283

Khanna-Chopra R (2012) Leaf senescence and abiotic stresses share reactive oxygen species-mediated chloroplast degradation. Protoplasma 249(3):469–481

Kim KS, Min JY, Dickman MB (2008) Oxalic acid is an elicitor of plant programmed cell death during *Sclerotinia sclerotiorum* disease development. Mol Plant-Microbe Interact 21(5): 605–612

Kim T, Kang K, Kim SH, An G, Paek NC (2019) OsWRKY5 promotes rice leaf senescence via senescence-associated NAC and abscisic acid biosynthesis pathway. Int J Mol Sci 20(18):4437

Kumar N, Ebel RC, Roberts PD (2011) H2O2 degradation is suppressed in kumquat leaves infected with *Xanthomonas axonopodis* pv. citri. Sci Hortic 130(1):241–247

Li Y, Liao S, Mei P, Pan Y, Zhang Y, Zheng X, Xie Y, Miao Y (2021) OsWRKY93 dually functions between leaf senescence and in response to biotic stress in rice. Front Plant Sci 12:327

Lim PO, Kim HJ, Gil Nam H (2007) Leaf senescence. Annu Rev Plant Biol 58:115–136

Masclaux C, Valadier MH, Brugière N, Morot-Gaudry JF, Hirel B (2000) Characterization of the sink/source transition in tobacco (*Nicotiana tabacum* L.) shoots in relation to nitrogen management and leaf senescence. Planta 211(4):510–518

Mukhtar MS, McCormack ME, Argueso CT, Pajerowska-Mukhtar KM (2016) Pathogen tactics to manipulate plant cell death. Curr Biol 26(13):R608–R619

Munné-Bosch S, Jubany-Marí T, Alegre L (2001) Drought-induced senescence is characterized by a loss of antioxidant defences in chloroplasts. Plant Cell Environ 24(12):1319–1327

Nguyen GN, Hailstones DL, Wilkes M, Sutton BG (2009) Drought-induced oxidative conditions in rice anthers leading to a programmed cell death and pollen abortion. J Agron Crop Sci 195(3): 157–164

Ougham HJ, Morris P, Thomas H (2005) The colors of autumn leaves as symptoms of cellular recycling and defenses against environmental stresses. Curr Top Dev Biol 66(1):135–160

Partelli FL, Batista-Santos P, Scotti-Campos P, Pais IP, Quartin VL, Vieira HD, Ramalho JC (2011) Characterization of the main lipid components of chloroplast membranes and cold induced changes in *Coffea* spp. Environ Exp Bot 74:194–204

Petrov V, Hille J, Mueller-Roeber B, Gechev TS (2015) ROS-mediated abiotic stress-induced programmed cell death in plants. Front Plant Sci 6:69

Qi Y, Wang H, Zou Y, Liu C, Liu Y, Wang Y, Zhang W (2011) Over-expression of mitochondrial heat shock protein 70 suppresses programmed cell death in rice. FEBS Lett 585(1):231–239

Qu GQ, Liu X, Zhang YL, Yao D, Ma QM, Yang MY, Zhu WH, Yu S, Luo YB (2009) Evidence for programmed cell death and activation of specific caspase-like enzymes in the tomato fruit heat stress response. Planta 229(6):1269–1279

Quan LJ, Zhang B, Shi WW, Li HY (2008) Hydrogen peroxide in plants: a versatile molecule of the reactive oxygen species network. J Integr Plant Biol 50(1):2–18

Raffetto JD, Leverkus M, Park HY, Menzoian JO (2001) Synopsis on cellular senescence and apoptosis. J Vasc Surg 34(1):173–177

Rasool S, Mir BA, Rehman MU, Amin I, Mir MUR, Ahmad SB (2019) Abiotic stress and plant senescence. In: Senescence signalling and control in plants. Academic Press, London, pp 15–27

Reape TJ, McCabe PF (2010) Apoptotic-like regulation of programmed cell death in plants. Apoptosis 15(3):249–256

Rogers HJ (2015) Senescence-associated programmed cell death. In: Plant programmed cell death. Springer, Cham, pp 203–233

Rosenvasser S, Mayak S, Friedman H (2006) Increase in reactive oxygen species (ROS) and in senescence-associated gene transcript (SAG) levels during dark-induced senescence of Pelargonium cuttings, and the effect of gibberellic acid. Plant Sci 170:873–879

Rubinstein B (2000) Regulation of cell death in flower petals. In: Programmed cell death in higher plants. Kluwer Academic Publishers, Boston, pp 59–74

Sachdev S, Ansari SA, Ansari MI, Fujita M, Hasanuzzaman M (2021) Abiotic stress and reactive oxygen species: generation, signaling, and defense mechanisms. Antioxidants 10(2):277

Sade N, del Mar Rubio-Wilhelmi M, Umnajkitikorn K, Blumwald E (2018) Stress-induced senescence and plant tolerance to abiotic stress. J Exp Bot 69(4):845–853

Salvesen GS, Hempel A, Coll NS (2016) Protease signaling in animal and plant-regulated cell death. FEBS J 283(14):2577–2598

Song C, Chen Q, Wu X, Zhang J, Huang C (2014) Heat stress induces apoptotic-like cell death in two *Pleurotus* species. Curr Microbiol 69(5):611–616

Thomas H (2013) Senescence, ageing and death of the whole plant. New Phytol 197(3):696–711

Tripathi SK, Tuteja N (2007) Integrated signaling in flower senescence: an overview. Plant Signal Behav 2(6):437–445

Tsiatsiani L, Van Breusegem F, Gallois P, Zavialov A, Lam E, Bozhkov PV (2011) Metacaspases. Cell Death Diff 18(8):1279–1288

Valandro F, Menguer PK, Cabreira-Cagliari C, Margis-Pinheiro M, Cagliari A (2020) Programmed cell death (PCD) control in plants: new insights from the *Arabidopsis thaliana* deathosome. Plant Sci 299:110603

Vanyushin BF, Bakeeva LE, Zamyatnina VA, Aleksandrushkina NI (2004) Apoptosis in plants: specific features of plant apoptotic cells and effect of various factors and agents. Int Rev Cytol 233:135–179

Watanabe M, Balazadeh S, Tohge T, Erban A, Giavalisco P, Kopka J, Mueller-Roeber B, Fernie AR, Hoefgen R (2013) Comprehensive dissection of spatiotemporal metabolic shifts in primary, secondary, and lipid metabolism during developmental senescence in Arabidopsis. Plant Physiol 162(3):1290–1310

Woltering EJ (2004) Death proteases come alive. Trends Plant Sci 9(10):469–472

Woo HR, Masclaux-Daubresse C, Lim PO (2018) Plant senescence: how plants know when and how to die. J Exp Bot 69(4):715–718

Woo HR, Kim HJ, Lim PO, Nam HG (2019) Leaf senescence: systems and dynamics aspects. Annu Rev Plant Biol 70:347–376

Yanık F, Çetinbaş-Genç A, Vardar F (2020) Abiotic stress–induced programmed cell death in plants. In: Plant life under changing environment. Academic Press, London, pp 1–24

Yao N, Tada Y, Sakamoto M, Nakayashiki H, Park P, Tosa Y, Mayama S (2002) Mitochondrial oxidative burst involved in apoptotic response in oats. Plant J 30(5):567–579

Zhao X, Nishimura Y, Fukumoto Y, Li J (2011) Effect of high temperature on active oxygen species, senescence and photosynthetic properties in cucumber leaves. Environ Exp Bot 70(2–3):212–216

Zimmermann P, Heinlein C, Orendi G, Zentgraf U (2006) Senescence-specific regulation of catalases in *Arabidopsis thaliana* (L.) Heynh. Plant Cell Environ 29(6):1049–1060

Photosynthetic Apparatus: Major Site of Oxidative Damage

5

5.1 Introduction

Plants are the most crucial living organisms which are instrumental in sustaining life on this planet. Plants are endowed with the magnificent ability to produce biomass by capturing solar energy through a process called photosynthesis. Photosynthesis is an essential physiological activity that converts photons into chemical energy and facilitates plant growth and development. Therefore, undisturbed photosynthetic activity is critical for maintaining plant health and productivity. Photosynthesis takes place in a specialized cellular compartment known as chloroplast and comprises two major phases, i.e., light and dark reactions. Both phases engage various pathways like electron chain reaction (ETC), generation of adenosine triphosphate (ATP), and nicotinamide adenine dinucleotide phosphate (NADPH), and assimilation of CO_2, ultimately generating plant biomass. The process of photosynthesis is influenced by unfavorable circumstances, which impinge plant growth and yield. The adverse conditions counting both biotic and abiotic factors such as pathogen attack, extreme temperature, water scarcity, CO_2 concentration, light, soil salinity, and nutrient deficiency affect photosynthetic performance (Reddy et al. 2010; Ansari et al. 2011; Gotoh et al. 2018). The onset of the stressful events triggers over-production of reactive oxygen species (ROS) in chloroplast by disrupting the photosynthetic pathways including over-reduction of ETC and reduced stomatal opening curtailing CO_2 intake. Accumulation of ROS in chloroplast beyond their threshold limit causes oxidation of cellular biomolecules resulting in photosynthetic pigments loss, eventually declining plant fitness.

Plants are sedentary organisms that complete their life journey in one particular space. Therefore, the plant cannot escape from prevailing adverse conditions. Harsh environmental cues threaten plant performance by modifying their metabolic, physiological, or molecular pathways. Moreover, the incessant changes in climatic conditions have amplified the incidences of stress such as increased CO_2 concentration in the atmosphere, extreme temperature, and prolonged drought conditions,

S. Sachdev et al., *Reactive Oxygen Species in Plants*,
https://doi.org/10.1007/978-981-19-9884-3_5

subjecting plants to more perilous conditions (Cassia et al. 2018; Bulgari et al. 2019). ROS are the by-product of aerobic activities, and nearly 1–2% of consumed oxygen is normally converted into different forms of ROS, namely superoxide radical ($O_2^{\bullet -}$), hydrogen peroxide (H_2O_2), singlet oxygen ($^1O_2$), and hydroxyl ion ($^{\bullet}OH$) (Cassia et al. 2018). The genesis of ROS in various cell organelles, however, is exacerbated under the isnfluence of stressful situations (Cassia et al. 2018). The organelles that show the presence of ROS include chloroplast, mitochondria, peroxisomes, plasma membrane, apoplast, cell wall, and endoplasmic reticulum. Among various cellular compartments, the chloroplast is the prime site for ROS production. The ROS produced in chloroplast either participate in the signaling pathway or induce oxidative stress depending upon the ROS scavenging efficiency of the photosynthetic apparatus (Sharma et al. 2012).

As photosynthesis is a vital activity of the plant, and any damage to photosynthetic apparatus may lead to plant termination, ultimately abridging food resources for other organisms. Production of ROS is an inevitable part of plant life. However, the exacerbated ROS production leads to significant damage to the chloroplast, dwindling plant-based food production. The ROS generate a specific signature depending upon the stress triggering their genesis (Choudhury et al. 2017; Cassia et al. 2018). For example, the drought conditions promote ROS production due to increased electron leakage from ETC to molecular oxygen that reduces activity of ribulose 1,5 bisphosphate carboxylase/oxygenase (Rubisco), fixation of CO_2, and $NADP^+$ regeneration by the Calvin cycle (Cruz 2008). Similarly, waterlogging stress accelerates ethylene-induced ROS production via increased activity of respiratory burst oxidases (RBOHs) and altered calcium flux (Voesenek and Bailey-Serres 2015). Therefore, it is paramount to understand the sequential events triggering ROS over-production in chloroplast under variable stress scenarios to elucidate mechanisms to minimize the impact on photosynthetic performance. The present chapter thus aims to provide a comprehensive overview of ROS production under adverse conditions that may impinge photosynthetic pathways, leading to plant death.

5.2 Chloroplast and Photosynthesis

Chloroplasts are important cellular organelles that partake in photosynthesis. They contain plenty of soluble protein (Rubisco), membrane protein (light-harvesting complex, LHC), and pigments (chlorophyll, carotenoid) (Kirchhoff 2019). Chloroplast is a double membrane-bounded structure forming an envelope enclosing space and other components (Cooper and Hausman 2000; Hameed et al. 2021). The two membranes are designated as inner and outer membranes and the space present between the two is called inter-membrane space. Inside the chloroplast envelope, disc-like thylakoids (forming the third internal membrane of the chloroplast) are found which remain arranged in a stack-like structure, known as grana (Cooper and Hausman 2000; Hameed et al. 2021). Each thylakoid is a membrane-enclosed structure (thylakoid membrane) with a space in the center, called lumen (Hameed et al. 2021). The space within the chloroplast envelope excluding grana is termed as stroma (Cooper and Hausman 2000). Grana and stroma are two important sites where light and dark reactions of photosynthesis take place, respectively. The thylakoids are the center for light reactions. In the thylakoid membrane, two photosystems (PSI and PSII) are found embedded. PSII is located in the appressed region and PSI is present in the stroma-exposed un-appressed region of the membrane (Joliot and Johnson 2011). The PS include reaction center complex and LHC (LHC-I and LHC-II). LHC contain chlorophyll molecules that act as an antenna to capture photon and transfer captured energy to the reaction center (Kouril et al. 2005). Both chlorophyll a and b are found in LHC in a ratio of 1.1–1.3 (Lichtenthaler and Babani 2004). The light absorption spectra of the reaction center of PSI and PSII are different (Kouril et al. 2005). The reaction center of PS contains only molecules of chlorophyll-a. The absorption peak of chlorophyll-a in the PSI reaction center is at 700 nm wavelength and is called P_{700}, whereas the absorption peak of chlorophyll-a in the PSII reaction center is at 680 nm wavelength and is known as P_{680}. In addition to chlorophyll-a, redox components present in the PSII reaction center include primary electron acceptor-pheophytin (chlorophyll-like molecule without Mg atom) and plastoquinone (Q_A and Q_B bounded with D2 and D1 protein, respectively). The final acceptor of the electron associated with PSI is ferredoxin. In the un-appressed region of thylakoid, ATPase is located and catalyzes the reaction leading to ATP production. The ultrastructure of chloroplast has been shown in Fig. 5.1.

Photosynthesis is a predominant and most significant physiological process that involves light and dark reactions to produce plant biomass. These reactions include a series of events taking place in the presence of some of the factors including light, CO_2, chlorophyll molecules, enzymes, electron acceptors, and proteins. The events that take place during light reactions have been illustrated in Fig. 5.1. The process of photosynthesis initiates with capturing of light (400–700 nm) by the pigments associated with PSI and PSII in the thylakoid membrane. The chlorophyll molecules associated with LHC absorb a photon and migrate excitation energy to the PSI and PSII reaction center via resonating chlorophyll molecules (Cruz 2008; Foyer et al. 2012). At PSII, the oxygen-evolving complex (OEC) site for water oxidation induces splitting of water molecules generating O_2 and proton (H^+), along with the generation of electrons (Gururani et al. 2015; Barber 2017). Generation of an electron at PS induces electron transport chain (ETC) which could be either linear or cyclic. The linear ETC includes the transfer of electrons from PSII to pheophytin, then to plastoquinone Q_A, followed by plastoquinone Q_B, and then to free plastoquinone. The free plastoquinone then subsequently transfers electron to cytochrome b_6/f complex and gets reduced. Further, electron transfer from Cyt b_6/f to PSI occurs via plastocyanin. PSI after accepting the electron, subsequently transfer it to ferredoxin, which acts as the final electron acceptor and reduces $NADP^+$, forming NADPH. The splitting of water and electron transport during the process results in proton accumulation in the thylakoid lumen which induces a positive charge. This develops an electrochemical proton gradient or proton motive force (pmf) across the thylakoid membrane, facilitating phosphorylation of ADP by ATP synthase, generating ATP

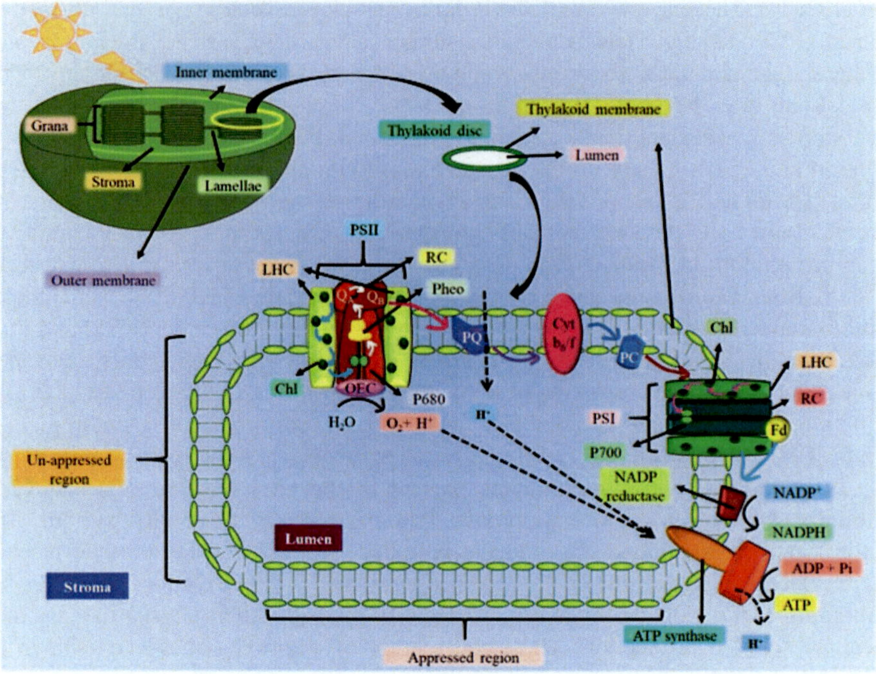

Fig. 5.1 Ultrastructure of Chloroplast and the mechanism of light reaction

(Foyer et al. 2012; Yamori 2016; Nawrocki et al. 2019). Analogous to linear electron transport, in cyclic ETC, an electron from PSI is transferred back to Cyt b_6/f via plastoquinone, increasing proton level within the lumen, thereby acidifying lumen and promoting ATP production (Nawrocki et al. 2019). The cyclic ETC takes place when demand for ATP increases in comparison to NADPH (Nawrocki et al. 2019). The steps involved in light reaction ultimately lead to the production of ATP and NADPH molecules that partake in CO_2 fixation in stroma through the carbon-reduction cycle known as the Calvin Benson cycle or Dark reaction (Yamori 2016). In the Calvin cycle, CO_2, ribulose 1,5 bisphosphate (RuBP), ribulose 1,5 bisphosphate carboxylase/oxygenase (Rubisco) enzyme, ATP, and NADPH are major components that through a series of steps results in the production of starch and other products. The Rubisco enzymes not only participate in carboxylation but also catalyze the oxygenation process during the presence of high O_2 level relative to CO_2 (Eisenhut et al. 2019; Timm and Hagemann 2020; Shi and Bloom 2021). The oxygenation of RuBP under light condition resulting in the consumption of O_2 and release of CO_2 is known as photorespiration (Eisenhut et al. 2019; Timm and Hagemann 2020; Shi and Bloom 2021). All the events or pathways leading to the generation of biological energy in the chloroplast occur in a coordinated manner. Disruption to the photosynthetic pathways at any particular instant due to unfavorable circumstances may hinder the process, resulting in the production of excess

ROS which induces oxidative stress and eventually declines plant productivity and survival status.

5.3 Photosynthesis and Stress

Plant encounters innumerable environmental exigencies like excess or low light, heat, flooding, drought, salinity, pollutants, and herbivore and pathogen attack, during life cycle which disrupt the photosynthetic pathways ensuing poor growth and development (Foyer and Shigeoka 2011; Hussain et al. 2018; Iqbal et al. 2020, 2021a, b; Humaira et al. 2020; Jalil and Ansari 2020; Misra et al. 2021; Sachdev et al. 2021). Chloroplasts are primary cell organelle which gets affected under stress, resulting in ROS over-production and photodamage due to impairment of photochemical reaction at PSII, over-reduction of ETC, and abridged Calvin cycle (Mittler 2006). The existence of surplus energy-intermediates, reductants, and O_2 level within chloroplast are certain factors that promote ROS genesis during unpredictable environmental conditions (Chen et al. 2012a). It has been reported that light reaction is primarily involved in the generation of ROS within photosynthetic apparatus (Khorobrykh et al. 2020). The ROS production is further aggravated under the condition of limited availability of CO_2 and impairment of ATP synthase activity (Gururani et al. 2015). In plants, under normal or favorable environmental conditions, LHC absorbs the photon and generates excited singlet chlorophyll ($^1Chl^*$), which partake in the conversion of excitation energy into electrochemical energy via charge separation (Sachdev et al. 2021). However, under stressful events, energy absorbed by LHC of PSII exceeds utilization limit that generates excited triplet chlorophyll ($^3Chl^*$) (Mattos and Moretti 2015; Pospisil 2016). LHC contains a carotenoid pigment which acts as a scavenger and quenches $^3Chl^*$. Insufficient quenching of $^3Chl^*$ results in its concentration built up in the chloroplast (Khorobrykh et al. 2020). The $^3Chl^*$ interacts with O_2 and triggers the production of $^1O_2$, an ROS (Khorobrykh et al. 2020). The highly energized chlorophyll molecules also excite O_2, resulting in the generation of $O_2^{\bullet-}$ (Maurya 2020). Another pathway leading to ROS production in photosynthetic apparatus is reduced availability of CO_2 to Calvin cycle due to condensed stomatal conductance under stress, causing over-reduction of plastoquinone Q_A and Q_B. This hampers charge separation between chlorophyll molecules of PSII reaction center (P680) and pheophytin, leading to the formation of triplet chlorophyll ($^3P680$) which also partake in the production of $^1O_2$ (Sharma et al. 2012). Moreover, reduced CO_2 level decreases $NADP^+$ availability that promotes leakage of an electron from ETC, ultimately reducing O_2 to $O_2^{\bullet-}$ at the acceptor side of PSI via reaction, known as Mehler's reaction (Cruz 2008; Sharma et al. 2012). The third route which induces ROS generation in the chloroplast is the over-reduction of ETC that favors electron leakage from Q_A and Q_B of plastoquinone to O_2, arising $O_2^{\bullet-}$ at reaction center of PSII (Xie et al. 2019). The $O_2^{\bullet-}$ radical formed further liberates the production of stable H_2O_2, through spontaneous or enzymatic dismutation. The thylakoid membrane-bounded, as well as stromal membrane-associated superoxide dismutase

(SOD), catalyzes the conversion of $O_2^{\bullet-}$ to H_2O_2 (Foyer and Shigeoka 2011). The formed H_2O_2 molecules undergo a Fenton reaction wherein the presence of reduced redox metals, particularly, Fe^{2+} or Cu^{1+}, gets converted into highly reactive and toxic $^{\bullet}OH$ radicals (Pospisil 2016). The ROS formed within photosynthetic apparatus rapidly react with crucial biomolecules, for instance, nucleic acids, lipids, proteins, and alter their structure and function, inducing photooxidative stress (Gururani et al. 2015). This causes membrane disruption and dysfunction of proteins and enzymes, affecting the ultrastructure of chloroplast, and may lead to plant cell death (PCD) (Miller et al. 2010; Das and Roychoudhury 2014). The impact of various environmental stress on chloroplast and photosynthesis processes has been listed in Table 5.1.

5.4 Light Stress and Photosynthesis

Light is the major driving factor that is obligatory for photosynthesis in green plants for their survival and production (Sharma et al. 2019). However, illumination exceeding photosynthetic capacity hinders photosynthesis by inducing light stress (Iqbal et al. 2021a), which leads to photoinhibition and photooxidation (Nouri et al. 2015; Sharma et al. 2019). Photoinhibition refers to the damage induced by high light intensity that inhibits the activity of PSII due to overloading of LHC, disrupting electron transfer flux, which ultimately interrupts CO_2 fixation (Murata et al. 2007; Takahashi and Murata 2008; Gururani et al. 2015; Maai et al. 2020; Sharma et al. 2019). The excess light absorbed by the PSII is generally dissipated as heat via a phenomenon called non-photochemical quenching (NPQ) (Gururani et al. 2015). However, light intensity exceeding the NPQ potential of chloroplast exacerbates oxidizing potential within the PSII reaction center that induces damage to D1 protein, which partake in PSII repair and lead to photo-inactivation of PSII (Sharma et al. 2019). The impact of photoinhibition is determined by the disequilibrium that arises due to the difference between the rate of photodamage that has occurred to the PS reaction center and the intrinsic rate of its repair (Murata et al. 2007). *Arabidopsis thaliana* exposed to high light intensity (photon flux density 1300 μmol photon m^{-2} s^{-1}) displays rapid PSII photoinhibition (Ksas et al. 2015). The PSII is the most susceptible component of chloroplast to high light, whereas the PSI displays mechanism protecting it from photoinhibition. However, extensive damage to PSII to the extent of loss of structural integrity prompts transfer of excitation energy toward PSI, exceeding their electron-accepting capacity that triggers photodamage to PSI (Gururani et al. 2015). During high light intensity, excess excitation energy leads to the production of ROS, which contributes to photooxidative damage (Triantaphylides et al. 2008). High light-induced ROS generation inhibits the translation of D1 protein and inactivates the repair process (Gururani et al. 2015). Moreover, $^1O_2$ has been reported to be the primary ROS generated under high irradiance that is responsible for cell death (Sharma et al. 2019). Excessive light also alleviates stomatal opening, limiting the inflow of CO_2 within the chloroplast. Restricted availability of CO_2 for the Calvin cycle reduces the rate of photosynthesis.

Table 5.1 The abiotic stress-triggered damage to chloroplast and photosynthetic activity

Stress	Effects on chloroplast/photosynthesis	References
High light stress	Induce photoinhibition due to overloading of light-harvesting complex (LHC); affect PSII reaction center; damage D1 protein; disturb electron transport flux; induce production of ROS; degrade pigments; reduce chlorophyll/carotenoid ratio, photosynthetic rate, and quantum efficiency	Triantaphylides et al. (2008), Chen et al. (2012b), Erickson et al. (2015), Gururani et al. (2015), Sato et al. (2015), Faseela and Puthur (2017), and Pascual et al. (2017)
Low-light stress	Reduce stomatal conductance, intercellular CO_2, photosynthetic efficiency; trigger oxidative stress; inhibit maximum quantum efficiency of PSII	Wang et al. (2013), Liu et al. (2014), Zhang et al. (2016), and Fu et al. (2017)
High-temperature stress	PSII, Rubisco enzyme, and plastoquinone are prime targets. The changes that occur are irreversible in severe conditions. Disruption of ETC induces ROS production; Stress results in the generation of heat shock proteins and secondary metabolites; decreases net photosynthetic rate; reduces stomatal opening, carbon assimilation, and apparent quantum yield	Sharkey and Zhang (2010), Tan et al. (2011), Mathur et al. (2014), and Abdulmajeed et al. (2017)
Low-temperature stress	Affect PSII due to alteration in chlorophyll antenna complex and reduce quantum yield; increase ROS production, oxidative damage, and MDA content; decrease photochemical efficiency and net photosynthetic rate; reduce chlorophyll content; decline chlorophyll fluorescence; inhibit photosynthesis; affect CO_2 fixation	Zhang et al. (2011), Adam and Murthy (2014), Li et al. (2014), Paredes and Quiles (2015), and Zhao et al. (2019)
Water-deficit stress	Increase stomatal closure; reduce intercellular CO_2; reduce ATP synthesis; inhibit Rubisco enzyme; reduce the quantum yield of PSII, affect the regeneration of $NADP^+$, PSII associated OEC and reaction center; trigger over-reduction of ETC and electron leakage; reduce chlorophyll fluorescence; degrade thylakoid membrane; negatively alter pigments and photosynthetic rate	Skotnica et al. (2000), Albert et al. (2011), Nishiyama and Murata (2014), Sevanto (2014), Tattini et al. (2014), Bertioli et al. (2016), Rivas et al. (2016), Chen et al. (2017), and Liu et al. (2018a)
Pesticide stress	Reduce chlorophyll and carotenoid content; alleviate CO_2 fixation; reduce PS; damage to OEC; restrict ETC from PSII to PSI, leading to ROS	Matouskova et al. (1999), Nadasy et al. (2000), Ralph (2000), Mishra et al. (2008), Parween et al. (2016), Fatma et al. (2018), Rusjan et al.

(continued)

Table 5.1 (continued)

Stress	Effects on chloroplast/photosynthesis	References
	generation; reduce maximum quantum yield; affect the stability of D1 protein; reduce ATP and NADPH production	(2018), and Yüzbaşıoğlu and Dalyan (2019)
Heavy metals stress	Induce ROS production, triggering oxidative damage to biomolecules and interrupt ETC; Damage ultrastructure of chloroplast, modifying the shape, size, and number of chloroplasts; thylakoid inflates; plastoglobule appearance, reduced stomatal conductance and CO_2 assimilation; inactivation of PSII reaction center; induce enzyme degradation; increase expression of chlorophyllase; reduce the content of chlorophyll and carotenoid	Appenroth et al. (2003), Najeeb et al. (2011), Wang et al. (2011), Parmar et al. (2013), Ali et al. (2018), Handa et al. (2018), Paunov et al. (2018), Boostani et al. (2019), Handa et al. (2019), and Kapoor et al. (2019)

Not only high light but low light also affects the photosynthesis process. Under low light, the activity of Rubisco declines dramatically which reduces carbon assimilation and net photosynthesis rate (Liu et al. 2014; Li et al. 2017; Lu et al. 2019). Low-light conditions increase the availability of an electron to O_2, elevating ROS accumulation and inducing oxidative stress (Li et al. 2017; Lu et al. 2019). Lu et al. (2019) have reported a significant decrease in chlorophyll a and b content, net photosynthetic rate, and maximal photochemical efficiency of PSII, in addition to inhibited linear electron transfer chain and increased cyclic electron flow in low-light-stressed tomato leaves. Similarly, pea leaves exposed to low-light stress demonstrate reduced PSII, ATP synthase, Cyt b_6/f complex, and Rubisco level, with an increased level of LHC-II (Leong and Anderson 1984a, b).

5.5 Salinity Stress and Photosynthesis

Salinity induces ion imbalance and water stress in plants which affect photosynthesis, resulting in decreased photosynthetic rate, stomatal conductance, and CO_2 concentration (Kwon et al. 2019). Salinity induces an increase in the concentration of salt in the chloroplast that affects pigments and reduces the photosynthetic capacity of the plant (Parihar et al. 2015). For instance, chlorophyll a, b, a+b, and carotenoid values were reduced by 12.88 and 27.09%, 21.96 and 23.38%, 16.43 and 72.44%, 7.29 and 27.19% in soybean on exposure to 75 and 150 mM NaCl stress, respectively (Osman et al. 2021). Analogously, barley grown under saline conditions was found to display damaging effects on photosynthesis (Tavakkoli et al. 2011). High Na^+ level inhibits K^+ and Ca^{2+} uptake abridging stomatal conductance, and excess Cl^- induced chlorophyll degradation and alleviated PSII quantum yield (Tavakkoli et al. 2011). Moreover, salinity also reduces photosynthetic efficiency due to the inhibition of PSII, stomatal conductance, and CO_2 level, which ultimately alleviate plant growth (Aldesuquy et al. 2014; Ahanger et al. 2017). The diminished stomatal conductance and CO_2 level within chloroplast were reported to decrease photosynthetic function in Carnation (Kwon et al. 2019). The production of ROS under salinity has also been reported to induce oxidative stress in plants. Salinity reduces the assimilation of CO_2, which disrupts the electron transfer chain and diverts electrons toward O_2, generating ROS (Hameed et al. 2021). For instance, deterioration to the ultrastructure of chloroplast in plants subjected to salinity stress has been attributed to ROS, in particular, $^1O_2$, $^{\bullet}OH$, H_2O_2, and $O_2^{\bullet-}$, which trigger damage to proteins, enzymes, and lipids (Yamane et al. 2012; Shu et al. 2013; Suo et al. 2017). The generation of H_2O_2 under salinity has been found to induce thylakoid swelling (Barhoumi et al. 2007; Yamane et al. 2012).

5.6 Temperature Stress and Photosynthesis

The extreme scenario of temperature, below or above ideal conditions (10–15 °C) (Wang et al. 2018), affects plant physiological activities. Photosynthesis is the prime function, which is altered by both high and low temperatures, decreasing plant growth and productivity (Liu et al. 2018b; Wang et al. 2018). PSII is the heat liable component of the thylakoid membrane and therefore high temperature affects the ETC and synthesis of ATP (Yamamoto et al. 2008; Wang et al. 2018). High temperature induces oxidative stress in chloroplast by inducing the production of ROS at PSI and PSII (Sarkar et al. 2018). The ROS generated in chloroplast trigger membrane lipid peroxidation, subsequently increasing electrolyte leakage that significantly damages chlorophyll content; membrane integrity; ultrastructure of chloroplast, in particular, distortion of PSI and PSII arrangement, loss of grana stacking; consequently abridging photosynthetic efficiency (Hasanuzzaman et al. 2013; Sarkar et al. 2018). In a study, high-temperature stress reduces chlorophyll content and damages chloroplast ultrastructure due to increased accumulation of ROS in the photosynthetic apparatus of the wheat plant (Sarkar et al. 2018). Exposure of *Arabidopsis* to high temperature reduces chlorophyll content (Todorov et al. 2003). ROS-induced oxidative stress under high temperature induces dissociation of OEC that inhibits the transfer of electron toward the acceptor side of PSII (Wang et al. 2018). In addition, the generation of ROS also affects protein molecules, particularly, D1 proteins associated with the repair of PSII (Allakhverdiev et al. 2008). Thus, high temperature or heat stress inhibits the repairing process of PSII, affecting linear electron flow (Allakhverdiev et al. 2008). Moreover, heat stress instigates cleavage of reaction center-binding proteins, followed by their aggregation (Allakhverdiev et al. 2008). High-temperature-induced cleavage of D1 and dissociation of manganese-stabilizing 33-kDa proteins of PSII in spinach have been observed (Yamane et al. 1998). In addition to damage induced to PSI and PSII, affecting light reaction, high temperature also disrupts dark reaction occurring in the stroma. High-temperature stress inactivates Rubisco activase that affects the activity of the Rubisco enzyme (Wang et al. 2018). Moreover, disturbed ETC and OEC inhibit the regeneration of Rubisco (Wang et al. 2018). These alterations impede the carbon assimilation process in the stroma of the chloroplast.

Similar to high-temperature stress, low temperature inhibits the photosynthesis process, damaging chloroplast ultrastructure and development (Liu et al. 2018b). Low temperature induces ROS production that makes the thylakoid membrane rigid as well as results in loss of enzymatic activity. The production of ROS in chloroplast under chilling stress occurs due to the reduced availability of $NADP^+$, which disturbs ETC (Hussain et al. 2018). The over-excitation of thylakoid membrane under low-temperature stress has been observed to induce photoinhibition of PS which eventually reduces the function of photosynthetic machinery (Hussain et al. 2018). Several studies have demonstrated that the low temperature triggers photoinhibition of PSI, which induces irreversible damage (Liu et al. 2018b). The cucumber leaves exposed to chilling stress exhibit low-light-induced photoinhibition of PSI and PSII, reduction in CO_2 assimilation and accumulation of ROS. However, the effect on PSI happens to be pronounced due to its sensitivity toward ROS (Zhang et al. 2014).

5.7 Water Stress and Photosynthesis

Water-deficit and waterlogging conditions, both impinge the photosynthetic activity of the plant. The production of ROS under stress has been reported to be a major cue that induces oxidative damages. During water-deficit conditions, the stomatal opening diminishes to reduce water loss through transpiration, depleting CO_2 level within the chloroplast (Ansari and Lin 2011; Dalal and Tripathy 2018; Iqbal et al. 2020). This reduces CO_2 fixation, and enhances photoreduction of O_2 leading to ROS over-production (Hussain et al. 2018). In sunflower, leakage of an electron to O_2 from ETC has been reported under water-deficit scenario (Sgherri et al. 1996). Moreover, water-deficit stress affects chlorophyll content and induces damage to the thylakoid membrane (Iqbal et al. 2020). The waterlogging condition is characterized by submergence of roots, which induces oxygen deficiency condition, i.e., anoxia. The onset of the anoxic condition disrupts ETC in chloroplast and triggers the production of ROS (Chang et al. 2012; Sasidharan et al. 2018), which subsequently induces oxidative damage. Under waterlogging conditions, ROS are responsible for the disruption of membrane integrity and abridged efficiency of PSII (Ashraf 2012). Mahmood et al. (2021) have reported waterlogging induced chlorophyll and carotenoid content reduction, diminished photosynthetic activity, and over-production of ROS, particularly, H_2O_2, in maize.

5.8 Toxic Compounds and Photosynthesis

Pesticides and heavy metals are certain toxic compounds omnipresent due to advances in industrialization. Contamination of soil with these toxic compounds instigates plant growth and productivity. Accumulation of heavy metals and pesticides within plant affect the photosynthetic process by inducing ROS production via over-reduction of ETC (Pandey et al. 2009; Trinh et al. 2014). For instance, pea plants exposed to Cr (VI) displays an elevated generation of ROS, including $O_2^{\bullet-}$, H_2O_2, and $^{\bullet}OH$ (Pandey et al. 2009). In another study, the exposure to metals such as Cd, Cu, Cr, Ni, and Zn inhibits photosynthetic components including PSII, ATP synthase, and enzymes of Calvin cycle; disrupts ETC and triggers the generation of ROS (Dietz et al. 1999; Shahzad et al. 2018). Moreover, Cd-induced oxidative stress in the wheat plant leads to the accumulation of high malondialdehyde (MDA) content and the promotion of electron leakage (Shakirova et al. 2016). Similar to heavy metals, pesticides also impede photosynthetic activity. For instance, herbicides particularly, Bentazon, 3-acetyl-5-isopropyltetramic acid, and Paraquat inhibit the process of photosynthesis either by blocking ETC from PSII by generating ROS which competes with the D1 binding site at PSII (Radwan et al. 2019) or by inhibiting $NADP^+$ (final electron acceptor) from accepting an electron from PS I, enhancing ROS production (Chen et al. 2012a).

5.9 Conclusion

Photosynthesis is the prime target for stress-induced alteration, which ultimately reduces agricultural yield. Environmental cues particularly, high light, extreme temperature, water deficiency, and excess organic and inorganic elements in soil, affect electron transport chain (ETC), light-harvesting complex (LHC), and oxygen-evolving complex (OEC), generating reactive oxygen species (ROS). Accumulation of ROS disrupts the thylakoid membrane, pigments, proteins, and enzymes, inducing oxidative damage. Due to oxidative stress, photosynthetic rate declines, abridging carbon product generation. The efficient scavenging by antioxidants can alleviate the ROS-induced oxidative damage. Several enzymatic and non-enzymatic antioxidants are found in the chloroplast that quenches excess ROS. Particular antioxidant scavenges specific ROS at a specific location; for instance, carotenoid quenches excess excited triplet chlorophyll produced at light-harvesting complex II. Moreover, the performance of antioxidants diminishes under extreme stress conditions. Thus, understanding the sites for the generation of ROS in chloroplast and fortifying antioxidant level could reduce the impact of oxidative damage on photosynthesis.

References

Abdulmajeed AM, Derby SR, Strickland SK, Qaderi MM (2017) Interactive effects of temperature and UVB radiation on methane emissions from different organs of pea plants grown in hydroponic system. J Photochem Photobiol B Biol 166:193–201

Adam S, Murthy SDS (2014) Effect of cold stress on photosynthesis of plants and possible protection mechanisms. In: Approaches to plant stress and their management. Springer, New Delhi, pp 219–226

Ahanger MA, Tomar NS, Tittal M, Argal S, Agarwal RM (2017) Plant growth under water/salt stress: ROS production; antioxidants and significance of added potassium under such conditions. Physiol Mol Biol Plants 23(4):731–744

Albert KR, Mikkelsen TN, Michelsen A, Ro-Poulsen H, van der Linden L (2011) Interactive effects of drought, elevated CO2 and warming on photosynthetic capacity and photosystem performance in temperate heath plants. J Plant Physiol 168(13):1550–1561

Aldesuquy H, Baka Z, Mickky B (2014) Kinetin and spermine mediated induction of salt tolerance in wheat plants: leaf area, photosynthesis and chloroplast ultrastructure of flag leaf at ear emergence. Egypt J Basic Appl Sci 1(2):77–87

Ali E, Hussain N, Shamsi IH, Jabeen Z, Siddiqui MH, Jiang LX (2018) Role of jasmonic acid in improving tolerance of rapeseed (*Brassica napus* L.) to Cd toxicity. J Zhejiang Univ Sci B 19(2):130–146

Allakhverdiev SI, Kreslavski VD, Klimov VV, Los DA, Carpentier R, Mohanty P (2008) Heat stress: an overview of molecular responses in photosynthesis. Photosynth Res 98(1):541–550

Ansari MI, Lin TP (2011) *Arabidopsis thaliana* thylakoid lumen 18.3 protein gene is up-regulated during dehydration. J Physiol Pathophysiol 2(2):18–24

Ansari MI, Lin TP, Liu MS (2011) Molecular characterization of TLP 18.3 gene of *Arabidopsis thaliana*. Int J Integr Biol 11:44–51

Appenroth KJ, Keresztes Á, Sárvári É, Jaglarz A, Fischer W (2003) Multiple effects of chromate on *Spirodela polyrhiza*: electron microscopy and biochemical investigations. Plant Biol 5(3): 315–323. https://doi.org/10.1055/s-2003-40801

Ashraf MA (2012) Waterlogging stress in plants: a review. Afr J Agric Res 7(13):1976–1981

Barber J (2017) A mechanism for water splitting and oxygen production in photosynthesis. Nat Plants 3(4):1–5

Barhoumi Z, Djebali W, Smaoui A, Chaïbi W, Abdelly C (2007) Contribution of NaCl excretion to salt resistance of *Aeluropus littoralis* (Willd) Parl. J Plant Physiol 164(7):842–850

Bertioli DJ, Cannon SB, Froenicke L, Huang G, Farmer AD, Cannon EK et al (2016) The genome sequences of *Arachis duranensis* and *Arachis ipaensis*, the diploid ancestors of cultivated peanut. Nat Genet 48(4):438–446

Boostani HR, Najaf-Ghiri M, Mirsoleimani A (2019) The effect of biochars application on reducing the toxic effects of nickel and growth indices of spinach (*Spinacia oleracea* L.) in a calcareous soil. Environ Sci Pollut Res 26(2):1751–1760

Bulgari R, Franzoni G, Ferrante A (2019) Biostimulants application in horticultural crops under abiotic stress conditions. Agronomy 9(6):306. https://doi.org/10.3390/agronomy9060306

Cassia R, Nocioni M, Correa-Aragunde N, Lamattina L (2018) Climate change and the impact of greenhouse gasses: CO2 and NO, friends and foes of plant oxidative stress. Front Plant Sci 9: 273

Chang R, Jang CJ, Branco-Price C, Nghiem P, Bailey-Serres J (2012) Transient MPK6 activation in response to oxygen deprivation and reoxygenation is mediated by mitochondria and aids seedling survival in *Arabidopsis*. Plant Mol Biol 78(1–2):109–122

Chen S, Yin C, Strasser RJ, Yang C, Qiang S (2012a) Reactive oxygen species from chloroplasts contribute to 3-acetyl-5-isopropyltetramic acid-induced leaf necrosis of *Arabidopsis thaliana*. Plant Physiol Biochem 52:38–51

Chen L, Jia H, Tian Q, Du L, Gao Y, Miao X, Liu Y (2012b) Protecting effect of phosphorylation on oxidative damage of D1 protein by down-regulating the production of superoxide anion in photosystem II membranes under high light. Photosynth Res 112(2):141–148. table

Chen YE, Zhang CM, Su YQ, Ma J, Zhang ZW, Yuan M et al (2017) Responses of photosystem II and antioxidative systems to high light and high temperature co-stress in wheat. Environ Exp Bot 135:45–55

Choudhury FK, Rivero RM, Blumwald E, Mittler R (2017) Reactive oxygen species, abiotic stress and stress combination. Plant J 90(5):856–867

Cooper GM, Hausman RE (2000) The cell: a molecular approach. Structure and organization of actin filaments, Sinauer Associates, Sunderland, MA

Cruz DCM (2008) Drought stress and reactive oxygen species: production, scavenging and signaling. Plant Signal Behav 3(3):156–165. https://doi.org/10.4161/psb.3.3.5536

Dalal VK, Tripathy BC (2018) Water-stress induced downsizing of light-harvesting antenna complex protects developing rice seedlings from photo-oxidative damage. Sci Rep 8(1):1–16

Das K, Roychoudhury A (2014) Reactive oxygen species (ROS) and response of antioxidants as ROS-scavengers during environmental stress in plants. Front Environ Sci 2:53. https://doi.org/10.3389/fenvs.2014.00053

Dietz KJ, Baier M, Kramer U (1999) Free radicals and reactive oxygen species as mediators of heavy metal toxicity in plants. In: Prasad MNV, Hagemeyer J (eds) Heavy metal stress in plants: from molecules to ecosystems. Springer, Berlin, pp 73–97. https://doi.org/10.1007/978-3-662-07745-0_4

Eisenhut M, Roell MS, Weber AP (2019) Mechanistic understanding of photorespiration paves the way to a new green revolution. New Phytol 223(4):1762–1769

Erickson E, Wakao S, Niyogi KK (2015) Light stress and photoprotection in *Chlamydomonas reinhardtii*. Plant J 82(3):449–465

Faseela P, Puthur JT (2017) Chlorophyll a fluorescence changes in response to short and long term high light stress in rice seedlings. Indian J Plant Physiol 22(1):30–33

Fatma F, Kamal A, Srivastava A (2018) Exogenous application of salicylic acid mitigates the toxic effect of pesticides in *Vigna radiata* (L.) Wilczek. J Plant Growth Regul 37(4):1185–1194

Foyer CH, Shigeoka S (2011) Understanding oxidative stress and antioxidant functions to enhance photosynthesis. Plant Physiol 155(1):93–100

Foyer CH, Neukermans J, Queval G, Noctor G, Harbinson J (2012) Photosynthetic control of electron transport and the regulation of gene expression. J Exp Bot 63(4):1637–1661

Fu Y, Li H, Yu J, Liu H, Cao Z, Manukovsky NS, Liu H (2017) Interaction effects of light intensity and nitrogen concentration on growth, photosynthetic characteristics and quality of lettuce (*Lactuca sativa* L. Var. youmaicai). Sci Hortic 214:51–57

Gotoh E, Suetsugu N, Yamori W, Ishishita K, Kiyabu R, Fukuda M, Higa T, Shirouchi B, Wada M (2018) Chloroplast accumulation response enhances leaf photosynthesis and plant biomass production. Plant Physiol 178(3):1358–1369

Gururani MA, Venkatesh J, Tran LSP (2015) Regulation of photosynthesis during abiotic stress-induced photoinhibition. Mol Plant 8(9):1304–1320

Hameed A, Ahmed MZ, Hussain T, Aziz I, Ahmad N, Gul B, Nielsen BL (2021) Effects of salinity stress on chloroplast structure and function. Cells 10(8):2023

Handa N, Kohli SK, Sharma A, Thukral AK, Bhardwaj R, Alyemeni MN, Wijaya L, Ahmad P (2018) Selenium ameliorates chromium toxicity through modifications in pigment system, antioxidative capacity, osmotic system, and metal chelators in *Brassica juncea* seedlings. S Afr J Bot 119:1–10

Handa N, Kohli SK, Sharma A, Thukral AK, Bhardwaj R, Abd Allah EF, Alqarawi AA, Ahmad P (2019) Selenium modulates dynamics of antioxidative defence expression, photosynthetic attributes and secondary metabolites to mitigate chromium toxicity in *Brassica juncea* L. plants. Environ Exp Bot 161:180–192

Hasanuzzaman M, Nahar K, Fujita M (2013) Extreme temperature responses, oxidative stress and antioxidant defense in plants. In: Abiotic stress-plant responses and applications in agriculture, vol 13. Intech, pp 169–205

Humaira R, Rahul G, Israil AM, Kumar SA (2020) Impact of bioengineered copper quantum dots on germination, photosynthetic pigment and antioxidant response in Chick-pea under Dark stress environment. Res J Biotechnol 15:12

Hussain HA, Hussain S, Khaliq A, Ashraf U, Anjum SA, Men S, Wang L (2018) Chilling and drought stresses in crop plants: implications, cross talk, and potential management opportunities. Front Plant Sci 9:393. https://doi.org/10.3389/fpls.2018.00393

Iqbal MS, Singh AK, Ansari MI (2020) Effect of drought stress on crop production. In: Rakshit A, Singh HB, Singh AK, Singh US, Fraceto L (eds) New frontiers in stress management for durable agriculture. Springer, Singapore, pp 35–47

Iqbal Z, Iqbal MS, Hashem A, Abd Allah EF, Ansari MI (2021a) Plant defense responses to biotic stress and its interplay with fluctuating dark/light conditions. Front Plant Sci 12:297

Iqbal Z, Iqbal MS, Khan MIR, Ansari MI (2021b) Toward integrated multi-omics intervention: rice trait improvement and stress management. Front Plant Sci 12:741419

Jalil SU, Ansari MI (2020) Physiological role of Gamma-aminobutyric acid in salt stress tolerance. In: Hasanuzzaman M, Tanveer M (eds) Salt and drought stress tolerance in plants. Springer, Cham, pp 337–350

Joliot P, Johnson GN (2011) Regulation of cyclic and linear electron flow in higher plants. Proc Natl Acad Sci 108(32):13317–13322

Kapoor D, Singh MP, Kaur S, Bhardwaj R, Zheng B, Sharma A (2019) Modulation of the functional components of growth, photosynthesis, and anti-oxidant stress markers in cadmium exposed *Brassica juncea* L. Plants 8(8):260

Khorobrykh S, Havurinne V, Mattila H, Tyystjärvi E (2020) Oxygen and ROS in photosynthesis. Plants 9(1):91

Kirchhoff H (2019) Chloroplast ultrastructure in plants. New Phytol 223(2):565–574

Kouril R, Zygadlo A, Arteni AA, de Wit CD, Dekker JP, Jensen PE, Scheller HV, Boekema EJ (2005) Structural characterization of a complex of photosystem I and light-harvesting complex II of *Arabidopsis thaliana*. Biochemistry 44(33):10935–10940

Ksas B, Becuwe N, Chevalier A, Havaux M (2015) Plant tolerance to excess light energy and photooxidative damage relies on plastoquinone biosynthesis. Sci Rep 5(1):1–16

Kwon OK, Mekapogu M, Kim KS (2019) Effect of salinity stress on photosynthesis and related physiological responses in carnation (*Dianthus caryophyllus*). Hortic Environ Biotechnol 60(6): 831–839

Leong TY, Anderson JM (1984a) Adaptation of the thylakoid membranes of pea chloroplasts to light intensities. I. Study on the distribution of chlorophyll-protein complexes. Photosynth Res 5(2):105–115

Leong TY, Anderson JM (1984b) Adaptation of the thylakoid membranes of pea chloroplasts to light intensities. II. Regulation of electron transport capacities, electron carriers, coupling factor (CF 1) activity and rates of photosynthesis. Photosynth Res 5(2):117–128

Li X, Cai J, Liu F, Dai T, Cao W, Jiang D (2014) Cold priming drives the sub-cellular antioxidant systems to protect photosynthetic electron transport against subsequent low temperature stress in winter wheat. Plant Physiol Biochem 82:34–43

Li Y, Fan Y, Ma Y, Zhang Z, Yue H, Wang L, Li J, Jiao Y (2017) Effects of exogenous γ-aminobutyric acid (GABA) on photosynthesis and antioxidant system in pepper (*Capsicum annuum* L.) seedlings under low light stress. J Plant Growth Regul 36(2):436–449

Lichtenthaler HK, Babani F (2004) Light adaptation and senescence of the photosynthetic apparatus. Changes in pigment composition, chlorophyll fluorescence parameters and photosynthetic activity. In: Chlorophyll a fluorescence. Springer, Dordrecht, pp 713–736

Liu QH, Xiu WU, Chen BC, Jie GAO (2014) Effects of low light on agronomic and physiological characteristics of rice including grain yield and quality. Rice Sci 21(5):243–251

Liu X, Li L, Li M, Su L, Lian S, Zhang B, Li X, Ge K, Li L (2018a) AhGLK1 affects chlorophyll biosynthesis and photosynthesis in peanut leaves during recovery from drought. Sci Rep 8(1): 1–11

Liu X, Zhou Y, Xiao J, Bao F (2018b) Effects of chilling on the structure, function and development of chloroplasts. Front Plant Sci 9:1715

Lu T, Yu H, Li Q, Chai L, Jiang W (2019) Improving plant growth and alleviating photosynthetic inhibition and oxidative stress from low-light stress with exogenous GR24 in tomato (*Solanum lycopersicum* L.) seedlings. Front Plant Sci 10:490

Maai E, Nishimura K, Takisawa R, Nakazaki T (2020) Light stress-induced chloroplast movement and midday depression of photosynthesis in sorghum leaves. Plant Product Sci 23(2):172–181

Mahmood U, Hussain S, Hussain S, Ali B, Ashraf U, Zamir S, Al-Robai SA, Alzahrani FO, Hano C, El-Esawi MA (2021) Morpho-physio-biochemical and molecular responses of maize hybrids to salinity and waterlogging during stress and recovery phase. Plants 10(7):1345

Mathur S, Agrawal D, Jajoo A (2014) Photosynthesis: response to high temperature stress. J Photochem Photobiol B Biol 137:116–126

Matouskova M, Naus J, Flasarova M (1999) A long-term response of chlorophyll fluorescence induction to one-shot application of cyanazine on barley plants and its relation to crop yield. Photosynthetica 37(2):281–294

Mattos LM, Moretti CL (2015) Oxidative stress in plants under drought conditions and the role of different enzymes. Enzym Eng 5:1–6

Maurya AK (2020) Oxidative stress in crop plants. In: Hasanuzzaman M (ed) Agronomic crops: stress responses and tolerance. Springer, Singapore, pp 349–380

Miller GAD, Suzuki N, Ciftci-Yilmaz SULTAN, Mittler RON (2010) Reactive oxygen species homeostasis and signalling during drought and salinity stresses. Plant Cell Environ 33(4): 453–467

Mishra V, Srivastava G, Prasad SM, Abraham G (2008) Growth, photosynthetic pigments and photosynthetic activity during seedling stage of cowpea (*Vigna unguiculata*) in response to UV-B and dimethoate. Pestic Biochem Physiol 92(1):30–37

Misra V, Mall AK, Ansari MI (2021) Physiological and molecular responses to salinity due to excessive Na+ in plants. In: Harsh environment and plant resilience: molecular and functional aspects. Springer, Cham, p 291

Mittler R (2006) Abiotic stress, the field environment and stress combination. Trends Plant Sci 11(1):15–19

Murata N, Takahashi S, Nishiyama Y, Allakhverdiev SI (2007) Photoinhibition of photosystem II under environmental stress. Biochim Biophys Acta Bioenerg 1767(6):414–421

Nadasy E, Lehoczky E, Lukacs P, Adam P (2000) Influence of different pre-emergent herbicides on the growth of soybean varieties. Z Pflanzenkr Pflanzenschutz Sonderheft 17:635–640

Najeeb U, Jilani G, Ali S, Sarwar M, Xu L, Zhou W (2011) Insights into cadmium induced physiological and ultra-structural disorders in *Juncus effusus* L. and its remediation through exogenous citric acid. J Hazard Mater 186(1):565–574

Nawrocki WJ, Bailleul B, Picot D, Cardol P, Rappaport F, Wollman FA, Joliot P (2019) The mechanism of cyclic electron flow. Biochim Biophys Acta Bioenerg 1860(5):433–438

Nishiyama Y, Murata N (2014) Revised scheme for the mechanism of photoinhibition and its application to enhance the abiotic stress tolerance of the photosynthetic machinery. Appl Microbiol Biotechnol 98(21):8777–8796

Nouri MZ, Moumeni A, Komatsu S (2015) Abiotic stresses: insight into gene regulation and protein expression in photosynthetic pathways of plants. Int J Mol Sci 16(9):20392–20416

Osman MS, Badawy AA, Osman AI, Abdel Latef AAH (2021) Ameliorative impact of an extract of the halophyte *Arthrocnemum macrostachyum* on growth and biochemical parameters of soybean under salinity stress. J Plant Growth Regul 40(3):1245–1256

Pandey V, Dixit V, Shyam R (2009) Chromium effect on ROS generation and detoxification in pea (*Pisum sativum*) leaf chloroplasts. Protoplasma 236:85–95. https://doi.org/10.1007/s00709-009-0061-8

Paredes M, Quiles MJ (2015) The effects of cold stress on photosynthesis in *Hibiscus* plants. PLoS One 10(9):e0137472

Parihar P, Singh S, Singh R, Singh VP, Prasad SM (2015) Effect of salinity stress on plants and its tolerance strategies: a review. Environ Sci Pollut Res 22(6):4056–4075

Parmar P, Kumari N, Sharma V (2013) Structural and functional alterations in photosynthetic apparatus of plants under cadmium stress. Bot Stud 54(1):1–6

Parween T, Jan S, Mahmooduzzafar S, Fatma T, Siddiqui ZH (2016) Selective effect of pesticides on plant—a review. Crit Rev Food Sci Nutr 56(1):160–179

Pascual J, Rahikainen M, Kangasjärvi S (2017) Plant light stress. In: eLS, pp 1–6

Paunov M, Koleva L, Vassilev A, Vangronsveld J, Goltsev V (2018) Effects of different metals on photosynthesis: cadmium and zinc affect chlorophyll fluorescence in durum wheat. Int J Mol Sci 19(3):787

Pospisil P (2016) Production of reactive oxygen species by photosystem II as a response to light and temperature stress. Front Plant Sci 7:1950. https://doi.org/10.3389/fpls.2016.01950

Radwan DEM, Mohamed AK, Fayez KA, Abdelrahman AM (2019) Oxidative stress caused by Basagran® herbicide is altered by salicylic acid treatments in peanut plants. Heliyon 5(5): e01791. https://doi.org/10.1016/j.heliyon.2019.e01791

Ralph PJ (2000) Herbicide toxicity of *Halophila ovalis* assessed by chlorophyll a fluorescence. Aquat Bot 66(2):141–152

Reddy AR, Rasineni GK, Raghavendra AS (2010) The impact of global elevated CO_2 concentration on photosynthesis and plant productivity. Curr Sci 99:46–57

Rivas R, Falcão HM, Ribeiro RV, Machado EC, Pimentel C, Santos MG (2016) Drought tolerance in cowpea species is driven by less sensitivity of leaf gas exchange to water deficit and rapid recovery of photosynthesis after rehydration. S Afr J Bot 103:101–107

Rusjan A, Dębski H, Koczkodaj D, Mitrus J, Horbowicz M (2018) Effect of Propaquizafop treatment on pigments content and growth of sweet maize seedlings. Acta Sci Pol Agric 16(4):227–234

Sachdev S, Ansari SA, Ansari MI, Fujita M, Hasanuzzaman M (2021) Abiotic stress and reactive oxygen species: generation, signaling, and defense mechanisms. Antioxidants 10(2):277

Sarkar J, Chakraborty B, Chakraborty U (2018) Plant growth promoting rhizobacteria protect wheat plants against temperature stress through antioxidant signalling and reducing chloroplast and membrane injury. J Plant Growth Regul 37(4):1396–1412

Sasidharan R, Hartman S, Liu Z, Martopawiro S, Sajeev N, van Veen H, Yeung E, Voesenek LA (2018) Signal dynamics and interactions during flooding stress. Plant Physiol 176(2): 1106–1117. https://doi.org/10.1104/pp.17.01232

Sato R, Ito H, Tanaka A (2015) Chlorophyll b degradation by chlorophyll b reductase under high-light conditions. Photosynth Res 126(2):249–259

Sevanto S (2014) Phloem transport and drought. J Exp Bot 65(7):1751–1759

Sgherri CL, Pinzino C, Navari-Izzo F (1996) Sunflower seedlings subjected to increasing stress by water deficit: changes in $O2^{\bullet -}$ production related to the composition of thylakoid membranes. Physiol Plant 96(3):446–452

Shahzad B, Tanveer M, Che Z, Rehman A, Cheema SA, Sharma A, Song H, ur Rehman S, Zhaorong D (2018) Role of 24-epibrassinolide (EBL) in mediating heavy metal and pesticide induced oxidative stress in plants: a review. Ecotoxicol Environ Saf 147:935–944

Shakirova FM, Allagulova CR, Maslennikova DR, Klyuchnikova EO, Avalbaev AM, Bezrukova MV (2016) Salicylic acid-induced protection against cadmium toxicity in wheat plants. Environ Exp Bot 122:19–28

Sharkey TD, Zhang R (2010) High temperature effects on electron and proton circuits of photosynthesis. J Integr Plant Biol 52(8):712–722

Sharma P, Jha AB, Dubey RS, Pessarakli M (2012) Reactive oxygen species, oxidative damage, and antioxidative defense mechanism in plants under stressful conditions. J Bot 2012:1–26

Sharma A, Kumar V, Shahzad B, Ramakrishnan M, Sidhu GPS, Bali AS, Handa N, Kapoor D, Yadav P, Khanna K, Bakshi P, Rehman A, Kohli SK, Khan EA, Parihar RD, Yuan H, Thukral AK, Bhardwaj R, Zheng B (2019) Photosynthetic response of plants under different abiotic stresses: a review. J Plant Growth Regul 39:1–23

Shi X, Bloom A (2021) Photorespiration: the futile cycle? Plants 10(5):908

Shu S, Yuan LY, Guo SR, Sun J, Yuan YH (2013) Effects of exogenous spermine on chlorophyll fluorescence, antioxidant system and ultrastructure of chloroplasts in *Cucumis sativus* L. under salt stress. Plant Physiol Biochem 63:209–216

Skotnica J, Matoušková M, Nauš J, Lazár D, Dvořák L (2000) Thermoluminescence and fluorescence study of changes in Photosystem II photochemistry in desiccating barley leaves. Photosynth Res 65(1):29–40

Suo J, Zhao Q, David L, Chen S, Dai S (2017) Salinity response in chloroplasts: insights from gene characterization. Int J Mol Sci 18(5):1011

Takahashi S, Murata N (2008) How do environmental stresses accelerate photoinhibition? Trends Plant Sci 13(4):178–182

Tan W, Wei Meng Q, Brestic M, Olsovska K, Yang X (2011) Photosynthesis is improved by exogenous calcium in heat-stressed tobacco plants. J Plant Physiol 168(17):2063–2071

Tattini M, Velikova V, Vickers C, Brunetti C, Di Ferdinando M, Trivellini A et al (2014) Isoprene production in transgenic tobacco alters isoprenoid, non-structural carbohydrate and phenylpropanoid metabolism, and protects photosynthesis from drought stress. Plant Cell Environ 37(8):1950–1964

Tavakkoli E, Fatehi F, Coventry S, Rengasamy P, McDonald GK (2011) Additive effects of Na+ and Cl–ions on barley growth under salinity stress. J Exp Bot 62(6):2189–2203

Timm S, Hagemann M (2020) Photorespiration—how is it regulated and how does it regulate overall plant metabolism? J Exp Bot 71(14):3955–3965

Todorov DT, Karanov EN, Smith AR, Hall MA (2003) Chlorophyllase activity and chlorophyll content in wild type and eti 5 mutant of *Arabidopsis thaliana* subjected to low and high temperatures. Biol Plant 46(4):633–636

Triantaphylides C, Krischke M, Hoeberichts FA, Ksas B, Gresser G, Havaux M, Breusegem FV, Mueller MJ (2008) Singlet oxygen is the major reactive oxygen species involved in photooxidative damage to plants. Plant Physiol 148(2):960–968

Trinh NN, Huang TL, Chi WC, Fu SF, Chen CC, Huang HJ (2014) Chromium stress response effect on signal transduction and expression of signaling genes in rice. Physiol Plant 150:205–224. https://doi.org/10.1111/ppl.12088

Voesenek LA, Bailey-Serres J (2015) Flood adaptive traits and processes: an overview. New Phytol 206(1):57–73

Wang F, Chen F, Cai Y, Zhang G, Wu F (2011) Modulation of exogenous glutathione in ultrastructure and photosynthetic performance against Cd stress in the two barley genotypes differing in Cd tolerance. Biol Trace Elem Res 144(1):1275–1288

Wang L, Deng F, Ren WJ, Yang WY (2013) Effects of shading on starch pasting characteristics of indica hybrid rice (*Oryza sativa* L.). PLoS One 8(7):e68220

Wang QL, Chen JH, He NY, Guo FQ (2018) Metabolic reprogramming in chloroplasts under heat stress in plants. Int J Mol Sci 19(3):849

Xie X, He Z, Chen N, Tang Z, Wang Q, Cai Y (2019) The roles of environmental factors in regulation of oxidative stress in plant. Biomed Res Int 2019:9732325. https://doi.org/10.1155/2019/9732325

Yamamoto Y, Aminaka R, Yoshioka M, Khatoon M, Komayama K, Takenaka D, Yamashita A, Nijo N, Inagawa K, Morita N, Sasaki T, Yamamoto Y (2008) Quality control of photosystem II: impact of light and heat stresses. Photosynth Res 98(1):589–608

Yamane Y, Kashino Y, Koike H, Satoh K (1998) Effects of high temperatures on the photosynthetic systems in spinach: oxygen-evolving activities, fluorescence characteristics and the denaturation process. Photosynth Res 57(1):51–59

Yamane K, Taniguchi M, Miyake H (2012) Salinity-induced subcellular accumulation of H_2O_2 in leaves of rice. Protoplasma 249(2):301–308

Yamori W (2016) Photosynthetic response to fluctuating environments and photoprotective strategies under abiotic stress. J Plant Res 129(3):379–395

Yüzbaşıoğlu E, Dalyan E (2019) Salicylic acid alleviates thiram toxicity by modulating antioxidant enzyme capacity and pesticide detoxification systems in the tomato (*Solanum lycopersicum* Mill.). Plant Physiol Biochem 135:322–330

Zhang Q, Zhang JZ, Chow WS, Sun LL, Chen JW, Chen YJ, Peng CL (2011) The influence of low temperature on photosynthesis and antioxidant enzymes in sensitive banana and tolerant plantain (*Musa* sp.) cultivars. Photosynthetica 49(2):201–208

Zhang ZS, Yang C, Gao HY, Zhang LT, Fan XL, Liu MJ (2014) The higher sensitivity of PSI to ROS results in lower chilling–light tolerance of photosystems in young leaves of cucumber. J Photochem Photobiol B Biol 137:127–134

Zhang H, Zhong H, Wang J, Sui X, Xu N (2016) Adaptive changes in chlorophyll content and photosynthetic features to low light in *Physocarpus amurensis* Maxim and *Physocarpus opulifolius* "Diabolo". PeerJ 4:e2125

Zhao M, Yuan L, Wang J, Xie S, Zheng Y, Nie L, Zhu S, Hou J, Chen G, Wang C (2019) Transcriptome analysis reveals a positive effect of brassinosteroids on the photosynthetic capacity of wucai under low temperature. BMC Genomics 20(1):1–19

6 Generation and Fate of ROS in Mitochondria

6.1 Introduction

The metabolic activity of aerobic organisms under biotic as well as abiotic stress generates reactive oxygen species (ROS) exorbitantly as a by-product in multiple cellular compartments including mitochondria (Jalil and Ansari 2018; Khan et al. 2021). Mitochondria, the chief metabolic organelle, perform cellular respiration via tricarboxylic acid (TCA) cycle and electron transport chain (ETC) integrated with oxidative phosphorylation (OXPHOS) (Keunen et al. 2011; Rao et al. 2017; Taylor 2018). The final output generated via these pathways includes CO_2, H_2O, and ATP (a form of energy) (Keunen et al. 2011; Rao et al. 2017; Taylor 2018). The ATP generated facilitates plant to perform metabolic activity and proliferate (Meyer et al. 2019). Plant mitochondria also display photorespiration process, alternative pathway of ETC, metabolism of biomolecules for "own biogenesis," retrograde signaling and induce programmed cell death (PCD) under adverse condition (Keunen et al. 2011; Rao et al. 2017). The alternative pathway of ETC does not yield ATP, however, has been reported to be a "target" as well as the "regulator" of stress encountered by the mitochondria (Keunen et al. 2011). The photorespiration and alternative pathway of ETC are unique features which occur only in plant mitochondria, due to photosynthetic functions; and presence of cyanide-insensitive alternate oxidase (AOX), rotenone-insensitive nicotinamide adenine dinucleotide (NAD(P)H) dehydrogenases (NDs), and uncoupling protein (UCP), respectively (Moller 2001; Keunen et al. 2011), formulating study of plant mitochondria different from its mammalian counterpart.

The production of ATP, the key operating task of mitochondria is dependent upon inner mitochondrial membrane (IMM) localized ETC and proton flux across the membrane generated by electrochemical gradient (Blokhina and Fagerstedt 2010). The intact functioning of ETC in conjugation with ATP synthase is essential for driving cellular activities, however, the altered physiological and pathological conditions cause over-reduction of ETC, eventually leaking electron from the

S. Sachdev et al., *Reactive Oxygen Species in Plants*,
https://doi.org/10.1007/978-981-19-9884-3_6

chain and producing ROS (Blokhina and Fagerstedt 2010). The generation of reactive nitrogen species (RNS) in mitochondria under unfavorable external stimuli elicits ROS formation by interfering with the activity of ETC (Navrot et al. 2007). ROS in mitochondria damage biomolecules inducing oxidative stress which impair its metabolic activities (Navrot et al. 2007). The mitochondrial DNA subjected to oxidative stress undergoes severe damages compared to nuclear DNA because of dearth of chromatin organization and low ability to repair DNA (Navrot et al. 2007). ROS in mitochondria though trigger oxidative stress, but also participate in signal transduction network (Navrot et al. 2007; Kowaltowski et al. 2009). Mitochondria under the state of dysfunctioning, induce retrograde signaling toward nucleus, which alter expression of nuclear genes (Rhoads et al. 2006). In signaling transduction pathway, ROS act as a messenger molecule, triggering transcription of genes and/or activation/deactivation of proteins, assisting in alleviation of stress-induced damages and/or PCD (Rhoads et al. 2006; Navrot et al. 2007).

ROS act as double-faced character, which induces oxidative damage as well as resilience against unfavorable stimuli. To alleviate the damaging impact of ROS, however, mitochondria recruit an arsenal of antioxidants, particularly, superoxide dismutase (SOD), ascorbate-glutathione (AsA-GSH) cycle enzymes, and peroxiredoxins (PRX) which scavenge and inhibit their accumulation reaching to the harmful level (Navrot et al. 2007; Blokhina and Fagerstedt 2010). Beside this, mitochondria regulate ROS through action of AOX, UCP, and ATP-sensitive K^+ channel (Blokhina and Fagerstedt 2010). The ultimate fate of ROS within cellular organelles can be deciphered through gaining deeper understanding of process and pathways leading to their generation, action, and maintenance within that particular organelle. This chapter aims to provide comprehensive knowledge and mechanisms pertaining to ROS generation and fate within plant mitochondria.

6.2 Structure and Functions of Plant Mitochondria

Mitochondria is a double membrane-bounded and spherical, filamentous or ellipsoidal-shaped organelle with size 2 μm in length (Douce 2012; Taylor 2018). The double-membrane of mitochondria comprises outer (OMM) and inner (IMM) membrane (Fig. 6.1). The two membranes enclose two compartments, (1) The intermembrane space (IMS), present between OMM and IMM; and (2) Matrix, the space in the center of the organelle, enveloped by the IMM (Douce 2012). The OMM serve as a barrier between mitochondria and the cytoplasm (Cogliati et al. 2016) and allow the movement of small molecules inside the mitochondria (Taylor 2018). The IMM is considered as the ultimate barrier of the organelle which allows entry of specific molecules (Taylor 2018). It has folded invaginations known as cristae, projected toward matrix (Cogliati et al. 2016), which increases membrane surface area (Taylor 2018). The five large complexes (I–V) viz., NADH: ubiquinone oxidoreductase (first complex), succinate dehydrogenase (SDH) (second complex), cytochrome c reductase (third complex), cytochrome c oxidase (fourth complex), and ATP synthase (fifth complex), found embedded in

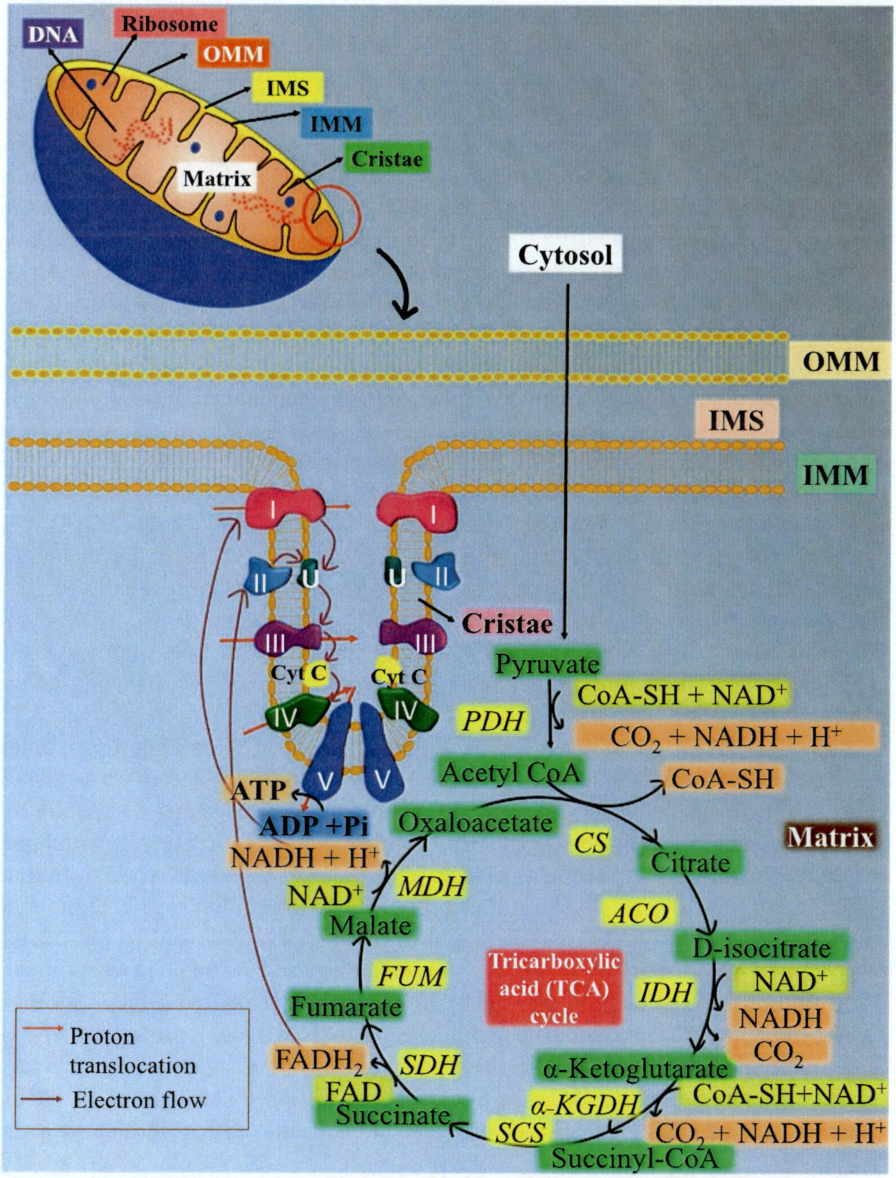

Fig. 6.1 Showing the simplified structural and functional features of mitochondria. Mitochondria, a double membrane-bound organelle, comprising outer mitochondrial membrane (OMM), inner mitochondrial membrane (IMM), and intermembrane space (IMS) between two membranes. The IMM shows inward invaginations called cristae which include enzymatic complexes I, II, III, IV and V; ubiquinone (U) and cytochrome c (cyt c) participating in electron transport chain (ETC) and oxidative phosphorylation. The area surrounded by IMM is called matrix, which contains DNA and ribosome and is a site where tricarboxylic acid (TCA) cycle occurs. The substrate pyruvate, entered in mitochondrial matrix from cytosol, is converted into acetyl Coenzyme A (Acetyl Co-A) on action of enzyme pyruvate dehydrogenase (PDH). Acetyl Co-A enters TCA cycle and through a series of eight reactions generates reduced nicotinamide adenine dinucleotide (NADH) and reduced flavin

IMM cristae, which perform ETC coupled with OXPHOS to generate ATP (Joubert and Puff 2021). The matrix of plant mitochondria is a protein-rich compartment, where protein molecules are found in closely packed form (Douce 2012). The matrix also contains ribosomes and some genetic material essentially required for the semi-autonomous functioning of the mitochondria (Douce 2012; Cupp and Nielsen 2014).

Mitochondria, the powerhouse of the cell, perform a catabolic process to generate energy (Taylor 2018) in multiple steps, beginning with the TCA cycle, subsequently followed by ETC associated with OXPHOS. The organic substrates, particularly, pyruvate is transferred from cytosol to mitochondria, where it changed into acetyl coenzyme A (acetyl-Co-A), which gets subsequently oxidized via TCA cycle (Taylor 2018). The reduced forms of NADH and flavin adenine dinucleotide ($FADH_2$) generated during the TCA cycle enters ETC and facilitates production of ATP, which then moves to the cytosol and utilized for the plant activity (Daloso et al. 2015; Taylor 2018). In ETC, NADH and $FADH_2$ are re-oxidized and again enter TCA cycle to continue the process. The CO_2 and water are liberated as by-products of TCA cycle and the ETC, respectively.

Tricarboxylic Acid Cycle Tricarboxylic acid (TCA) cycle is the major metabolic process occurring in aerobic organisms, which ensue in series of steps to oxidize biochemical compounds to generate energy (Gasmi et al. 2021). TCA cycle generates nearly 70% of ATP to fulfill energy demand of a living body (Gasmi et al. 2021). It occurs within mitochondrial matrix, where pyruvate, the end product of glycolysis, is imported from the cytosol and acts as a substrate to generate ATP (Gasmi et al. 2021). The process begins with the decarboxylation of pyruvate in the presence of *pyruvate dehydrogenase* (PDH) enzyme, forming acetyl-CoA (Gasmi et al. 2021) (Fig. 6.1). The acetyl-CoA react with oxaloacetate (OAA) in the presence of citrate synthase (CS) to form citrate, which subsequently undergoes reversible isomerization in the presence of aconitase (ACO) to form isocitrate (Mongin et al. 2012). The cycle further proceeded by formation of α-ketoglutarate in the presence of enzyme isocitrate dehydrogenase (IDH), followed by production of succinyl CoA, succinate, fumarate, and malate in each subsequent event of the cycle on action of α-ketoglutarate dehydrogenase complex (α-KGDH), succinyl CoA synthetase (SCS), succinate dehydrogenase (SDH), and fumarase (FUM), respectively, which concludes with the regeneration of OAA in a reaction catalyzed by malate dehydrogenase (MDH) (Sweetlove et al. 2010; Mongin et al. 2012) (Fig. 6.1). The TCA cycle engages eight reactions catalyzed by eight enzymes to

Fig. 6.1 (continued) adenine dinucleotide ($FADH_2$), which supply electron to ETC at complex I and complex III, respectively. The electron flow from one complex to another ultimately generates ATP at complex V due to proton gradient. *Co-A-SH* Coenzyme A with sulfhydryl group, *CS* citrate synthase, *ACO* aconitase, *IDH* Isocitrate dehydrogenase, *α-KGDH* α-ketoglutarate dehydrogenase, *SCS* succinyl Coenzyme A synthetase, *SDH* succinate dehydrogenase, *FUM* fumarase, *MDH* malate dehydrogenase, *FAD* oxidized flavin adenine dinucleotide, *NAD^+* oxidized nicotinamide adenine dinucleotide produce three NADH and one $FADH_2$ in a cycle with CO_2 (Mongin et al. 2012; Daloso et al. 2015; Gasmi et al. 2021). The NADH and $FADH_2$ drive the respiratory ETC, which eventually generates ATP through oxidative phosphorylation (Gasmi et al. 2021).

Respiratory Electron Transport Chain and Oxidative Phosphorylation The classical respiratory electron transport chain (ETC) (cytochrome pathway) tightly coupled with oxidative phosphorylation (OXPHOS) serves the function of energy production by utilizing reducing equivalents generated during TCA cycle (Mongin et al. 2012; Nolfi-Donegan et al. 2020). A simplified structure of ETC and OXPHOS has been demonstrated in Fig. 6.1. The ETC comprises five major multiprotein enzyme complexes I–V along with mobile ubiquinone and mobile cytochrome c (Mongin et al. 2012; Schertl and Braun 2014; Sachdev et al. 2021). The electron required for the functioning of ETC is supplied by NADH and $FADH_2$ at complex I and II, respectively (Mongin et al. 2012; Nolfi-Donegan et al. 2020). These electrons via ubiquinone and mobile cytochrome c are transported to complex III and IV, respectively. Complex IV ultimately transfer electron to oxygen, the final acceptor, to form water molecule. The transfer of electron through complexes establishes electrochemical gradient due to translocation of protons from matrix to IMS at complex I, III, and IV (Vanlerberghe 2013). The gradient development is eventually employed by complex V to generate ATP through phosphorylation of ADP (Vanlerberghe 2013) (Fig. 6.2).

Alternative Respiratory Pathway Plant mitochondria in addition to discussed "classical ETC" or "cytochrome pathway" also demonstrate an "alternative pathway" for respiration, which is non-phosphorylating (Keunen et al. 2011; Rasmusson and Møller 2011; Smith et al. 2011; Sweetman et al. 2020) (Fig. 6.2). The alternative pathway comprised type II NAD(P)H dehydrogenases (NDs) and AOX (Schertl and Braun 2014). The type II or alternative NDs are rotenone-insensitive proteins, localized on both (internal and external) sides of IMM (Smith et al. 2011). These alternative NDs do not participate in proton translocation and on linking with AOX, enable transfer of electron directly to O_2, skipping complex III and IV, which prevent over-reduction of ETC and maintain cell redox homeostasis under unfavorable circumstances (Smith et al. 2011; Vanlerberghe et al. 2016; Czarnocka and Karpinski 2018).

6.3 ROS Production in Mitochondria Under Stress

Mitochondrion is one of the major organelles for cellular ROS genesis in non-photosynthetic cells and photosynthetic cells during dark conditions (Moller 2001; Plaxton and Podestá 2006; Zorov et al. 2014; Nolfi-Donegan et al. 2020). The classical ETC pathway in mitochondria contribute to cellular ROS in the presence of stress stimuli (Zhao et al. 2020) (Fig. 6.2). The respiratory ETC yields small titer of ROS during normal conditions, however, their volume increases significantly in

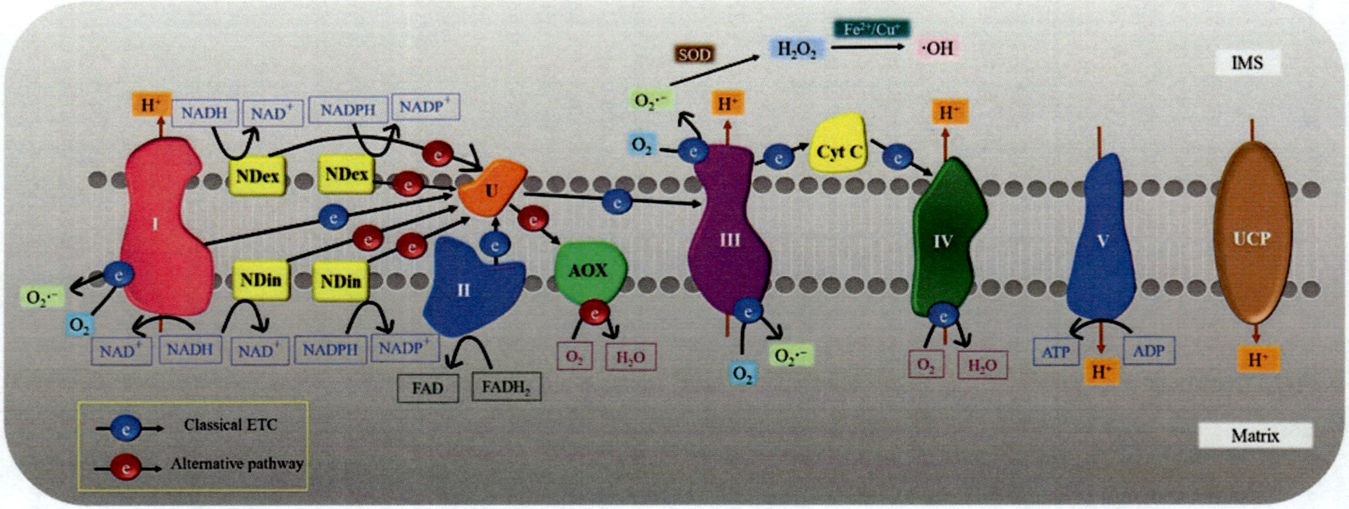

Fig. 6.2 Mitochondrial classical electron transport chain (ETC) coupled with oxidative phosphorylation; and alternative pathway. In classical ETC, tricarboxylic (TCA) cycle derived reduced nicotinamide adenine dinucleotide (NADH) and reduced flavin adenine dinucleotide ($FADH_2$) transfer electron to ETC at complex I and complex II, respectively. At complex I, NADH gets converted into oxidized nicotinamide adenine dinucleotide (NAD^+) and proton (H^+) with the generation of electron (e). At complex II, $FADH_2$ gets changed into oxidized flavin adenine dinucleotide (FAD) with electron generation. From both complex I and II, electron gets transferred to ubiquinone (U) and then to complex III. From complex III via cytochrome c (cyt c), (e) flow toward complex IV, where it reduces O_2 to water molecule (H_2O). During electron flow, H^+ flows across inner mitochondrial membrane (IMM) from matrix to inner mitochondrial space (IMS), generating proton gradient. This proton gradient facilitates phosphorylation of ADP to ATP at complex V (ATP synthase). Under stress, due to over-reduction of complex I and III of ETC, single electron leakage occurs which reduces O_2 into superoxide radical ($O_2^{\bullet-}$). $O_2^{\bullet-}$ undergoes dismutation and yields hydrogen peroxide (H_2O_2) on action of superoxide dismutase (SOD). H_2O_2 further undergoes Fenton reaction in the presence of reduced iron (Fe^{2+}) and copper (Cu^+) to form hydroxyl radical ($^{\bullet}OH$). In alternative pathway, NAD(P)H acts as electron carrier, and supplies electron to rotenone-insensitive nicotinamide adenine dinucleotide dehydrogenase (NDex or NDin), instead of complex I and complex II, from where electron is transferred to ubiquinone (U) and then directly to alternative oxidase (AOX). At AOX, electron reduces O_2 into H_2O, thereby eliminating formation of ROS. Moreover, the uncoupling protein (UCP), dissipates excess energy by pumping proton from intermembrane space (IMS) to matrix

response to unfavorable cues (Czarnocka and Karpinski 2018). The generation of ROS in mitochondria is triggered due to highly reduced state of the components of the respiratory ETC (Czarnocka and Karpinski 2018; Zhao et al. 2020). The over-reduction of ETC triggers electron leakage from the components of the chain to O_2, generating ROS (Rhoads et al. 2006; Miller et al. 2010; Gill and Tuteja 2010; Jalil and Ansari 2020). The over-reduction of ETC is induced due to surplus input of electron to the ETC in comparison to the rate of electron leaving the respiratory pathway at the terminal complex, either due to reduced ATP turnover, and/or presence of inhibitor or stress affecting ETC components (Rhoads et al. 2006; Cvetkovska et al. 2014).

Superoxide radical ($O_2^{\bullet -}$), the primary ROS generates through monovalent reduction of O_2 on single electron leakage from ETC components (Kowaltowski et al. 2009; Czarnocka and Karpinski 2018; Ansari et al. 2021). At the complex I, over-reduction causes leakage of an electron to O_2, eventually producing $O_2^{\bullet -}$ (Czarnocka and Karpinski 2018; Poór 2020). The flavin mononucleotide subunit and iron-sulphur cluster are the prime sites associated with complex I, participating in ROS production (Minibayeva et al. 2012). Moreover, leakage of electron from ubiquinone and complex III to O_2 also forms $O_2^{\bullet -}$ (Jacoby et al. 2012). Nearly 2–4% of the mitochondrial membrane absorbed O_2 is reduced to $O_2^{\bullet -}$ (Kohli et al. 2019). The over-reduction of ETC and ubiquinone under the influence of salinity, chilling, and drought facilitates production of $O_2^{\bullet -}$ due to electron leakage (Hu et al. 2008; Miller et al. 2010; Yamane et al. 2012; Bose et al. 2014; Das and Roychoudhury 2014; Sako et al. 2020). The complex II, itself does not generate ROS, however, could be potential source of ROS due to reverse transfer of electron to ubiquinone and then to complex I (Kowaltowski et al. 2009; Jacoby et al. 2012). Complex II displays reverse electron flow in the condition of scarcity of NAD^+-linked substrate (Turrens 2003). The $O_2^{\bullet -}$ formed undergoes dismutation on subsequent action of mitochondrial manganese superoxide dismutase (Mn-SOD) and/or ascorbate peroxidase (APX) enzyme, resulting in production of hydrogen peroxide (H_2O_2) (Sharma et al. 2012; Czarnocka and Karpinski 2018; Maurya 2020; Ansari et al. 2021). H_2O_2 is comparatively a stable and long-lived molecule which can travel to greater distance and can participate in signaling, formation of toxic hydroxyl radical ($^{\bullet}OH$) and PCD (Hossain et al. 2015; Jalil and Ansari 2018). Excess production of H_2O_2 under stress due to over-reduction of ETC, triggers opening of mitochondrial permeability transition pore (MTP) and releases cytochrome C from the mitochondria to cytosol, inducing PCD (Balk et al. 1999; Tiwari et al. 2002). The presence of redox metal in mitochondria promotes formation of ROS inducing oxidative damage (Keunen et al. 2011). The H_2O_2 formed reacts with redox metal (Fe^{2+}, Cu^+) via the Fenton reaction and produces highly toxic hydroxyl radical ($^{\bullet}OH$) (Gill and Tuteja 2010; Rasmusson and Møller 2011; Kohli et al. 2019). Parallelly, non-redox active metals have been found contributing to ROS production in mitochondria (Keunen et al. 2011). Exposure of pea root mitochondria to Cr has been observed to promote the generation of $O_2^{\bullet -}$ and inhibit respiratory ETC under in vivo conditions (Dixit et al. 2002). Similarly, Cd exposure induces ROS production via respiratory ETC and disrupts mitochondrial functions (Heyno et al. 2008; Bi et al. 2009; Smiri et al. 2010).

6.4 Oxidative Damage to Mitochondrial Biomolecules

The mitochondria may not be the foremost site for ROS origin in plants, however, are the prime target of ROS-mediated damage and oxidative stress. Stressful environment enhances the demand for ATP, eventually accelerating the activity of ETC, mitochondrial membrane potential and ROS production (Minibayeva et al. 2012). The amplified level of ROS under stressful situations causes oxidation of mitochondrial biomolecules, in particular, proteins, enzymes, lipid, and DNA, altering their structure and function (Hernandez et al. 1993; Minibayeva et al. 2012) (Fig. 6.3). For instance, under saline stress, generation of H_2O_2 in mitochondria prompts

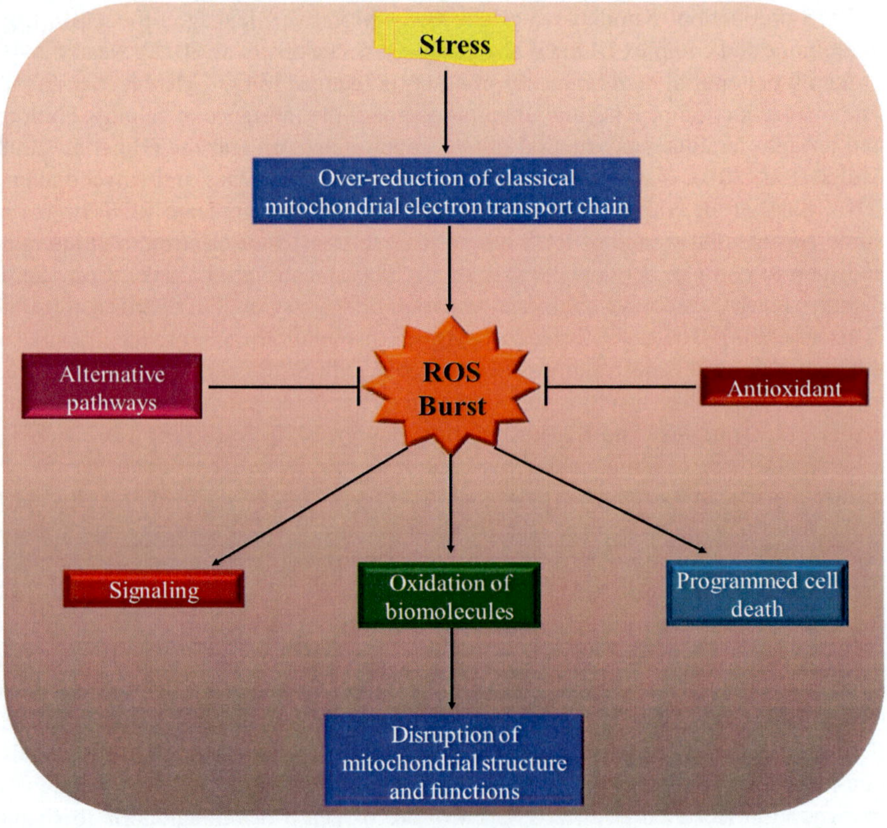

Fig. 6.3 Impact of ROS production in mitochondria on the perception of stress stimulus and mitochondrial recruited defense mechanism for prevention of accumulation of reactive oxygen species (ROS)

degradation of mitochondrial cristae (Yamane et al. 2012). The impairment of mitochondrial function due to oxidative damage further accelerates ROS formation, which may severely affect cell and trigger PCD (Minibayeva et al. 2012). ROS affect synthesis and inhibit function of multiple proteins and enzymes in mitochondria, particularly, enzymes participating in TCA cycle and components of the ETC such as ATP synthase and pyruvate decarboxylase subunits (Rhoads et al. 2006; Taylor and Millar 2007). Evidently, the enzymes containing Fe–S cluster and related to TCA cycle are highly susceptible and primarily affected by the oxidative stress (Minibayeva et al. 2012). H_2O_2 inactivates aconitase (ACO) of TCA cycle in potato (Verniquet et al. 1991). The ACO inactivation by H_2O_2 arises from 4Fe–4S cluster alteration (Jacoby et al. 2012; Ansari et al. 2021). Moreover, inactivation of enzymes, particularly ACO, releasing redox metal can result in production of highly toxic $^{\bullet}OH$ radical via the Fenton reaction (Dumont and Rivoal 2019). The ROS can induce oxidative damage to mitochondrial DNA, altering their repair and replication process (Kohli et al. 2019). The mitochondrial DNA shows high oxidation incidence comparative to nuclear DNA owing to low repair system and lack of chromatin organization, enhancing mutation rate 10–20 times (Navrot et al. 2007; Minibayeva et al. 2012). The ROS cause peroxidation of polyunsaturated fatty acids (PUFAs) of mitochondrial membrane through H^+ abstraction, generating lipid peroxidation products, which are considered as the major markers of cell death (Poór 2020). Peroxidation of lipids and oxidative modification of protein, alters membrane potential, which opens membrane permeability transition pore and eases relocation of cytochrome c from mitochondria to cytosol, initiating PCD (Vacca et al. 2006; Rhoads et al. 2006; Gao et al. 2008; Poór 2020). Under heat shock, oxidation of mitochondrial proteins and membranes orchestrates the primary event of PCD (Vacca et al. 2004). The relocation of cytochrome c to cytosol, eliciting PCD in cucumber under heat stress has been observed by Balk et al. (1999). DNA laddering, nucleus fragmentation, activation of caspase-like protease in cytosol, and PCD in *Arabidopsis thaliana* occurs under excess UV radiation due to ROS formation and loss of mitochondrial membrane potential (Gao et al. 2008).

6.5 Defense Against Oxidative Damage in Mitochondria

Plant mitochondria show three different approaches to reduce or manage oxidative damage under unfavorable circumstances. (1) Avoiding production of ROS by keeping ETC in oxidized condition, (2) Maintaining redox hemostasis by detoxifying ROS, and (3) Repairing ROS-mediated damage (Pastore et al. 2007) (Fig. 6.3). The alternative respiratory pathway and uncoupling proteins (UCP) avoid ROS generation in mitochondria. The alternative pathway prevents over-reduction of ETC by transferring electron from ubiquinone to AOX directly and then to O_2, eventually forming H_2O and alleviating ROS production (Poór 2020; Zhao et al. 2020) (Fig. 6.2). Thus, the pathway regulates O_2 homeostasis and halt accelerating production of ROS, under stressful circumstances when over-reduction of O_2 is frequent (Minibayeva et al. 2012). The overexpression of AOX reduces mitochondrial-mediated PCD significantly (Liu et al. 2014). The UCP dissipates excess energy released by the ETC by transporting H^+ back to the mitochondrial matrix across the IMM and reduces membrane potential as well as ROS formation induced by over-reduced ETC (Pastore et al. 2007; Minibayeva et al. 2012; Poór 2020). However, for the activation of UCP, the presence of ROS is required (Poór 2020).

Plants recruit enzymatic as well as non-enzymatic antioxidants to quench excess ROS and balancing redox status (Minibayeva et al. 2012; Sachdev et al. 2021). Matrix-localized Mn-SOD quenches $O_2^{\bullet-}$ and converts it into H_2O_2. The accumulation of H_2O_2 in mitochondria is maintained by ascorbate peroxidase (APX), glutathione peroxidases (GPXs), and peroxiredoxins (PRX) (Rhoads et al. 2006; Poór 2020). APX, GPX, and PRX reduce H_2O_2 to water molecule (Rhoads et al. 2006; Poór 2020). Other compounds such as lipoic acid, a coenzyme of mitochondrial enzymes, in its reduced form are pivotal for chelation of transition metals and can reduce $O_2^{\bullet-}$ and $^{\bullet}OH$ accumulation. Additionally, a non-enzymatic antioxidant or osmoprotectant proline can scavenge $O_2^{\bullet-}$ and $^{\bullet}OH$, and prevent lipid peroxidation and PCD (Poór 2020). Moreover, glutathione transferases (GST) participate in detoxification of array of biomolecules oxidation by-products such as lipid peroxides, and reactive aldehydes. For instance, lipid peroxidation products-4-hydroxynonenal (HNE) is converted into glutathione conjugate (GS-HNE) on action of GST (Rhoads et al. 2006; Poór 2020). The type II peroxiredoxin/thioredoxin system could alleviate lipid peroxides (Rhoads et al. 2006). The oxidation of proteins is often irreversible and results in protein degradation, however, oxidized proteins have been found to be reduced into disulfides by *Arabidopsis* mitochondria localized protein disulfide isomerase in conjugation with thioredoxin and NADPH (Sweetlove et al. 2002; Rhoads et al. 2006).

6.6 Conclusion

ROS in mitochondria are the major outcome of energy generation via ETC-associated oxidative phosphorylation. The genesis of ROS in mitochondria is unavoidable, nevertheless, limiting their accumulation is essential for the plant fitness. ROS formed in mitochondria severely affect biomolecules and trigger retrograde signaling and PCD in response to stress stimuli. To curb accelerated ROS formation and incurred damage on cell, mitochondria recruit various pathways, including alternative pathway to prevent over-reduction of ETC and recruitment of antioxidants and osmolytes to scavenge ROS and/or repair induced damages. These pathways facilitate management of ROS in mitochondria, but sometimes occur at the cost of reduced energy production and/or cell death, which may affect cellular performance. Thus, there is a need to explore approaches that can fortify intrinsic defense system and decipher other defensive mechanisms, which could significantly aid in the improvement of organelle redox status under stress.

References

Ansari MI, Jalil SU, Ansari SA, Hasanuzzaman M (2021) GABA shunt: a key-player in mitigation of ROS during stress. Plant Growth Regul 94(2):131–149

Balk J, Leaver CJ, McCabe PF (1999) Translocation of cytochrome c from the mitochondria to the cytosol occurs during heat-induced programmed cell death in cucumber plants. FEBS Lett 463(1–2):151–154

Bi Y, Chen W, Zhang W, Zhou Q, Yun L, Xing D (2009) Production of reactive oxygen species, impairment of photosynthetic function and dynamic changes in mitochondria are early events in cadmium-induced cell death in *Arabidopsis thaliana*. Biol Cell 101(11):629–643

Blokhina O, Fagerstedt KV (2010) Reactive oxygen species and nitric oxide in plant mitochondria: origin and redundant regulatory systems. Physiol Plant 138(4):447–462

Bose J, Rodrigo-Moreno A, Shabala S (2014) ROS homeostasis in halophytes in the context of salinity stress tolerance. J Exp Bot 65(5):1241–1257

Cogliati S, Enriquez JA, Scorrano L (2016) Mitochondrial cristae: where beauty meets functionality. Trends Biochem Sci 41(3):261–273

Cupp JD, Nielsen BL (2014) Minireview: DNA replication in plant mitochondria. Mitochondrion 19:231–237

Cvetkovska M, Dahal K, Alber NA, Jin C, Cheung M, Vanlerberghe GC (2014) Knockdown of mitochondrial alternative oxidase induces the 'stress state' of signaling molecule pools in Nicotiana tabacum, with implications for stomatal function. New Phytol 203(2):449–461

Czarnocka W, Karpinski S (2018) Friend or foe? Reactive oxygen species production, scavenging and signaling in plant response to environmental stresses. Free Radic Biol Med 122:4–20

Daloso DM, Müller K, Obata T, Florian A, Tohge T, Bottcher A, Riondet C, Bariet L, Carrari F, Nunes-Nesi A, Buchanan BB, Reichheld JP, Araujo WL, Fernie AR (2015) Thioredoxin, a master regulator of the tricarboxylic acid cycle in plant mitochondria. Proc Natl Acad Sci 112(11):E1392–E1400

Das K, Roychoudhury A (2014) Reactive oxygen species (ROS) and response of antioxidants as ROS-scavengers during environmental stress in plants. Front Environ Sci 2:53

Dixit V, Pandey V, Shyam R (2002) Chromium ions inactivate electron transport and enhance superoxide generation in vivo in pea (*Pisum sativum* L. cv. Azad) root mitochondria. Plant Cell Environ 25(5):687–693

Douce R (2012) Mitochondria in higher plants: structure, function, and biogenesis. Elsevier, Amsterdam

Dumont S, Rivoal J (2019) Consequences of oxidative stress on plant glycolytic and respiratory metabolism. Front Plant Sci 10:166

Gao C, Xing D, Li L, Zhang L (2008) Implication of reactive oxygen species and mitochondrial dysfunction in the early stages of plant programmed cell death induced by ultraviolet-C overexposure. Planta 227(4):755–767

Gasmi A, Peana M, Arshad M, Butnariu M, Menzel A, Bjørklund G (2021) Krebs cycle: activators, inhibitors and their roles in the modulation of carcinogenesis. Arch Toxicol 95(4):1161–1178

Gill SS, Tuteja N (2010) Reactive oxygen species and antioxidant machinery in abiotic stress tolerance in crop plants. Plant Physiol Biochem 48(12):909–930

Hernandez JA, Corpas FJ, Gomez M, del Rio LA, Sevilla F (1993) Salt-induced oxidative stress mediated by activated oxygen species in pea leaf mitochondria. Physiol Plant 89(1):103–110

Heyno E, Klose C, Krieger-Liszkay A (2008) Origin of cadmium-induced reactive oxygen species production: mitochondrial electron transfer versus plasma membrane NADPH oxidase. New Phytol 179(3):687–699

Hossain MA, Bhattacharjee S, Armin SM, Qian P, Xin W, Li HY, Burritt DJ, Fujita M, Tran LSP (2015) Hydrogen peroxide priming modulates abiotic oxidative stress tolerance: insights from ROS detoxification and scavenging. Front Plant Sci 6:420

Hu WH, Song XS, Shi K, Xia XJ, Zhou YH, Yu JQ (2008) Changes in electron transport, superoxide dismutase and ascorbate peroxidase isoenzymes in chloroplasts and mitochondria of cucumber leaves as influenced by chilling. Photosynthetica 46(4):581–588

Jacoby RP, Li L, Huang S, Pong Lee C, Millar AH, Taylor NL (2012) Mitochondrial composition, function and stress response in plants F. J Integr Plant Biol 54(11):887–906

Jalil SU, Ansari MI (2018) Plant microbiome and its functional mechanism in response to environmental stress. Int J Green Pharm 12(01):S81–S92

Jalil SU, Ansari MI (2020) Stress implications and crop productivity. In: Plant ecophysiology and adaptation under climate change: mechanisms and perspectives I. Springer, Singapore, pp 73–86

Joubert F, Puff N (2021) Mitochondrial cristae architecture and functions: lessons from minimal model systems. Membranes 11(7):465

Keunen E, Remans T, Bohler S, Vangronsveld J, Cuypers A (2011) Metal-induced oxidative stress and plant mitochondria. Int J Mol Sci 12(10):6894–6918

Khan MIR, Jalil SU, Chopra P, Chhillar H, Ferrante A, Khan NA, Ansari MI (2021) Role of GABA in plant growth, development and senescence. Plant Gene 26:100283

Kohli SK, Khanna K, Bhardwaj R, Abd Allah EF, Ahmad P, Corpas FJ (2019) Assessment of subcellular ROS and NO metabolism in higher plants: multifunctional signaling molecules. Antioxidants 8(12):641

Kowaltowski AJ, de Souza-Pinto NC, Castilho RF, Vercesi AE (2009) Mitochondria and reactive oxygen species. Free Radic Biol Med 47(4):333–343

Liu J, Li Z, Wang Y, Xing D (2014) Overexpression of ALTERNATIVE OXIDASE1a alleviates mitochondria-dependent programmed cell death induced by aluminium phytotoxicity in *Arabidopsis*. J Exp Bot 65(15):4465–4478

Maurya AK (2020) Oxidative stress in crop plants. In: Agronomic crops. Springer, Singapore, pp 349–380

Meyer EH, Welchen E, Carrie C (2019) Assembly of the complexes of the oxidative phosphorylation system in land plant mitochondria. Annu Rev Plant Biol 70:23–50

Miller GAD, Suzuki N, Ciftci-Yilmaz SULTAN, Mittler RON (2010) Reactive oxygen species homeostasis and signalling during drought and salinity stresses. Plant Cell Environ 33(4): 453–467

Minibayeva F, Dmitrieva S, Ponomareva A, Ryabovol V (2012) Oxidative stress-induced autophagy in plants: the role of mitochondria. Plant Physiol Biochem 59:11–19

Moller IM (2001) Plant mitochondria and oxidative stress: electron transport, NADPH turnover, and metabolism of reactive oxygen species. Annu Rev Plant Biol 52(1):561–591

Mongin AA, Dohare P, Jourd'heuil D (2012) Selective vulnerability of synaptic signaling and metabolism to nitrosative stress. Antioxid Redox Signal 17(7):992–1012

Navrot N, Rouhier N, Gelhaye E, Jacquot JP (2007) Reactive oxygen species generation and antioxidant systems in plant mitochondria. Physiol Plant 129(1):185–195

Nolfi-Donegan D, Braganza A, Shiva S (2020) Mitochondrial electron transport chain: oxidative phosphorylation, oxidant production, and methods of measurement. Redox Biol 37:101674

Pastore D, Trono D, Laus MN, Di Fonzo N, Flagella Z (2007) Possible plant mitochondria involvement in cell adaptation to drought stress: a case study: durum wheat mitochondria. J Exp Bot 58(2):195–210

Plaxton WC, Podestá FE (2006) The functional organization and control of plant respiration. Crit Rev Plant Sci 25(2):159–198

Poór P (2020) Effects of salicylic acid on the metabolism of mitochondrial reactive oxygen species in plants. Biomolecules 10(2):341

Rao RSP, Salvato F, Thal B, Eubel H, Thelen JJ, Møller IM (2017) The proteome of higher plant mitochondria. Mitochondrion 33:22–37

Rasmusson AG, Møller IM (2011) Mitochondrial electron transport and plant stress. In: Plant mitochondria. Springer, New York, NY, pp 357–381

Rhoads DM, Umbach AL, Subbaiah CC, Siedow JN (2006) Mitochondrial reactive oxygen species. Contribution to oxidative stress and interorganellar signaling. Plant Physiol 141(2):357–366

Sachdev S, Ansari SA, Ansari MI, Fujita M, Hasanuzzaman M (2021) Abiotic stress and reactive oxygen species: generation, signaling, and defense mechanisms. Antioxidants 10(2):277

Sako K, Futamura Y, Shimizu T, Matsui A, Hirano H, Kondoh Y, Muroi M, Aono H, Tanaka M, Honda K, Shimizu K, Kawatani M, Nakano T, Osada H, Noguchi K, Seki M (2020) Inhibition of mitochondrial complex I by the novel compound FSL0260 enhances high salinity-stress tolerance in Arabidopsis thaliana. Sci Rep 10(1):1–12

Schertl P, Braun HP (2014) Respiratory electron transfer pathways in plant mitochondria. Front Plant Sci 5:163

Sharma P, Jha AB, Dubey RS, Pessarakli M (2012) Reactive oxygen species, oxidative damage, and antioxidative defense mechanism in plants under stressful conditions. J Bot 2012:217037

Smiri M, Chaoui A, Rouhier N, Kamel C, Gelhaye E, Jacquot JP, El Ferjani E (2010) Cadmium induced mitochondrial redox changes in germinating pea seed. Biometals 23(6):973–984

Smith C, Barthet M, Melino V, Smith P, Day D, Soole K (2011) Alterations in the mitochondrial alternative NAD (P) H dehydrogenase NDB4 lead to changes in mitochondrial electron transport chain composition, plant growth and response to oxidative stress. Plant Cell Physiol 52(7): 1222–1237

Sweetlove LJ, Heazlewood JL, Herald V, Holtzapffel R, Day DA, Leaver CJ, Millar AH (2002) The impact of oxidative stress on *Arabidopsis mitochondria*. Plant J 32(6):891–904

Sweetlove LJ, Beard KF, Nunes-Nesi A, Fernie AR, Ratcliffe RG (2010) Not just a circle: flux modes in the plant TCA cycle. Trends Plant Sci 15(8):462–470

Sweetman C, Miller TK, Booth NJ, Shavrukov Y, Jenkins CL, Soole KL, Day DA (2020) Identification of alternative mitochondrial electron transport pathway components in chickpea indicates a differential response to salinity stress between cultivars. Int J Mol Sci 21(11):3844

Taylor NL (2018) Editorial for special issue "plant mitochondria". Int J Mol Sci 19(12):3849

Taylor NL, Millar AH (2007) Oxidative stress and plant mitochondria. In: Mitochondria. Humana Press, Clifton, pp 389–403

Tiwari BS, Belenghi B, Levine A (2002) Oxidative stress increased respiration and generation of reactive oxygen species, resulting in ATP depletion, opening of mitochondrial permeability transition, and programmed cell death. Plant Physiol 128(4):1271–1281

Turrens JF (2003) Mitochondrial formation of reactive oxygen species. J Physiol 552(2):335–344

Vacca RA, de Pinto MC, Valenti D, Passarella S, Marra E, De Gara L (2004) Production of reactive oxygen species, alteration of cytosolic ascorbate peroxidase, and impairment of mitochondrial metabolism are early events in heat shock-induced programmed cell death in tobacco Bright-Yellow 2 cells. Plant Physiol 134(3):1100–1112

Vacca RA, Valenti D, Bobba A, Merafina RS, Passarella S, Marra E (2006) Cytochrome c is released in a reactive oxygen species-dependent manner and is degraded via caspase-like proteases in tobacco Bright-Yellow 2 cells en route to heat shock-induced cell death. Plant Physiol 141(1):208–219

Vanlerberghe GC (2013) Alternative oxidase: a mitochondrial respiratory pathway to maintain metabolic and signaling homeostasis during abiotic and biotic stress in plants. Int J Mol Sci 14(4):6805–6847

Vanlerberghe GC, Martyn GD, Dahal K (2016) Alternative oxidase: a respiratory electron transport chain pathway essential for maintaining photosynthetic performance during drought stress. Physiol Plant 157(3):322–337

Verniquet F, Gaillard J, Neuburger M, Douce R (1991) Rapid inactivation of plant aconitase by hydrogen peroxide. Biochem J 276(3):643–648

Yamane K, Taniguchi M, Miyake H (2012) Salinity-induced subcellular accumulation of H2O2 in leaves of rice. Protoplasma 249(2):301–308

Zhao Y, Yu H, Zhou JM, Smith SM, Li J (2020) Malate circulation: linking chloroplast metabolism to mitochondrial ROS. Trends Plant Sci 25(5):446–454

Zorov DB, Juhaszova M, Sollott SJ (2014) Mitochondrial reactive oxygen species (ROS) and ROS-induced ROS release. Physiol Rev 94(3):909–950

Peroxisomes and ROS Under Stress Conditions

7

7.1 Introduction

Peroxisomes are small cellular organelle, enclosed with single lipid bilayer membrane, ubiquitously found in eukaryotes. These are small subcellular compartments with diameters ranging from 0.1 to 1.5 μm (Walker et al. 2018; Oikawa et al. 2019). The shape of peroxisomes varies from sphere to ellipsoidal depending upon the developmental stage and persisting environmental cues with respect to metabolic and physiological performance (Oikawa et al. 2019). These are major cellular organelles with simple structure, constitute an amorphous matrix, housing numerous enzymatic and non-enzymatic components, and are devoid of genetic material (Corpas et al. 2020; Sachdev et al. 2021). The metabolism of peroxisome is an oxidative type and shows plasticity due to variation in the enzymatic pool, depending upon multiple factors including the type of organism, type of cell or tissue, and prevailing environmental scenario (del Río et al. 2006). The plasticity of peroxisomes facilitates remodeling of their shape, size as well as enzymatic content (Nyathi and Baker 2006). This could be advocated by the ability of peroxisomes to proliferate under the influence of biotic and abiotic stress (Nyathi and Baker 2006). These organelles are known with different names depending on the developmental stage of the organism and/or organ with which they are associated (Corpas et al. 2020). For instance, peroxisomes associated with seed germination are known as glyoxysomes, whereas those found in leaves and roots are called the leaf and root peroxisomes, respectively (Corpas et al. 2020). The reason for this nomenclature is the specific metabolic pathway associated with each type of peroxisome (Corpas et al. 2020). Nevertheless, owing to the existence of an array of common metabolites and enzymes, in general, all kinds are preferred to be known as "Peroxisomes," irrespective of their specific metabolic pathways (Corpas et al. 2019, 2020). In peroxisomes, hydrogen peroxide (H_2O_2) and catalase (CAT) are the plenteous and frequently occurring metabolites and enzymes, respectively, which partake in ROS metabolism (Sousa et al. 2019; Corpas et al. 2020). Despite being small in size,

S. Sachdev et al., *Reactive Oxygen Species in Plants*,
https://doi.org/10.1007/978-981-19-9884-3_7

peroxisomes are morphologically and metabolically versatile organelle and partake in various key metabolic processes, in particular, breakdown of fatty acid via β-oxidation; acetate-malate shunt; photorespiration; catabolism of toxic or bioactive metabolites such as polyamines, sulfite, urate, and methylglyoxal; hormones and cofactors biosynthesis; nicotinamide adenine dinucleotide phosphate (NADPH) regeneration; and metabolism of amino acid, reactive species of oxygen (ROS) and nitrogen (RNS) (Kunze and Hartig 2013; Fahy et al. 2017; Oikawa et al. 2019; Su et al. 2019; Pan et al. 2020). The disruption of its functions and biogenesis could induce lethal effects (Oikawa et al. 2019; Pan et al. 2020).

Peroxisome is the key organelle for ROS production, distinctly under stressful circumstances (Petrov et al. 2015; Ansari et al. 2021; Jalil and Ansari 2019). The oxidation of glycolate to glyoxylate in peroxisome lumen induces overproduction and accumulation of ROS (Fahy et al. 2017). Moreover, photorespiration leads to the production of H_2O_2 in bulk on exposure to adverse stimuli (Petrov et al. 2015; Iqbal et al. 2021). Due to the overloading of ROS, peroxisomes are highly susceptible to oxidative damage (Fahy et al. 2017). The other ROS, singlet oxygen ($^1O_2$) and hydroxyl radical ($^{\bullet}OH$) formed in peroxisome, trigger peroxidation of lipids and induce oxidation, nitrosylation, as well as carbonylation of amino acids (Fahy et al. 2017). These changes cause irreversible damage to peroxisomal membranes, protein molecules, and enzymes, leading to altered membrane permeability and fluidity, eventually resulting in dysfunction of peroxisomes (Fahy et al. 2017). The severe condition of oxidative stress may alter peroxisome biogenesis and induce its degradation (Su et al. 2019). Peroxisomes participate in the genesis of ROS. However, they also rescue cells from oxidative damage (Tiew et al. 2015). Peroxisomes contain an array of antioxidants, which maintain ROS equilibrium (Tiew et al. 2015). Peroxisomes show crosstalk with other cellular organelles particularly, chloroplast, mitochondria, endoplasmic reticulum, and lipid bodies through physiological and metabolic processes (Sinclair et al. 2009; Cui et al. 2016; Oikawa et al. 2019; Su et al. 2019; Corpas et al. 2020; Pan et al. 2020). The ROS, particularly, H_2O_2 produced in peroxisome is instrumental in organelles crosstalk and co-ordinate several functions such as redox metabolism, photorespiration, and others (Su et al. 2019; Corpas et al. 2020). The ROS formed within peroxisomes on the onset of stress activate retrograde signaling, which induces adaptive response including regulation of defense machinery (Su et al. 2019). The chapter provides a recent update on ROS metabolism in peroxisomes and their implication on cellular structure and function to delineate the challenges and opportunities faced by plants under changing environmental scenarios.

7.2 Peroxisomes: Source of ROS Under Stress

Peroxisomes are the notable location for genesis of H_2O_2 in a cell (del Río et al. 2006; Ansari 2018). Moreover, superoxide radicals ($O_2^{\bullet -}$), $^1O_2$, and $^{\bullet}OH$ are also generated in peroxisomes (del Río et al. 2006; Fahy et al. 2017; Pan et al. 2020) (Table 7.1). Photorespiration, enzymatic activity of flavin oxidases, catabolism of

Table 7.1 Metabolic activities participating in the generation of ROS in peroxisomes

Metabolic activities	Enzyme involved	Substrate	Peroxisomal location	ROS generated
Photorespiration	Glycolate oxidase (GOX)	Glycolate	Matrix	H_2O_2
β-oxidation of Fatty acid	Acyl-CoA oxidase (ACX)	Fatty acid	Matrix	H_2O_2
Enzymatic action of Xanthine oxidase	Xanthine oxidase (XOD)	Xanthine/ Hypoxanthine	Matrix	$O_2^{\bullet -}$
Metabolism of urate	Urate oxidase (UO)	Uric acid	Matrix	H_2O_2
Catabolism of polyamines	Polyamine oxidase (PAO), Diamine oxidase (DAO), Copper-containing amino oxidase (CuAO)	Polyamines	Matrix	H_2O_2
Sulfite oxidation	Sulfite oxidase (SO)	Sulfite	Matrix	H_2O_2
Demethylation of sarcosine	Sarcosine oxidase (SOX)	Sarcosine	Matrix	H_2O_2
Dismutation of superoxide radical	Superoxide dismutase (SOD)	Superoxide radical	Matrix/ Membrane	H_2O_2
NADPH-dependent small ETC	–	–	Membrane	$O_2^{\bullet -}$
Haber–Weiss reaction	–	–	Matrix	$^1O_2$; $^{\bullet}OH$

fatty acid via β-oxidation, sulfite oxidation, catabolism of polyamines by polyamine oxidase, dismutation of $O_2^{\bullet -}$, are some major peroxisomal metabolic pathways, inducing intracellular H_2O_2 formation (del Río et al. 2006; Corpas et al. 2020). In peroxisomes, genesis of $O_2^{\bullet -}$ takes place in the organelle matrix and at the membrane, which after production is transferred to the cytosol (Ansari 2018; Ansari et al. 2021; Sachdev et al. 2021). In the matrix, the action of xanthine oxidase (XOD) enzyme catalyzes the formation of $O_2^{\bullet -}$ and at the membrane, NAD(P)H-dependent small electron transport chain (ETC) is responsible for the production of $O_2^{\bullet -}$. The reaction of H_2O_2 and $O_2^{\bullet -}$ via Haber–Weiss has been suggested as the potential pathway for $^1O_2$ production in peroxisomes (Mor et al. 2014). In peroxisomes, redox metals such as iron and copper are found abundantly in complex-form. Under adverse conditions, these metals are released and catalyze site-specific Haber–Weiss reaction leading to the generation of $^{\bullet}OH$ from H_2O_2 (del Río et al. 1992). The stress conditions including heavy metals, salinity, pesticides, and others, exacerbate the production of ROS in peroxisomes, altering redox homeostasis (Demidchik 2012). For instance, in pea leaves, overproduction of H_2O_2 was reported to occur in the presence of toxic compounds (cadmium and 2,4-Dichlorophenoxy acetic acid) (Ortega-Galisteo et al. 2012). Correspondingly, in peroxisomes of *Arabidopsis thaliana* seedlings, cadmium reportedly induces overproduction of

$O_2^{\bullet -}$ (Corpas and Barroso 2014). The amplified accumulation of $^1O_2$ in root peroxisomes of *Arabidopsis* occurs when subjected to an array of biotic and abiotic stress in combination with dark (Mor et al. 2014). The stressful events, in particular, salinity, drought, ground-level ozone, high irradiance, heavy metals, and others, trigger proliferation of peroxisomes, due to ROS generation (Rodríguez-Serrano et al. 2009; del Río and López-Huertas 2016; Sandalio et al. 2021). The production of ROS and peroxisomal proliferation take place in pea plant subjected to heavy metal (cadmium) stress (Romero-Puertas et al. 2002).

7.3 Photorespiration Mediated ROS Generation in Peroxisomes

Photorespiration is a process, which takes place in peroxisomes in coordination with chloroplast and mitochondria (Voss et al. 2013) (Fig. 7.1). Chloroplast is the first site, where the process begins in the presence of light, and involves the utilization of O_2, releasing CO_2 (Voss et al. 2013; Su et al. 2019). Plant exposed to harsh conditions, particularly, salinity, water-logging, drought, elevated temperature, metal toxicity, and others, alleviates the availability of CO_2 for the enzyme Ribulose-1,5-bisphosphate carboxylase/oxygenase (Rubisco) due to stomatal

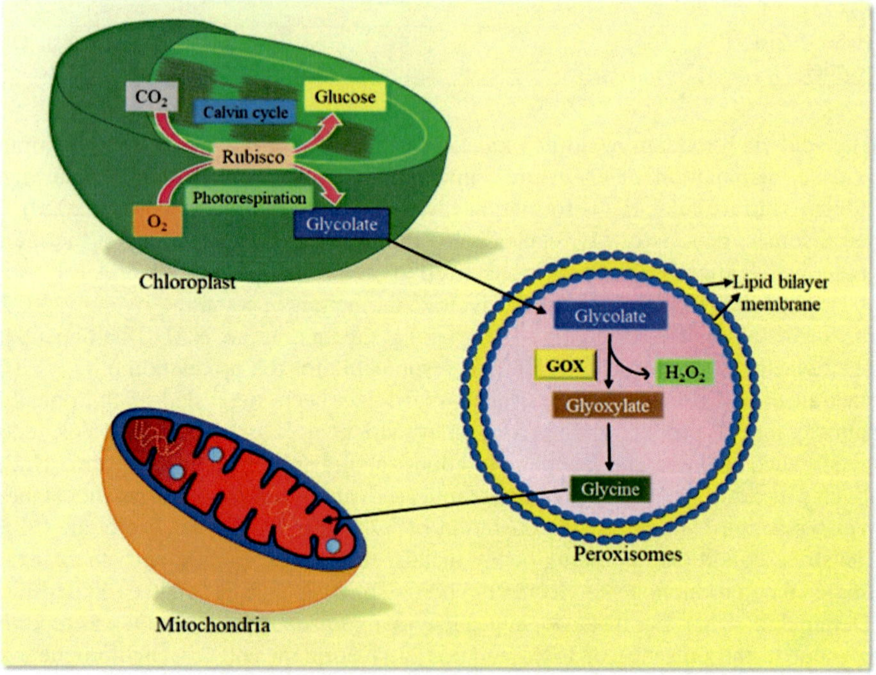

Fig. 7.1 Production of hydrogen peroxide in peroxisomes through photorespiration

closure and/or increased solubility of O_2 in the chloroplast (Voss et al. 2013; Suo et al. 2017; Sachdev et al. 2021). This changes the specificity of the Rubisco enzyme more for O_2 than CO_2, eventually increasing oxygenase activity of the enzyme relatively higher than carboxylase, which leads to the formation of peroxisomal H_2O_2 (Petrov et al. 2015). In chloroplast, photorespiration generates glycolate, which gets transported to the peroxisome, where on the action of glycolate oxidase (GOX) it produces H_2O_2 (Cruz 2008; Sandalio et al. 2013; Das and Roychoudhury 2014) (Fig. 7.1). The generation of H_2O_2 in peroxisomes, however, exceeds exorbitantly under adverse situations (Petrov et al. 2015; Suo et al. 2017; Sachdev et al. 2021). The ROS generated via the action of GOX trigger oxidative damage under adverse condition induced by excess salt (Voss et al. 2013). For instance, under saline stress, peroxisomal GOX activity accelerates ROS production, triggering oxidative damage in the tomato plants by enhancing peroxidation of lipid molecules and abridging antioxidant concentration such as reduced ascorbate (AsA) and glutathione (GSH) (Mittova et al. 2003). Analogously, Yamane et al. (2012) have observed an increase in photorespiration-induced H_2O_2 accumulation in the peroxisomes under salinity stress. It has been documented that photorespiration stands alone for 70% H_2O_2 genesis in photosynthetic tissues (Sandalio et al. 2021). The genesis of H_2O_2 via photorespiration is 2 and 50 times more in peroxisomes than in chloroplast and mitochondria, respectively (Foyer and Noctor 2003; Sandalio et al. 2013). The shift in H_2O_2 production from chloroplast to peroxisomes is considered relatively less damaging for the plant, as it protects chloroplast from photoinhibition and the peroxisomes can more effectively quench ROS by recruiting antioxidants (Petrov et al. 2015). The mutants with disabled photorespiratory pathway display higher rate of photoinhibition under light stress, demonstrating instrumental role of photorespiration in the prevention of photoinhibition under CO_2 scarce conditions (Takahashi and Badger 2011; Phua et al. 2021).

β-Oxidation of Fatty Acid Under stress conditions, increased availability of oxygen (O_2) or reduced availability of carbon dioxide (CO_2) in chloroplast, shifts carboxylase activity of Rubisco to oxygenase, inducing photorespiration process. The photorespiration results in the production of glycolate in chloroplast, which gets transferred to the peroxisomes and thereafter on action of enzyme glycolate oxidase (GOX), it generates H_2O_2 (hydrogen peroxide) and glyoxylate. The glyoxylate formed is converted in glycine, which is then transported to the mitochondria.

Catabolism of fatty acids via β-oxidation is an additional route for the formation of peroxisomal H_2O_2. β-oxidation of fatty acid is an important metabolic process, required by plants to produce energy, carbon skeleton, and phytohormones for growth and development (Yu et al. 2019; Pan et al. 2020; Plett et al. 2020). In plants, peroxisomes are the major site where the β-oxidation process occurs (Plett et al. 2020). This process is more commonly evident in germinating seeds due to the presence of stored lipids, which are utilized for seed development and proliferation (Sandalio et al. 2013). The fatty acids are imported to the peroxisomes, where its breakdown begins with the esterification, yielding CoA, which further oxidized to form trans-2-enoyl-CoA on the action of acyl-CoA oxidase (ACX) enzyme along with the generation of H_2O_2 (Kaur et al. 2009; Sandalio et al. 2013; Yu et al. 2019; Pan et al. 2020). β-oxidation in plant peroxisomes is affected by stress conditions, in particular, darkness; excess light; organic pollutants like heavy metals; clofibrate; salinity; and herbicides, inducing oxidative stress and peroxisomes proliferation (Palma et al. 1991; Castillo et al. 2008; Sandalio et al. 2013; Su et al. 2019; Yu et al. 2019; Pan et al. 2020). The generation of H_2O_2 via peroxisomal β-oxidation of fatty acid under salinity stress produces oxidative stress (Yu et al. 2019). Analogously, the oxidative stress triggered in the presence of cadmium, induces peroxisome proliferation and causes formation of peroxules (Rodríguez-Serrano et al. 2016). Under high illumination, the formation of peroxules is accompanied by amplified H_2O_2 production (Rodríguez-Serrano et al. 2016). Moreover, excess salt reportedly accelerates peroxisome proliferation in *Arabidopsis* (Fahy et al. 2017).

7.4 Peroxisomal Enzymes Generating ROS

Besides GOX and ACX, other peroxisomal enzymes including XOD, sarcosine oxidase (SOX), superoxide dismutase (SOD), urate oxidase (UO), polyamine oxidases (PAO) and other amino oxidase (AO) like copper-containing amine oxidases (CuAOs), catalyze reaction, which subsequently generate ROS (Ortega-Galisteo et al. 2012; Sandalio et al. 2013; del Río and López-Huertas 2016). The enzymes XOD and xanthine dehydrogenase (XDH) are two isoforms of xanthine oxidoreductase (XOR) found in peroxisomes and partake in the generation of ROS (Phua et al. 2021). The XOD and XDH can be interconverted via proteolysis and/or by regulating the sulfuration state of manganese cofactor in the protein (Sandalio et al. 2013). The xanthine/hypoxanthine undergoes oxidation on action of XOD in the peroxisomal matrix, forming uric acid along with the generation of nicotinamide adenine dinucleotide (NADH) and $O_2^{\bullet-}$ (Das and Roychoudhury 2014; Corpas et al. 2020; Ansari et al. 2021). The uric acid formed undergoes a reaction, catalyzed by the enzyme UO and converted into allantoin with subsequent production of H_2O_2 (Sandalio et al. 2013; Corpas et al. 2020). The two enzymes XOD and UO, localized in peroxisomes, are instrumental in nucleic acid catabolism (Werner and Witte 2011; Sandalio et al. 2013). In peroxisomes, the metabolism of sulfur is catalyzed by sulfite oxidase (SO)-a molybdenum dependent non-heme enzyme, generating H_2O_2 (Nyathi and Baker 2006; Phua et al. 2021). The reaction involves the conversion of sulfite into sulfate in the presence of SO, leading to the production of H_2O_2 (Phua et al. 2021). The catalytic reaction of SO in peroxisomes is functionally relevant as it protects the activity of CAT enzyme, which is known to be affected by sulfite even at low concentrations (Corpas et al. 2020), however, this reaction contributes to ROS accumulation in peroxisomes. Accumulation of H_2O_2 under such circumstances facilitates detoxification of sulfite by oxidizing it into sulphate (Schrader and Fahimi 2006; Hänsch et al. 2007; Sandalio et al. 2013; Phua et al. 2021). The antioxidant SOD, possessing ability to scavenge ROS, generates H_2O_2 on dismutation of the $O_2^{\bullet-}$ (Corpas et al. 2020). The two types of SOD viz., Cu/Zn-SOD with copper-zinc cofactor and Mn-SOD with manganese cofactor, have been reported from various plant peroxisomes, in particular, watermelon, rice, and pea plant, catalyzing dismutation reactions (Nyathi and Baker 2006; Ueda et al. 2013; Corpas et al. 2017a). The catabolism of polyamines by AO enzymes such as flavin adenine dinucleotide (FAD) dependent-PAO, diamine oxidase (DAO), and CuAOs also yields H_2O_2 in plant peroxisome (Sandalio et al. 2013; Kusano et al. 2015; Corpas et al. 2020; Phua et al. 2021). The presence of DAO and PAO in peroxisomes of *Arabidopsis* and rice plants is instrumental in the catabolism of polyamines (Kamada-Nobusada et al. 2008; Ono et al. 2012; Sandalio et al. 2013). Another enzyme SOX, dependent on FAD for its function, mediates demethylation of sarcosine, leading to the generation of H_2O_2 in peroxisome (Nyathi and Baker 2006; Sandalio et al. 2013; Corpas et al. 2017a).

7.5 Peroxisomal Electron Transport Chain

The ETC operating at the membrane of peroxisome generates $O_2^{\bullet-}$ (Fig. 7.2). This chain involves the NAD^+ and $NADP^+$ regeneration to maintain the peroxisomal oxidative metabolism (Nyathi and Baker 2006; Phua et al. 2021). The peroxisomal ETC in pea leaves is composed of three peroxisome membrane polypeptides (PMPs). The two NADH dependent PMP viz., cytochrome (cyt) b and flavoprotein NADH: ferricyanide reductase has a molecular mass of 18 kDa and

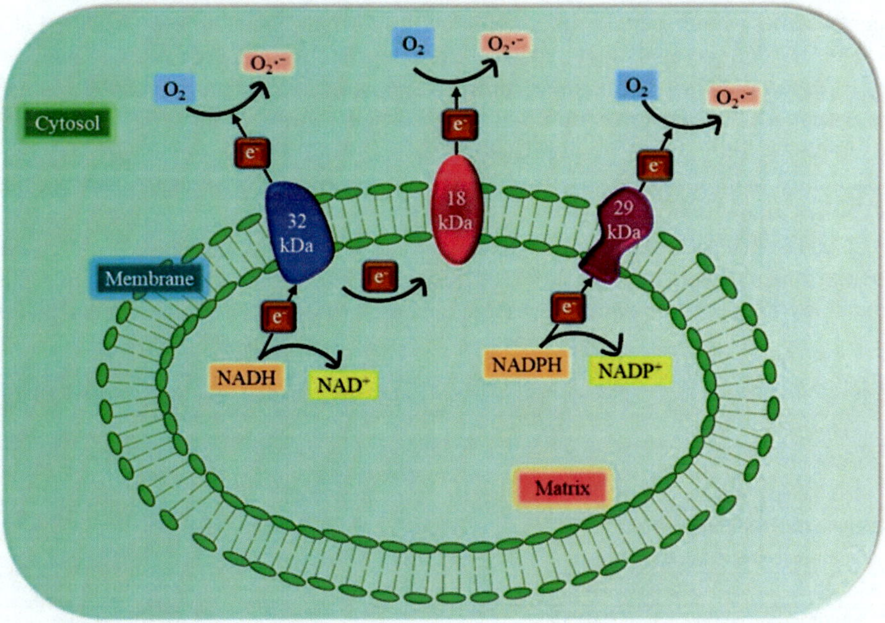

Fig. 7.2 ROS generation through NAD(P)H-dependent small electron transport chain

32 kDa and are designated as PMP18 and PMP32, respectively (Camejo et al. 2016). The third NADPH-dependent PMP of 29 kDa associated with pea leaf peroxisome ETC has been designated as PMP29 (Karuppanapandian et al. 2011). PMP18 mainly participates in $O_2^{\bullet -}$ production, which is released to the cytosol (Das and Roychoudhury 2014). It mediates the transfer of an electron from monodehydroascorbate reductase (MDAR) to O_2, generating $O_2^{\bullet -}$ as a resultant (Nyathi and Baker 2006). Additionally, PMP29 displaying cyt c reductase activity can induce $O_2^{\bullet -}$ production by reducing O_2 or cyt c (Sandalio et al. 2013; Das and Roychoudhury 2014; Del Río and López-Huertas 2016; Phua et al. 2021). Moreover, PMP32 also generates $O_2^{\bullet -}$ on subsequent reduction of O_2 (Nyathi and Baker 2006). This reduction occurs either directly by transferring an electron to O_2 or indirectly through PMP18, which involves transfer of an electron from PMP32 to PMP18 and then to O_2 (Nyathi and Baker 2006). The $O_2^{\bullet -}$ formed is either enzymatically or spontaneously converted into H_2O_2 (Nyathi and Baker 2006).

7.6 Regulation of ROS in Peroxisomes

Peroxisomes have a dynamic system to regulate organelle redox homeostasis, comprising a large range of antioxidants. Ascorbate-glutathione (AsA-GSH) cycle (Foyer-Halliwell-Asada cycle) enzymes like ascorbate peroxidase (APX), monodehydroascorbate reductase (MDHAR), dehydroascorbate reductase (DHAR), and glutathione reductase (GR); and other enzymes such as peroxiredoxins (Prx), CAT, and SOD are indispensable arsenal of antioxidants found in peroxisomes which partake in neutralization of ROS (Sandalio et al. 2013; Fahy et al. 2017; Corpas et al. 2019, 2020) (Table 7.2). The presence of extensive network of ROS scavengers in peroxisomes makes peroxisomes a regulating organelle, which is essential for plant protection and survival under stress (Fahy et al. 2017). For example, salinity stress up-regulated SOD and CAT expression in peroxisomes efficiently neutralizes the $O_2^{\bullet -}$ and H_2O_2, respectively, in several plant species (Mittova et al. 2003; Amor et al. 2019).

Catalase (CAT) is the key enzymatic antioxidant found abundantly in all peroxisomes (Sandalio et al. 2013; Corpas et al. 2020; Palma et al. 2020). CAT is localized in the peroxisomal matrix, constituting 10–25% of total protein in the

Table 7.2 Antioxidants in peroxisomes involved in reactive oxygen species (ROS) scavenging

Antioxidant	Peroxisomal site	Target ROS
Catalase (CAT)	Matrix	H_2O_2
Ascorbate peroxidase (APX)	Membrane	H_2O_2
Superoxide dismutase (SOD)	Matrix/Membrane	$O_2^{\bullet -}$
Glutathione reductase (GR)	Matrix	H_2O_2
Monodehydroascorbate reductase (MDHAR)	Matrix/Membrane	H_2O_2
Dehydroascorbate reductase (DHAR)	Matrix	H_2O_2
Peroxiredoxins (Prxs)	Matrix	H_2O_2

peroxisome and facilitates the elimination of H_2O_2 by degrading or reducing it into H_2O (Sandalio et al. 2013; Su et al. 2019; Pan et al. 2020). Though CAT is a potent quencher of H_2O_2, it has weak affinity toward it and therefore, for efficient performance of CAT, the presence of H_2O_2 in high concentration is needed (Mhamdi et al. 2012; Sandalio et al. 2013). Moreover, in plants, the activation of CAT is mediated by Ca^{2+} and calmodulin, which stimulate the catalytic function of the enzyme (Yang and Poovaiah 2002). The in vivo increase in intraperoxisomal Ca^{2+} showed a stimulatory effect on CAT activity, enhancing peroxisomal H_2O_2 quenching potential (Costa et al. 2010). In *Arabidopsis thaliana* peroxisomes, proteins encoding CAT partake in H_2O_2 neutralization under adverse circumstances (Su et al. 2018). The expression of CAT encoding genes viz., *CAT1*, *CAT2*, and *CAT3* in *Arabidopsis* have been reported by Mhamdi et al. (2012) to be associated with the fatty acid β-oxidation, photorespiration pathway and senescence pathway, respectively. The CAT-deficient mutant of *A. thaliana* displays reduced activity of peroxisomal CAT enzyme, enhanced ozone sensitivity, and photorespiratory H_2O_2 triggered cell death (Vandenabeele et al. 2004). Moreover, under adverse environmental cues, the interaction between peroxisomal CAT gene *CAT2* and small heat shock protein (HSP) Hsp17.6CII occurs and enhances the activity of *CAT2* to improve protection against stress (Li et al. 2017).

Superoxide dismutase (SOD) is another important peroxisomal antioxidant, localized both in matrix and on organellar membrane (Corpas et al. 2020). SOD prompts disproportionation of $O_2^{\bullet-}$ into H_2O_2 and prevents oxidative damage induced by $O_2^{\bullet-}$ on biomolecules such as CAT enzyme (Sandalio et al. 2013). The activity of SOD in peroxisomes has initially been reported in the tissues of pea leaves in the early 1980s (del Río et al. 1983; Sandalio et al. 2013; Corpas et al. 2019). Three different types of SOD isoforms based on metal(s) present in their prosthetic group have been documented in plant peroxisomes viz., Cu/Zn-SOD, Fe-SOD, and Mn-SOD (Sandalio et al. 2013; Corpas et al. 2020). The type of SOD isoenzymes that exist in peroxisomes is determined by the organ or the plant species (del Río et al. 2018; Corpas et al. 2019, 2020). For example, two isoforms of SOD, i.e., Cu/Zn-SOD as well as Mn-SOD have been reported from watermelon cotyledons, localized in matrix and bound to the membrane, respectively (Rodríguez-Serrano et al. 2007; Corpas et al. 2020) whereas in olive fruit peroxisomes, four SOD isoforms have been reported comprising single Fe-SOD, single Mn-SOD and two Cu/Zn-SOD isoforms (Lopez-Huertas and del Río 2014). Similarly, Cu/Zn-SOD have been identified in *Arabidopsis* peroxisomes (Pan and Hu 2018).

Concomitant to CAT, AsA-GSH cycle is pivotal in regulating H_2O_2 in peroxisomes (Sandalio et al. 2013). Its associated enzymes like APX is found bounded with peroxisomal membrane on the cytosolic side, GR has been observed in matrix only and MDHAR is found in membrane as well as in matrix (Leterrier et al. 2005; Romero-Puertas et al. 2006; Eastmond 2007; Sandalio et al. 2013; Corpas et al. 2020). Membrane-bound APX shows more affinity for H_2O_2 compared to CAT, which assists in management of H_2O_2 (Ishikawa et al. 1998; Pan et al. 2020). APX localization on peroxisomal membrane facilitates quenching of H_2O_2

formed on spontaneous dismutation of $O_2^{\bullet -}$, which is generated via membrane-associated small ETC (Lopez-Huertas et al. 1999; Sandalio et al. 2013). Thus, it has been proposed that APX functions primarily to prevent leakage of H_2O_2 out from the peroxisomes (Kaur et al. 2009; Pan et al. 2020). The two APX viz., APX3 and APX5 have been documented from *Arabidopsis* peroxisomes (Narendra et al. 2006; Pan et al. 2018). The level of peroxisomal APX and their expression is up-regulated by unfavorable events in various plant species, triggering tolerance against the stress (Anjum et al. 2016). Similar to APX, the MDHAR found associated with the peroxisomal membranes assists in H_2O_2 scavenging generated through small ETC (Lopez-Huertas et al. 1999; Sandalio et al. 2013). Peroxisomal matrix has also been reported to harbor protein, which is immunologically related to plant Peroxiredoxins (Prxs) and shows capacity to scavenge H_2O_2 (Bonekamp et al. 2009; Mhamdi et al. 2010; Corpas et al. 2017b; Su et al. 2019). Prxs are the thioredoxin-dependent peroxidases, which partake in the alleviation of H_2O_2 (Sandalio et al. 2013; Del Río and López-Huertas 2016). Moreover, Prxs in peroxisomes could improve the antioxidant potential of the organelle (Sandalio et al. 2013).

7.7 Oxidative Stress in Peroxisomes

Plants subjected to either biotic or abiotic stress, eventually lead to the excess generation of ROS. The exorbitant generation of ROS is catalyzed by the surplus performance of pathways mediating ROS generation such as photorespiration, and others, and/or abridged activity of antioxidants (Gupta et al. 2018). For instance, tomato plants affected by the salt stress underwent increased lipid peroxidation and displayed diminished activity of antioxidants including AsA and GSH (Mittova et al. 2003). The leaf peroxisomes exposed to cadmium elevate production of H_2O_2, antioxidant enzymes dysfunctioning, oxidative alteration to exogenous proteins with increment in the number of peroxisomes (Romero-Puertas et al. 1999). Cadmium also brings changes in pea leaf peroxisomes such as the development of senescence-like symptoms and metabolic shift from leaf peroxisomes to glyoxysomes (McCarthy et al. 2001). The generation of ROS via β-oxidation can induce oxidative stress and affect plant survival and their ability to tolerate stressful events (Fan et al. 2017; Yu et al. 2019). The exposure of CAT-deficient tobacco plants to excess light accelerates H_2O_2 content in the peroxisome, leading to leaf bleaching indicating oxidative stress generation (Willekens et al. 1997). The ROS generated in peroxisome induces post-translational modification (PTMs) of proteins, such as carbonylation and *S*-sulfenylation, which can change protein function and stability (Waszczak et al. 2018; Corpas et al. 2020). The proteomic techniques have revealed that nearly 2% of peroxisomal proteins are susceptible to *S*-sulfenylation (Akter et al. 2017; Huang et al. 2019; Corpas et al. 2020). The thiol group associated with cysteine amino acid residues is the primary target of H_2O_2-induced oxidation via a process called *S*-sulfenylation, resulting in enzymatic inactivation (Corpas et al. 2020). The carbonylation of the peroxisomal protein carbon groups due to overproduction of ROS under adverse circumstances, accelerates incidence of protein oxidation (Sandalio et al. 2013). The major proteins related to β-oxidation of fatty acids, malate synthase, multifunctional protein AIM1, and enzyme like GR, Mn-SOD and CAT have been found to experience carbonylation (Nguyen and Donaldson 2005; Sandalio et al. 2013; Mano et al. 2014; Rodríguez-Ruiz et al. 2019; Corpas et al. 2020). Additionally, the reaction of $O_2^{\bullet-}$ with nitric oxide (NO) results in the generation of peroxynitrite ($ONOO^-$), a powerful oxidizing agent that induces PTM like tyrosine (protein amino acid) nitration, which is irreversible and can inhibit protein molecules, eventually modifying their function (Mata-Pérez et al. 2016; Corpas et al. 2020). For instance, in *Arabidopsis*, $ONOO^-$ production in peroxisomes affects Cu/Zn-SOD, causing 65% inhibition of the enzyme function (Holzmeister et al. 2015). In peroxisomes of *Arabidopsis* seedlings, increased accumulation of NO, accompanied by $O_2^{\bullet-}$ content accelerates the generation of $ONOO^-$ under cadmium, and lead stress (Corpas and Barroso 2014, 2017; Corpas et al. 2017a).

7.8 Peroxisomal ROS-Mediated Signaling

The retrograde signaling induced from organelle to nucleus can alter the nuclear genes expression, mediating stress tolerance (Su et al. 2019). The ROS generated in plants not only induce oxidative stress, but significantly take part in signaling pathway which trigger responses like stress resistance (Sandalio et al. 2013). For instance, suppression of *CAT2* by salicylic acid has been found to increase H_2O_2 accumulation in plants and trigger resistance toward biotrophic bacteria (Yuan et al. 2017). The transient buildup of ROS in particular organelle under stress activates specific defense responses (Sandalio et al. 2013; Sewelam et al. 2014; Su et al. 2019). H_2O_2 production from chloroplast of GOX overexpressing *A. thaliana* triggers expression of genes participating in signaling process, including transcription factors, defense proteins, and protein/receptor kinases encoding genes, and secondary signaling messengers biosynthesis, whereas H_2O_2 generated from peroxisomes in CAT-deficient mutant of *Arabidopsis* induces genes coding for HSP and other proteins required for the degradation of ubiquitin-based proteins, which trigger responses mediating stress recovery (Sewelam et al. 2014). Nevertheless, studies also suggest that in some cases H_2O_2 mediated signaling is induced irrespective of the site of its origin (Sewelam et al. 2014).

Peroxisomes, which is one of the important sites of ROS generation, are considered as a key player in the generation of signaling molecules, particularly, H_2O_2 (Sandalio et al. 2013; Waszczak et al. 2018; Su et al. 2019). Many studies have evidently shown that the reduced accumulation of H_2O_2 under stressed conditions, abridges plants' ability to cope up with the adverse circumstances. For instance, the peroxisomal GOX-deficient mutants of *Nicotiana benthamiana* and *A. thaliana* become susceptible to non-host pathogens. The silencing of GOX genes: *HAOX2* and *GOX3*, decreases H_2O_2 generation via photorespiration, attenuating expression of many defense-related genes, which reduce resistance capacity of plants to hypersensitive response-inducing non-host pathogenic *Pseudomonas* (Rojas et al. 2012).

The reduced activity of peroxisomal antioxidant enzymes, particularly, CAT assists transient increase in H_2O_2 concentration buildup which mediates signaling process (Inzé et al. 2012). The adverse environmental conditions, promoting photorespiration, elevates ROS generation. These elevated ROS, in particular, H_2O_2, are quenched by the CAT antioxidants (Mateo et al. 2004; Vanderauwera et al. 2005). However, mutants lacking CAT activity result in the accumulation of photorespiratory ROS (Vanderauwera et al. 2005; Inzé et al. 2012). The diffusion of H_2O_2 from peroxisomes to cytosol shows inter-organellar communication, which facilitates activation of nuclear gene expression, leading to extensive transcriptional response (Corpas et al. 2001; Vanderauwera et al. 2005; Inzé et al. 2012). For instance, *A. thaliana* mutant lacking CAT activity displays modulated nuclear gene expression under high light stress (Vandenabeele et al. 2004). Moreover, due to the loss of peroxisomal CAT function, photorespiratory H_2O_2 accumulation enhances and mediates signal by up-regulating the expression of HSP (Vanderauwera et al. 2005). The *CAT2* knock-out mutant of *A. thaliana* shows oxidative signaling under photorespiratory circumstances (Queval et al. 2007). Similarly, the G-Box Binding Factor 1 (GBF1), suppressing expression of *CAT2* in *Arabidopsis*, increases the accumulation of H_2O_2, which promotes plant resistance against pathogens (Giri et al. 2017).

7.9 Conclusion

Peroxisome metabolic activities, as a by-product, generate reactive oxygen species (ROS), which exacerbates under stress conditions. To regulate redox homeostasis, peroxisomes recruit many antioxidants or undergo proliferation, and to improve plants' resilience toward stress, ROS mediate retrograde signaling. However, undue increase in ROS, disrupts intrinsic defense mechanisms leading to severe oxidative damage. To improve crop performance under a variety of stressful circumstances, fortifying plant-defensive machinery or up-regulating the expression of genes orchestrating resistance is imperative. Thus there is a need to enhance understanding on metabolism of ROS in peroxisomes and explore pathways as well as genes involved in the elicitation of resistance under varying adverse conditions to improve food security.

References

Akter S, Carpentier S, Van Breusegem F, Messens J (2017) Identification of dimedone-trapped sulfenylated proteins in plants under stress. Biochem Biophys Rep 9:106–113

Amor N, Jimenez A, Boudabbous M, Sevilla F, Abdelly C (2019) Implication of peroxisomes and mitochondria in the halophyte *Cakile maritima* tolerance to salinity stress. Biol Plant 63:113–121

Anjum NA, Sharma P, Gill SS, Hasanuzzaman M, Khan EA, Kachhap K, Mohamed AA, Thangavel P, Devi GD, Vasudhevan P, Sofo A, Khan NA, Misra AN, Lukatkin AS, Singh

HP, Pereira E, Tuteja N (2016) Catalase and ascorbate peroxidase—representative H2O2-detoxifying heme enzymes in plants. Environ Sci Pollut Res 23(19):19002–19029
Ansari MI (2018) Plant microbiome and its functional mechanism in response to environmental stress. Int J Green Pharm 12(01):S81
Ansari MI, Jalil SU, Ansari SA, Hasanuzzaman M (2021) GABA shunt: a key-player in mitigation of ROS during stress. Plant Growth Regul 94(2):131–149
Bonekamp NA, Völkl A, Fahimi HD, Schrader M (2009) Reactive oxygen species and peroxisomes: struggling for balance. Biofactors 35(4):346–355
Camejo D, Guzmán-Cedeño Á, Moreno A (2016) Reactive oxygen species, essential molecules, during plant–pathogen interactions. Plant Physiol Biochem 103:10–23
Castillo MC, Sandalio LM, Del Rio LA, Leon J (2008) Peroxisome proliferation, wound-activated responses and expression of peroxisome-associated genes are cross-regulated but uncoupled in *Arabidopsis thaliana*. Plant Cell Environ 31(4):492–505
Corpas FJ, Barroso JB (2014) Peroxynitrite (ONOO−) is endogenously produced in *Arabidopsis* peroxisomes and is overproduced under cadmium stress. Ann Bot 113(1):87–96
Corpas FJ, Barroso JB (2017) Lead-induced stress, which triggers the production of nitric oxide (NO) and superoxide anion ($O2^{\bullet -}$) in *Arabidopsis* peroxisomes, affects catalase activity. Nitric Oxide 68:103–110
Corpas FJ, Barroso JB, del Rio LA (2001) Peroxisomes as a source of reactive oxygen species and nitric oxide signal molecules in plant cells. Trends Plant Sci 6(4):145–150
Corpas FJ, Barroso JB, Palma JM, Rodriguez-Ruiz M (2017a) Plant peroxisomes: a nitro-oxidative cocktail. Redox Biol 11:535–542
Corpas FJ, Pedrajas JR, Palma JM, Valderrama R, Rodríguez-Ruiz M, Chaki M, del Rio LA, Barroso JB (2017b) Immunological evidence for the presence of peroxiredoxin in pea leaf peroxisomes and response to oxidative stress conditions. Acta Physiol Plant 39(2):1–12
Corpas FJ, Del Río LA, Palma JM (2019) Impact of nitric oxide (NO) on the ROS metabolism of peroxisomes. Plants 8(2):37
Corpas FJ, González-Gordo S, Palma JM (2020) Plant peroxisomes: a factory of reactive species. Front Plant Sci 11:853
Costa A, Drago I, Behera S, Zottini M, Pizzo P, Schroeder JI, Pozzan T, Schiavo FL (2010) H2O2 in plant peroxisomes: an in vivo analysis uncovers a Ca2+-dependent scavenging system. Plant J 62(5):760–772
Cruz DCM (2008) Drought stress and reactive oxygen species: production, scavenging and signaling. Plant Signal Behav 3(3):156–165
Cui S, Hayashi Y, Otomo M, Mano S, Oikawa K, Hayashi M, Nishimura M (2016) Sucrose production mediated by lipid metabolism suppresses the physical interaction of peroxisomes and oil bodies during germination of *Arabidopsis thaliana*. J Biol Chem 291(38):19734–19745
Das K, Roychoudhury A (2014) Reactive oxygen species (ROS) and response of antioxidants as ROS-scavengers during environmental stress in plants. Front Environ Sci 2:53
Del Río LA, López-Huertas E (2016) ROS generation in peroxisomes and its role in cell signaling. Plant Cell Physiol 57(7):1364–1376
del Río LA, Lyon DS, Olah I, Glick B, Salin ML (1983) Immunocytochemical evidence for a peroxisomal localization of manganese superoxide dismutase in leaf protoplasts from a higher plant. Planta 158(3):216–224
del Río LA, Sandalio LM, Palma J, Bueno P, Corpas FJ (1992) Metabolism of oxygen radicals in peroxisomes and cellular implications. Free Radic Biol Med 13(5):557–580
del Río LA, Sandalio LM, Corpas FJ, Palma JM, Barroso JB (2006) Reactive oxygen species and reactive nitrogen species in peroxisomes. Production, scavenging, and role in cell signaling. Plant Physiol 141(2):330–335
del Río LA, Corpas FJ, López-Huertas E, Palma JM (2018) Plant superoxide dismutases: function under abiotic stress conditions. In: Antioxidants and antioxidant enzymes in higher plants. Springer, Cham, pp 1–26

Demidchik V (2012) Reactive oxygen species and oxidative stress in plants. In: Plant stress physiology, vol 8. Delve Publishing, Ashland, pp 24–58
Eastmond PJ (2007) MONODEHYROASCORBATE REDUCTASE4 is required for seed storage oil hydrolysis and postgerminative growth in *Arabidopsis*. Plant Cell 19(4):1376–1387
Fahy D, Sanad MN, Duscha K, Lyons M, Liu F, Bozhkov P, Kunz HH, Hu J, Neuhaus HE, Steel PG, Smertenko A (2017) Impact of salt stress, cell death, and autophagy on peroxisomes: quantitative and morphological analyses using small fluorescent probe N-BODIPY. Sci Rep 7(1):1–18
Fan J, Yu L, Xu C (2017) A central role for triacylglycerol in membrane lipid breakdown, fatty acid β-oxidation, and plant survival under extended darkness. Plant Physiol 174(3):1517–1530
Foyer CH, Noctor G (2003) Redox sensing and signalling associated with reactive oxygen in chloroplasts, peroxisomes and mitochondria. Physiol Plant 119(3):355–364
Giri MK, Singh N, Banday ZZ, Singh V, Ram H, Singh D, Chattopadhyay S, Nandi AK (2017) GBF 1 differentially regulates CAT 2 and PAD 4 transcription to promote pathogen defense in *Arabidopsis thaliana*. Plant J 91(5):802–815
Gupta DK, Palma JM, Corpas FJ (eds) (2018) Antioxidants and antioxidant enzymes in higher plants. Springer International Publishing, Berlin
Hänsch R, Lang C, Rennenberg H, Mendel RR (2007) Significance of plant sulfite oxidase. Plant Biol 9(05):589–595
Holzmeister C, Gaupels F, Geerlof A, Sarioglu H, Sattler M, Durner J, Lindermayr C (2015) Differential inhibition of *Arabidopsis* superoxide dismutases by peroxynitrite-mediated tyrosine nitration. J Exp Bot 66(3):989–999
Huang J, Willems P, Wei B, Tian C, Ferreira RB, Bodra N, Ache SAM, Wahni K, Liu K, Vertommen D, Gevaert K, Carroll KS, Van Montagu M, Yang J, Van Breusegem F, Messens J (2019) Mining for protein S-sulfenylation in *Arabidopsis* uncovers redox-sensitive sites. Proc Natl Acad Sci 116(42):21256–21261
Inzé A, Vanderauwera S, Hoeberichts FA, Vandorpe M, Van Gaever TIM, Van Breusegem F (2012) A subcellular localization compendium of hydrogen peroxide-induced proteins. Plant Cell Environ 35(2):308–320
Iqbal Z, Iqbal MS, Hashem A, Abd Allah EF, Ansari MI (2021) Plant defense responses to biotic stress and its interplay with fluctuating dark/light conditions. Front Plant Sci 12:297
Ishikawa T, Yoshimura K, Sakai K, Tamoi M, Takeda T, Shigeoka S (1998) Molecular characterization and physiological role of a glyoxysome-bound ascorbate peroxidase from spinach. Plant Cell Physiol 39(1):23–34
Jalil SU, Ansari MI (2019) Nanoparticles and abiotic stress tolerance in plants: synthesis, action, and signaling mechanisms. In: Plant signaling molecules. Woodhead Publishing, Duxford, pp 549–561
Kamada-Nobusada T, Hayashi M, Fukazawa M, Sakakibara H, Nishimura M (2008) A putative peroxisomal polyamine oxidase, AtPAO4, is involved in polyamine catabolism in *Arabidopsis thaliana*. Plant Cell Physiol 49(9):1272–1282
Karuppanapandian T, Moon JC, Kim C, Manoharan K, Kim W (2011) Reactive oxygen species in plants: their generation, signal transduction, and scavenging mechanisms. Aust J Crop Sci 5(6): 709–725
Kaur N, Reumann S, Hu J (2009) Peroxisome biogenesis and function. In: The *Arabidopsis* book. American Society of Plant Biologists, Rockville, MD, p 7
Kunze M, Hartig A (2013) Permeability of the peroxisomal membrane: lessons from the glyoxylate cycle. Front Physiol 4:204
Kusano T, Kim DW, Liu T, Berberich T (2015) Polyamine catabolism in plants. In: Polyamines. Springer, Tokyo, pp 77–88
Leterrier M, Corpas FJ, Barroso JB, Sandalio LM, del Río LA (2005) Peroxisomal monodehydroascorbate reductase. Genomic clone characterization and functional analysis under environmental stress conditions. Plant Physiol 138(4):2111–2123

Li G, Li J, Hao R, Guo Y (2017) Activation of catalase activity by a peroxisome-localized small heat shock protein Hsp17. 6CII. J Genet Genomics 44(8):395–404

Lopez-Huertas E, del Río LA (2014) Characterization of antioxidant enzymes and peroxisomes of olive (*Olea europaea* L.) fruits. J Plant Physiol 171(16):1463–1471

Lopez-Huertas E, Corpas FJ, Sandalio LM, del Rio LA (1999) Characterization of membrane polypeptides from pea leaf peroxisomes involved in superoxide radical generation. Biochem J 337(3):531–536

Mano JI, Nagata M, Okamura S, Shiraya T, Mitsui T (2014) Identification of oxidatively modified proteins in salt-stressed *Arabidopsis*: a carbonyl-targeted proteomics approach. Plant Cell Physiol 55(7):1233–1244

Mata-Pérez C, Begara-Morales JC, Chaki M, Sánchez-Calvo B, Valderrama R, Padilla MN, Corpas FJ, Barroso JB (2016) Protein tyrosine nitration during development and abiotic stress response in plants. Front Plant Sci 7:1699

Mateo A, Mühlenbock P, Rustérucci C, Chang CCC, Miszalski Z, Karpinska B, Parker JE, Mullineaux PM, Karpinski S (2004) LESION SIMULATING DISEASE 1 is required for acclimation to conditions that promote excess excitation energy. Plant Physiol 136(1): 2818–2830

McCarthy I, Romero-Puertas MC, Palma JM, Sandalio LM, Corpas FJ, Gómez M, Del Rio LA (2001) Cadmium induces senescence symptoms in leaf peroxisomes of pea plants. Plant Cell Environ 24(10):1065–1073

Mhamdi A, Queval G, Chaouch S, Vanderauwera S, Van Breusegem F, Noctor G (2010) Catalase function in plants: a focus on *Arabidopsis* mutants as stress-mimic models. J Exp Bot 61(15): 4197–4220

Mhamdi A, Noctor G, Baker A (2012) Plant catalases: peroxisomal redox guardians. Arch Biochem Biophys 525(2):181–194

Mittova V, Tal M, Volokita M, Guy M (2003) Up-regulation of the leaf mitochondrial and peroxisomal antioxidative systems in response to salt-induced oxidative stress in the wild salt-tolerant tomato species *Lycopersicon pennellii*. Plant Cell Environ 26(6):845–856

Mor A, Koh E, Weiner L, Rosenwasser S, Sibony-Benyamini H, Fluhr R (2014) Singlet oxygen signatures are detected independent of light or chloroplasts in response to multiple stresses. Plant Physiol 165(1):249–261

Narendra S, Venkataramani S, Shen G, Wang J, Pasapula V, Lin Y, Kornyeyev D, Holaday AS, Zhang H (2006) The *Arabidopsis* ascorbate peroxidase 3 is a peroxisomal membrane-bound antioxidant enzyme and is dispensable for *Arabidopsis* growth and development. J Exp Bot 57(12):3033–3042

Nguyen AT, Donaldson RP (2005) Metal-catalyzed oxidation induces carbonylation of peroxisomal proteins and loss of enzymatic activities. Arch Biochem Biophys 439(1):25–31

Nyathi Y, Baker A (2006) Plant peroxisomes as a source of signalling molecules. Biochim Biophys Acta Mol Cell Res 1763(12):1478–1495

Oikawa K, Hayashi M, Hayashi Y, Nishimura M (2019) Re-evaluation of physical interaction between plant peroxisomes and other organelles using live-cell imaging techniques. J Integr Plant Biol 61(7):836–852

Ono Y, Kim DW, Watanabe K, Sasaki A, Niitsu M, Berberich T, Kusano T, Takahashi Y (2012) Constitutively and highly expressed *Oryza sativa* polyamine oxidases localize in peroxisomes and catalyze polyamine back conversion. Amino Acids 42(2):867–876

Ortega-Galisteo AP, Rodríguez-Serrano M, Pazmiño DM, Gupta DK, Sandalio LM, Romero-Puertas MC (2012) S-Nitrosylated proteins in pea (*Pisum sativum* L.) leaf peroxisomes: changes under abiotic stress. J Exp Bot 63(5):2089–2103

Palma J, Garrido M, Rodriguez-Garcia MI, Luis A (1991) Peroxisome proliferation and oxidative stress mediated by activated oxygen species in plant peroxisomes. Arch Biochem Biophys 287(1):68–74

Palma JM, Mateos RM, López-Jaramillo J, Rodríguez-Ruiz M, González-Gordo S, Lechuga-Sancho AM, Corpas FJ (2020) Plant catalases as NO and H2S targets. Redox Biol 34:101525

Pan R, Hu J (2018) Proteomics of peroxisomes. Subcell Biochem 89:3–45

Pan R, Reumann S, Lisik P, Tietz S, Olsen LJ, Hu J (2018) Proteome analysis of peroxisomes from dark-treated senescent *Arabidopsis* leaves. J Integr Plant Biol 60(11):1028–1050

Pan R, Liu J, Wang S, Hu J (2020) Peroxisomes: versatile organelles with diverse roles in plants. New Phytol 225(4):1410–1427

Petrov V, Hille J, Mueller-Roeber B, Gechev TS (2015) ROS-mediated abiotic stress-induced programmed cell death in plants. Front Plant Sci 6:69

Phua SY, De Smet B, Remacle C, Chan KX, Van Breusegem F (2021) Reactive oxygen species and organellar signaling. J Exp Bot 72(16):5807–5824

Plett A, Charton L, Linka N (2020) Peroxisomal cofactor transport. Biomolecules 10(8):1174

Queval G, Issakidis-Bourguet E, Hoeberichts FA, Vandorpe M, Gakière B, Vanacker H, Miginiac-Maslow M, Breusegem FV, Noctor G (2007) Conditional oxidative stress responses in the *Arabidopsis* photorespiratory mutant cat2 demonstrate that redox state is a key modulator of daylength-dependent gene expression, and define photoperiod as a crucial factor in the regulation of H2O2-induced cell death. Plant J 52(4):640–657

Rodríguez-Ruiz M, González-Gordo S, Cañas A, Campos MJ, Paradela A, Corpas FJ, Palma JM (2019) Sweet pepper (*Capsicum annuum* L.) fruits contain an atypical peroxisomal catalase that is modulated by reactive oxygen and nitrogen species. Antioxidants 8(9):374

Rodríguez-Serrano M, Romero-Puertas MC, Pastori GM, Corpas FJ, Sandalio LM, del Río LA, Palma JM (2007) Peroxisomal membrane manganese superoxide dismutase: characterization of the isozyme from watermelon (*Citrullus lanatus* Schrad.) cotyledons. J Exp Bot 58(10): 2417–2427

Rodríguez-Serrano M, Romero-Puertas MC, Sparkes I, Hawes C, Luis A, Sandalio LM (2009) Peroxisome dynamics in *Arabidopsis* plants under oxidative stress induced by cadmium. Free Radic Biol Med 47(11):1632–1639

Rodríguez-Serrano M, Romero-Puertas MC, Sanz-Fernández M, Hu J, Sandalio LM (2016) Peroxisomes extend peroxules in a fast response to stress via a reactive oxygen species-mediated induction of the peroxin PEX11a. Plant Physiol 171(3):1665–1674

Rojas CM, Senthil-Kumar M, Wang K, Ryu CM, Kaundal A, Mysore KS (2012) Glycolate oxidase modulates reactive oxygen species–mediated signal transduction during nonhost resistance in *Nicotiana benthamiana* and *Arabidopsis*. Plant Cell 24(1):336–352

Romero-Puertas MC, McCarthy I, Sandalio LM, Palma JM, Corpas FJ, Gómez M, Del Rio LA (1999) Cadmium toxicity and oxidative metabolism of pea leaf peroxisomes. Free Radic Res 31 (sup1):25–31

Romero-Puertas MC, Palma JM, Gómez M, Del Rio LA, Sandalio LM (2002) Cadmium causes the oxidative modification of proteins in pea plants. Plant Cell Environ 25(5):677–686

Romero-Puertas MC, Corpas FJ, Sandalio LM, Leterrier M, Rodríguez-Serrano M, Del Río LA, Palma JM (2006) Glutathione reductase from pea leaves: response to abiotic stress and characterization of the peroxisomal isozyme. New Phytol 170(1):43–52

Sachdev S, Ansari SA, Ansari MI, Fujita M, Hasanuzzaman M (2021) Abiotic stress and reactive oxygen species: generation, signaling, and defense mechanisms. Antioxidants 10(2):277

Sandalio LM, Rodríguez-Serrano M, Romero-Puertas MC, Río LAD (2013) Role of peroxisomes as a source of reactive oxygen species (ROS) signaling molecules. In: Peroxisomes and their key role in cellular signaling and metabolism. Springer, Dordrecht, pp 231–255

Sandalio LM, Peláez-Vico MA, Molina-Moya E, Romero-Puertas MC (2021) Peroxisomes as redox-signaling nodes in intracellular communication and stress responses. Plant Physiol 186(1):22

Schrader M, Fahimi HD (2006) Peroxisomes and oxidative stress. Biochim Biophys Acta Mol Cell Res 1763(12):1755–1766

Sewelam N, Jaspert N, Van Der Kelen K, Tognetti VB, Schmitz J, Frerigmann H, Stahl E, Zeier J, Van Breusegem F, Maurino VG (2014) Spatial H2O2 signaling specificity: H2O2 from chloroplasts and peroxisomes modulates the plant transcriptome differentially. Mol Plant 7(7): 1191–1210

Sinclair AM, Trobacher CP, Mathur N, Greenwood JS, Mathur J (2009) Peroxule extension over ER-defined paths constitutes a rapid subcellular response to hydroxyl stress. Plant J 59(2): 231–242

Sousa RH, Carvalho FE, Lima-Melo Y, Alencar VT, Daloso DM, Margis-Pinheiro M, Komatsu S, Silveira JA (2019) Impairment of peroxisomal APX and CAT activities increases protection of photosynthesis under oxidative stress. J Exp Bot 70(2):627–639

Su T, Wang P, Li H, Zhao Y, Lu Y, Dai P, Ren T, Wang X, Li X, Shao Q, Zhao D, Zhao Y, Ma C (2018) The *Arabidopsis* catalase triple mutant reveals important roles of catalases and peroxisome-derived signaling in plant development. J Integr Plant Biol 60(7):591–607

Su T, Li W, Wang P, Ma C (2019) Dynamics of peroxisome homeostasis and its role in stress response and signaling in plants. Front Plant Sci 10:705

Suo J, Zhao Q, David L, Chen S, Dai S (2017) Salinity response in chloroplasts: insights from gene characterization. Int J Mol Sci 18(5):1011

Takahashi S, Badger MR (2011) Photoprotection in plants: a new light on photosystem II damage. Trends Plant Sci 16(1):53–60

Tiew TWY, Sheahan MB, Rose RJ (2015) Peroxisomes contribute to reactive oxygen species homeostasis and cell division induction in *Arabidopsis* protoplasts. Front Plant Sci 6:658

Ueda Y, Uehara N, Sasaki H, Kobayashi K, Yamakawa T (2013) Impacts of acute ozone stress on superoxide dismutase (SOD) expression and reactive oxygen species (ROS) formation in rice leaves. Plant Physiol Biochem 70:396–402

Vandenabeele S, Vanderauwera S, Vuylsteke M, Rombauts S, Langebartels C, Seidlitz HK, Zabeau M, Montagu MV, Inze D, Van Breusegem F (2004) Catalase deficiency drastically affects gene expression induced by high light in *Arabidopsis thaliana*. Plant J 39(1):45–58

Vanderauwera S, Zimmermann P, Rombauts S, Vandenabeele S, Langebartels C, Gruissem W, Inze D, Van Breusegem F (2005) Genome-wide analysis of hydrogen peroxide-regulated gene expression in *Arabidopsis* reveals a high light-induced transcriptional cluster involved in anthocyanin biosynthesis. Plant Physiol 139(2):806–821

Voss I, Sunil B, Scheibe R, Raghavendra AS (2013) Emerging concept for the role of photorespiration as an important part of abiotic stress response. Plant Biol 15(4):713–722

Walker CL, Pomatto LC, Tripathi DN, Davies KJ (2018) Redox regulation of homeostasis and proteostasis in peroxisomes. Physiol Rev 98(1):89–115

Waszczak C, Carmody M, Kangasjärvi J (2018) Reactive oxygen species in plant signaling. Annu Rev Plant Biol 69:209–236

Werner AK, Witte CP (2011) The biochemistry of nitrogen mobilization: purine ring catabolism. Trends Plant Sci 16(7):381–387

Willekens H, Chamnongpol S, Davey M, Schraudner M, Langebartels C, Van Montagu M, Inze D, Van Camp W (1997) Catalase is a sink for H2O2 and is indispensable for stress defence in C3 plants. EMBO J 16(16):4806–4816

Yamane K, Taniguchi M, Miyake H (2012) Salinity-induced subcellular accumulation of H2O2 in leaves of rice. Protoplasma 249(2):301–308

Yang TPBW, Poovaiah BW (2002) Hydrogen peroxide homeostasis: activation of plant catalase by calcium/calmodulin. Proc Natl Acad Sci 99(6):4097–4102

Yu L, Fan J, Xu C (2019) Peroxisomal fatty acid β-oxidation negatively impacts plant survival under salt stress. Plant Signal Behav 14(2):1561121

Yuan HM, Liu WC, Lu YT (2017) CATALASE2 coordinates SA-mediated repression of both auxin accumulation and JA biosynthesis in plant defenses. Cell Host Microbe 21(2):143–155

ROS Production and Function at Plasma Membrane and Apoplast

8

8.1 Introduction

The plasma membrane, apoplast, and cell wall are three major extracellular compartments, delimiting the intracellular components of the plant cell to safeguard their integrity, which is essential for maintaining metabolic activities and communication with external stimuli such as environmental cues and microbial interactions (Janku et al. 2019). The plasma membrane or cytoplasmic membrane is the first extracellular compartment, present in direct contact with the cytoplasmic content of the plant cell, acting as an interface between intracellular components and the external environment (Furt et al. 2011; Takahashi et al. 2013). Its composition is highly diverse, constituting different types of lipids and proteins, depending on cell type, stage of development, and environmental conditions (Furt et al. 2011). Both proteins and lipids content of the plasma membrane is essential for mediating key functions of the membrane such as signal transmission, transportation via the membrane, activities of membrane-bound enzymes, and regulation of membrane fluidity (Furt et al. 2011; Takahashi et al. 2013). The plasma membrane is a selectively permeable membrane, which regulates the movement of metal ions, solutes, and macromolecules in and out of the cytoplasm (Furt et al. 2011; Takahashi et al. 2013). Moreover, it is the primary site for signal recognition and transduction to the intracellular region, thus facilitating nutrients uptake and their distribution, response to environmental stimuli, and developmental signaling (Furt et al. 2011). Apoplast, the extraprotoplasmic matrix, extends out from the plasma membrane and encompasses the cell wall and intercellular space (Horemans et al. 2000; Farvardin et al. 2020). Apoplast serves several functions in plants including transportation of nutrients and water, synthesis of defense-related molecules as well as cell wall components (Farvardin et al. 2020). The external stimuli are primarily detected at the apoplast, as it is the first contact site for environmental signals or substances invading plant cells, and thus mediate cell-to-cell or plant-to-plant communication (Vaahtera et al. 2014; Farvardin et al. 2020). Both plasma membrane and apoplast

S. Sachdev et al., *Reactive Oxygen Species in Plants*,
https://doi.org/10.1007/978-981-19-9884-3_8

are important sites for extracellular reactive oxygen species production (ROS) in plants.

The genesis of ROS in the plasma membrane and apoplast takes place as a result of the activity of multiple enzymes located in the apoplast, plasma membrane, and/or cell wall such as NADPH (nicotinamide adenine dinucleotide phosphate) oxidases (NOXs), peroxidases (POXs), amine oxidases (AO), and others (Janku et al. 2019; Sachdev et al. 2021). The extracellular ROS constitute only a small fraction of ROS generated in plant cell. However, this small titer is extremely and extraordinarily important (Podgórska et al. 2017). The accumulation of ROS in apoplast induces oxidative burst, which either leads to cellular damage or signaling mediating tolerance against unfavorable circumstances and also participates in physiological functions like cell growth and cell wall lignification (Janku et al. 2019). For instance, ROS in apoplasts are found to be involved in the extension of the cell wall during leaf growth and elongation of root hairs and pollen tubes (Janku et al. 2019). Moreover, the apoplastic ROS play a noteworthy role in the elimination of invaded pathogens and tolerance to abiotic stresses (Podgórska et al. 2017; Qi et al. 2017; Li et al. 2019). Though extracellular ROS cater to multiple beneficial roles for plant cells, their extreme accumulation could be toxic and hence their level is maintained through the recruitment of various antioxidants (Kukavica et al. 2005).

8.2 Sources of ROS Generation in Extracellular Compartments

ROS production (Fig. 8.1) in apoplast takes place as a result of enzymatic activity of plasma membrane-bound NOX, POX, AO, germin-like oxalate oxidases (OXO), and non-heme iron-containing lipoxygenase (LOX) (Cona et al. 2006; Minibayeva et al. 2009; Qi et al. 2017; Hasanuzzaman et al. 2020). In addition, the activity of plasma membrane-localized quinone reductase, the reaction between superoxide radicals ($O_2^{\bullet-}$) and nitric oxide (NO) in the apoplast, metal-catalyzed Fenton or Haber–Weiss reaction and sequential non-enzymatic degradation of ascorbate also induce the formation of ROS in the apoplast (Kärkönen and Kuchitsu 2015; Podgórska et al. 2017). The reaction of NO with $O_2^{\bullet-}$ and Fenton/Haber–Weiss reaction generate highly toxic hydroxyl radical ($^{\bullet}OH$) in the apoplast (Podgórska et al. 2017). The LOX enzyme participates in ROS generation by catalyzing hydroperoxidation of polyunsaturated fatty acid (Das and Roychoudhury 2014; Podgórska et al. 2017). The phylloquinone mediates the generation of apoplastic ROS by transferring an electron from cytosolic NADPH-dependent quinone reductase to NOX (Bridge et al. 2000). The OXOs are the oligomeric proteins containing manganese in their structure (Podgórska et al. 2017; Janku et al. 2019). These OXOs can mediate oxidative breakdown of oxalic acid, resulting in CO_2 and hydrogen peroxide (H_2O_2) generation as the end product (Podgórska et al. 2017). In maize plants, suffering from water-deficit stress, displayed increased accumulation of apoplastic H_2O_2, particularly in the apical portion of root growth zone, due to accelerated activity of OXO (Voothuluru and Sharp 2013). The H_2O_2 formed via OXOs evokes defense responses in the sunflower against a fungal pathogen

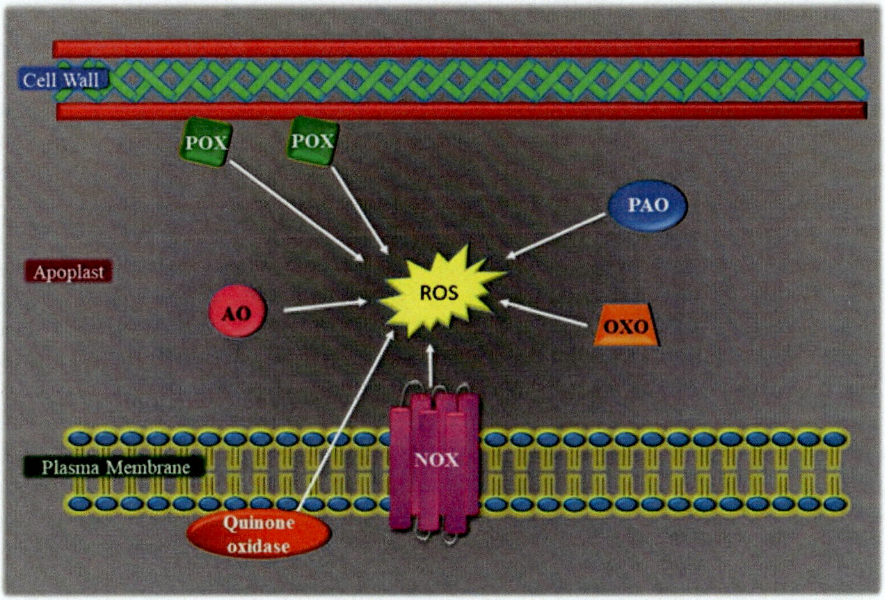

Fig. 8.1 Sources of extracellular ROS generation

Sclerotinia sclerotiorum (Hu et al. 2003). The other sources of ROS in the apoplast, particularly NOX and POX, which are the most prominent contributors, have been discussed in detail.

8.2.1 NADPH Oxidases

Plant NOX localized in the plasma membrane is also termed respiratory burst oxidase homologs (RBOHs) (Podgórska et al. 2017; Sachdev et al. 2021). It is structurally as well as functionally homologous to NOX found in mammalian cells (Qi et al. 2017). The plant NOX (RBOH) is a six transmembrane domain structure with C- and N-terminals lying in the cytosol (Marino et al. 2012; Podgórska et al. 2017; Chang et al. 2020). The C-terminal consists of two heme groups, FAD and NADPH-binding domains, which facilitate the movement of an electron from the cytosol to the apoplast (Marino et al. 2012; Kadota et al. 2015; Podgórska et al. 2017; Chang et al. 2020; Ogboo et al. 2022). The hydrophilic N-terminal region of NOX positioned in cytosol consists of two highly conserved Ca^{2+} binding EF-hand motifs and phosphorylation sites, which regulate the enzymatic activity of NOX (Kaur et al. 2014; Podgórska et al. 2017; Chang et al. 2020). One of the two EF-hand motifs, which has α-helix "E" can bind with the Ca^{2+} and mediate conformational changes in NOX (RBOH) (Oda et al. 2010; Kawasaki and Kretsinger 2017; Podgórska et al. 2017). Moreover, its activity is activated by the symplastic signals such as protein phosphorylation and interaction of NOX with specific proteins and/or lipids (Zhang et al. 2009; Kimura et al. 2012; Kaur et al. 2014; Vaahtera et al. 2014; Podgórska et al. 2017). In contrast to animal RBOH, the N-terminal of plant NOX can interact only with a few proteins, which regulates its activity, either positively or negatively (Podgórska et al. 2017). The conformational alterations in phosphorylated RBOH proteins allow their interaction with the Rac GTPases (Wong et al. 2007), which partake in various signaling pathways (Nagawa et al. 2010).

NOX mediates a pathway for the production of apoplastic ROS using cytosolic NADPH as an electron source (Kaur et al. 2014; Podgórska et al. 2017; Yu et al. 2020). The electron from cytosol is transported to the apoplast via FAD and heme cofactors of NOX, where it reduces O_2 to $O_2^{\bullet-}$ (Kaur et al. 2014; Vaahtera et al. 2014; Arthikala et al. 2017; Arnaud et al. 2021; Sachdev et al. 2021; Ogboo et al. 2022). The $O_2^{\bullet-}$ further undergoes dismutation either enzymatically or non-enzymatically, resulting in the formation of H_2O_2 (Chapman et al. 2019; Chang et al. 2020). The activity of NOX is encoded by the number of *Rboh* genes (Zurbriggen et al. 2010). In *Arabidopsis* and *Nicotiana tabacum*, 10 and 20 *Rboh* genes designated as *AtRbohA-AtRbohJ* and *NtabRbohA-T*, respectively, have been documented (Zurbriggen et al. 2010; Chang et al. 2020; Yu et al. 2020). These *Rboh* genes are upregulated transcriptionally on the perception of stress like wounding, pathogen invasion or during plant development (Marino et al. 2012; Zurbriggen et al. 2010; Yu et al. 2020). Among various known *Rboh* genes, *RbohD* is predominant, which partakes in apoplastic ROS burst and stomatal defense response (Chang et al. 2020; Arnaud et al. 2021). For instance, under low temperatures, the inducible protein AtSRC2 has been observed to activate *AtRbohD* (Kawarazaki et al. 2013). Similarly, in *Arabidopsis*, phosphorylation of NOX RbohD intensified ROS production, inducing stomatal closure and increased resistance to pathogenic bacteria *Pseudomonas syringae* (Chen et al. 2017).

8.2.2 Peroxidases

The class III peroxidases (POXs) are heme-containing enzymes, which significantly contribute to ROS production in apoplasts (Podgórska et al. 2017; Janku et al. 2019). In *Arabidopsis*, POXs have been reported to account for the generation of 50% ROS during an oxidative burst in pathogen defense (O'Brien et al. 2012). The POXs in plants are localized in the cell wall and are encoded by a large multigene family (Janku et al. 2019). In *Arabidopsis thaliana*, POX consists of 73 genes (Welinder et al. 2002; Survila et al. 2016; Podgórska et al. 2017). It has a wide range of substrate specificity and is involved in both ROS production and elimination depending on the reaction mechanism involving the hydroxylation (as well as oxidative) cycle and peroxidation cycle, respectively (Podgórska et al. 2017; Janku et al. 2019). For instance, in the oxidative cycle, POX utilizes NADH as a substrate. In the peroxidation cycle, phenolics are used as a substrate (Maksimović et al. 2008; Podgórska et al. 2017). The POX in the oxidative cycle uses apoplastic reductants to reduce O_2 into $O_2^{\bullet-}$ or H_2O_2 and in the hydroxylation cycle, they mediate reaction generating $^{\bullet}OH$ from $O_2^{\bullet-}$ and H_2O_2 (Podgórska et al. 2017). The transient alkalinization of the apoplast has been observed to be instrumental in the generation of ROS by POXs (Podgórska et al. 2017). Moreover, the generation of ROS in apoplast by POX triggers *Rbohs* transcription, enhancing apoplastic ROS accumulation under stress as a response mechanism (Podgórska et al. 2017). For instance, the *A. thaliana* overexpressing gene *RCI3*, encoding POX activity displayed elevated tolerance toward dehydration and salt stress, whereas its antisense suppression resulted in a phenotype showing sensitivity to similar stress conditions (Llorente et al. 2002). During pathogen infection, POX belonging to pathogenesis-related (PR) proteins are being induced, which limits pathogen growth by creating a barrier via increased cross-linking of cell wall components, leading to wall rigidification or by inducing oxidative burst (van Loon et al. 2006; Survila et al. 2016; Farvardin et al. 2020).

8.2.3 Amine Oxidases

Amino oxidases (AO) activity is the other source for the production of extracellular ROS in plants. Two classes of AO viz., copper-containing amine oxidase (CuAO) and flavin-containing polyamine oxidases (PAO) localized in plant cells, actively participate in the oxidative deamination of aliphatic polyamines at a primary and secondary amino group, respectively, generating H_2O_2 as a product (Angelini et al. 2010; Pottosin et al. 2014; Podgórska et al. 2017). CuAOs are loosely attached to the cell wall and are commonly found in dicots, particularly in plants belonging to the Fabaceae family (Pottosin et al. 2014). CuAOs catalyze the oxidation of diamines like putrescine (Put) and cadaverine (Cad) and to a lesser extent spermine (Spm) and spermidine (Spd), inducing oxidation (Angelini et al. 2010; Pottosin et al. 2014; Podgórska et al. 2017). Likewise, PAO targets polyamines such as Spm, Sdp, and/or their acetyl derivatives (Angelini et al. 2010; Podgórska et al. 2017). The localization of AO in apoplasts has been reported in a large number of plant taxa. For instance, in *Arabidopsis* two genes encoding CuAOs, i.e., AtCuAOβ and AtCuAOγ1 have been reported in the extracellular compartment (Planas-Portell et al. 2013; Tavladoraki et al. 2016). The presence of PAOs has been reported in apoplasts of maize and barley (Cona et al. 2006; Podgórska et al. 2017). The generation of H_2O_2 by AO takes place on the catabolism of polyamine substrates under both stress and plant physiological conditions (Angelini et al. 2010; Janku et al. 2019). Under salt stress, Moschou et al. (2008) have observed the secretion of Spd in the apoplast of a tobacco plant cell, which apoplastic PAO oxidizes to produce extracellular H_2O_2 and induce programmed cell death (PCD). Analogously, Rodríguez et al. (2009) have reported the instrumental role of apoplastic ROS (H_2O_2, $^{\bullet}OH$, and $O_2^{\bullet-}$), generated through oxidation of free polyamine substrates by PAO, in the elongation of maize leaf blade under salinity conditions.

8.3 Apoplastic ROS in Cell Growth

The metabolism of ROS in apoplast facilitates plant cell division and cell elongation, thereby promoting plant growth and development (Podgórska et al. 2017). In *Arabidopsis*, ROS generation via pollen tube plasma membrane-localized NOX *RbohH* and *RbohJ* promote tip growth of the pollen tube and trigger polar cell growth (Kaya et al. 2014). The plant cell growth by elongation depends on cell wall extension, which is the outcome of the delicate balance between loosening and stiffening of the cell wall (Wolf et al. 2012; Schmidt et al. 2016). The cellulose matrix of the primary cell wall is cross-linked by hemicellulose-like xyloglucans, and pectin molecules in such a manner that it provides both mechanical strength and expansion capacity to the cell (Schmidt et al. 2016). Loosening of the cell wall occurs due to cleavage of the cross-linking bonds of wall polysaccharides, facilitating relaxation of wall stress, cell enlargement, and water uptake (Cosgrove 2016). In plants, apoplastic ROS mediate oxidative loosening and stiffening of the cell wall, facilitating its extension (Müller et al. 2009; Schmidt et al. 2016; Podgórska et al. 2017). The apoplastic $^{\bullet}OH$ promotes cell elongation by triggering oxidative cleavage of pectins and xyloglucans, which results in cell wall loosening (Podgórska et al. 2017; Janku et al. 2019). In apoplast, accumulation of $O_2^{\bullet-}$ and H_2O_2 results in the formation of $^{\bullet}OH$ in the presence of transition metals via the Fenton/Haber–Weiss reaction, which induces cell wall loosening and cell expansion (Richards et al. 2015; Schmidt et al. 2016). However, the production of $^{\bullet}OH$ under stress conditions abridges leaf growth. For instance, the alleviated $^{\bullet}OH$ generation from $O_2^{\bullet-}$ hinders the expansion of maize leaf and leaf growth under saline stress conditions (Rodríguez et al. 2004). The process of cell elongation is also affected by the presence of H_2O_2 in the apoplast. H_2O_2 restricts the elongation process by initiating disulfide bond formation between polysaccharide cross-linking with protein (Podgórska et al. 2017; Janku et al. 2019). Thus, H_2O_2, which is specifically relevant for the genesis of $^{\bullet}OH$ may inhibit the growth process, when its concentration excessively built up in the apoplast, particularly under stress condition (Kravchik and Bernstein 2013; Avramova et al. 2015; Schmidt et al. 2016). Moreover, the generation of ROS by POX and NOX induces polymerization of monolignols, which promote cell wall lignification (Vaahtera et al. 2014). Under stress conditions, accelerated lignification and rigidification of cell walls due to the exacerbated activity of POX restrict leaf growth (Lee et al. 2007). In *Arabidopsis*, the expression of *POX71* increases the accumulation of H_2O_2, which arrests cell expansion (Raggi et al. 2015). The activity of POX in apoplast, not always, generates H_2O_2 but can induce $^{\bullet}OH$ production from H_2O_2 and $O_2^{\bullet-}$ under low pH condition, resulting in cell extension (Chen and Schopfer 1999; Liszkay et al. 2003; Schmidt et al. 2016).

8.4 Oxidative Damage Induced by Extracellular ROS Under Stress

Ineffective functioning of antioxidants results in the accumulation of ROS in the apoplast. The concentration of ROS exceeding threshold level might have toxic impact and induce oxidative damage to plasma membrane (Podgórska et al. 2017) (Table 8.1). ROS catalyzes oxidation of protein and lipids localized in the plasma membrane (Idänheimo et al. 2014; Podgórska et al. 2017). The excess production of ROS in an oxidative burst often occurs on recognition of biotic or abiotic stress cues (Souza-Santos et al. 2001; Łukawska-Kuźma et al. 2012; Janku et al. 2019). For instance, the plants subjected to salinity stress often show plasma membrane permeability disruption due to oxidation of proteins and peroxidation of lipids (Mansour 2013).

8.4.1 Biotic Stress

During biotic stress, the apoplast is the first interaction site of a plant with pathogens. The plasma membrane has some specific receptors that assist in recognizing pathogen-associated molecular patterns (PAMPs). After recognizing pathogenic invasion, plasma membrane-localized NOX and cell wall-associated POX trigger ROS production, resulting in transient apoplastic oxidative burst (Bolwell et al.

Table 8.1 Generation of extracellular ROS and their response under stress

Stress	Plant	Effects	References
Salinity	Rice seedling	The increased activity of cell wall-associated POX under stress elevates H_2O_2 accumulation, which reduces root growth	Lin and Kao (2001)
Salt stress	Tobacco	Secretion and extrusion of spermidine into apoplastic space on the perception of stress, increased activity of polyamine oxidases (PAO) generate H_2O_2, which induces programmed cell death (PCD)	Moschou et al. (2008)
Avirulent *Xanthomonas campestris* pv. *malvacearum*	Cotton cotyledons	Stress-induced activation of apoplastic POX generates $O_2^{\bullet -}$ in apoplast, triggering hypersensitive response (HR) cell death	Martinez et al. (1998)
Ozone	–	Generation of apoplastic H_2O_2 and $O_2^{\bullet -}$ triggers oxidative burst and HR cell death	Langebartels et al. (2002)
Drought stress	Grapevine	Abscisic acid promotes production of polyamines and their extrusion in apoplast elevates activity of amine oxidases, resulting in the generation of H_2O_2, which induces signaling or PCD	Toumi et al. (2010)

2001; Janku et al. 2019). NOX is the primary source of ROS in the presence of biotic elicitors (Iqbal et al. 2021). The ROS in apoplast, particularly H_2O_2, diffuses into the cytoplasm via aquaporins in plasma membranes and activates downstream signals, inducing defense response (Podgórska et al. 2017; Janku et al. 2019). The most common defensive response of plants to pathogen invasion is hypersensitivity, which results in death of cells lying in vicinity of the site of infection (Podgórska et al. 2017). The production of ROS under biotic stress also leads to oxidative stress in extracellular compartments, resulting in plasma membrane lipid peroxidation. The inoculation of Plum Pox Virus (PPV) to the susceptible cultivars of apricot and peach plants promotes increased accumulation of H_2O_2 in the apoplast accompanied by an elevated electrolyte leakage (Díaz-Vivancos et al. 2006). The AK toxin 1 produced by the pathogen *Alternaria alternata* modifies plasma membrane and induces H_2O_2 generation in the tissues of sensitive-Japanese Pear, eventually resulting in plasma membrane lipid peroxidation (Shimizu et al. 2006).

8.4.2 Abiotic Stress

Heavy Metals Heavy metals are very reactive and toxic elements, which affect physiological and metabolic activities of plants. Plants possess a defensive mechanism against heavy metals, which involve their binding to the cell wall or inhibition of long distance movement, thereby restricting them in apoplast, which subsequently reduces their uptake by the plant (Podgórska et al. 2017). However, this mechanism leads to the accumulation of heavy metals in apoplast, which may render toxic effects to extracellular biomolecules. The accumulation of heavy metals, particularly, transition metals like Cu^+, Mn^{2+}, and Fe^{2+} partakes in Fenton/Haber Weiss reaction, resulting in the generation of ROS in the apoplast (Pourrut et al. 2008; Źróbek-Sokolnik et al. 2009; Podgórska et al. 2017). The exposure of plants to heavy metals such as Cu, Cd, and Zn at stress-inducing levels often accompanies with elevated expression of NOX and LOXs (Remans et al. 2010, 2012; Cuypers et al. 2011; Podgórska et al. 2017). Exposure of wheat seedling roots to nickel elicits production of H_2O_2 and $O_2^{\bullet-}$ via NOX, which leads to membrane lipid peroxidation (Hao et al. 2006). Besides, heavy metals can lead to enzyme inactivation, eclipsing activity of apoplastic enzymatic antioxidants like SODs, POXs, or CATs by binding to their cysteine residue or displacing their cofactors (Bhaduri and Fulekar 2012; Kimura et al. 2014; Podgórska et al. 2017). These heavy metals can trigger membrane lipids peroxidation and even displace Ca^{2+} from the cell wall, affecting cellular structure (Podgórska et al. 2017). Iron-induced ROS induces oxidative damage to corn root plasma membrane H^+-ATPase, inhibiting the enzyme activity irreversibly due to peroxidation of membrane phospholipid generating lipoperoxide, which eventually oxidizes sulfhydryl groups of the enzyme (Souza-Santos et al. 2001).

Temperature Stress The alteration in plasma membrane stability also brings the fluctuation in the ambient temperature conditions. Heat stress elevates the membrane fluidity, which activates plasma membrane-localized Ca^{2+} channel, eventually increasing cytosolic Ca^{2+} level (Saidi et al. 2009; Podgórska et al. 2017). The increased influx of Ca^{2+} in cytosol leads to the activation of NOX by inducing phosphorylation, which promotes production of ROS in the apoplast (Saidi et al. 2010; Podgórska et al. 2017). The increased accumulation of ROS under heat stress results in oxidative damage to plasma membrane and cell wall. For instance, forage rape seeds exposed to heat stress display ruptured plasma membrane and cell wall due to accumulation of ROS along with peroxidation of membrane lipids. The lipid peroxidation reduces the membrane stability and accelerates electrolyte leakage (Rashid et al. 2020). Chilling stress also affects membrane stability and causes membrane dysfunctioning (Podgórska et al. 2017). Under unfavorable cold conditions, NOX favors ROS generation in the apoplast, inducing oxidative damage to biomolecules including proteins and lipids (Zhou et al. 2012). The exposure of wild-type winter wheat to cold stress results in the accumulation of ROS, which induces lipid peroxidation, reducing membrane integrity and H^+-ATPase activity (Zhang et al. 2010).

Ground-Level Ozone The ground-level ozone is considered toxic for plants due to its strong oxidizing potential, which causes cellular injuries and affects plant productivity (Podgórska et al. 2017). Due to reactive nature, the first site of its action in plants is apoplast, where it results in production of ROS. Ozone undergoes spontaneous decomposition, generating ROS and thus induces oxidative burst in the apoplast space. The sunflower exposed to ozone for two hours accumulates H_2O_2 in the apoplast (Ranieri et al. 2003). The generation of ROS in apoplastic space probably is a manifestation of NOX and POX activity. The unsaturated fatty acids of the plasma membrane are the primary components to get oxidized due to ozonolysis, resulting in lipid peroxidation, which alters membrane permeability and increases electrolyte leakage. The products formed as a result of lipid peroxidation, further decompose to generate organic radicals with ROS as a byproduct, aggravating the oxidative damage (Podgórska et al. 2017).

8.5 Antioxidants Maintaining Redox Homeostasis in Extracellular Compartments

The generation and accumulation of ROS in extracellular compartments possess several beneficial attributes; however, their abundance and/or generation of highly toxic reactive intermediates can be harmful for plants (Kukavica et al. 2005; Farvardin et al. 2020). To regulate the accumulation of ROS in apoplast and maintain redox status, cells recruit several enzymatic antioxidants such as superoxide dismutase (SOD), catalases (CAT), POX, ascorbate peroxidase (APX), glutathione reductase (GR), monodehydroascorbate reductase (MDHAR) and dehydroascorbate reductase (DHAR); and low molecular weight antioxidants like ascorbate (AsA), glutathione (GSH), phenolics, and polyamine (Podgórska et al. 2017; Farvardin et al. 2020).

Superoxide Dismutase Superoxide dismutase (SOD) is an enzymatic antioxidant, which catalyzes the dismutation of $O_2^{\bullet -}$, resulting in the generation of H_2O_2 (Kukavica et al. 2005). Earlier SOD has been considered as an intracellular enzyme. The recent findings have confirmed its presence in the extracellular compartments, including the apoplast and plasma membrane (Ogawa et al. 1996; Karpinska et al. 2001; Vuletić et al. 2003; Kukavica et al. 2005). In plants, Cu/Zn isoform of SOD has been localized in apoplasts of several number of plant species (Kukavica et al. 2005). Moreover, FeSOD isoform associated with a plasma membrane has been documented in *Arabidopsis* (Marmagne et al. 2007). The Cu/ZnSOD has been proposed to be closely attached with the NOX. Therefore, it acts as a major source for H_2O_2 production in apoplast as it dismutates the $O_2^{\bullet -}$ generated by NOX (Ogawa et al. 1997).

Catalases Catalase (CAT) mediates enzymatic conversion of H_2O_2 into H_2O and O_2 and this decomposition does not require any reductant (Podgórska et al. 2017). The activity of CAT has been reported from cell walls and the apoplast of various plants (Podgórska et al. 2017). The apoplast of barley and oats accounts for 0.2–2% of the total foliar CAT activity (Vanacker et al. 1998a, b).

Peroxidases Peroxidases (POXs) in addition to generation, partake in the elimination of ROS. POX uses H_2O_2 as a substrate, thereby reducing its level (Podgórska et al. 2017). The soluble isoforms of cell wall-bound POX have been reported to be released into the apoplast under stress, or oxidative burst conditions that prevent accumulation of H_2O_2 and thereby protect cellular damage (Bolwell et al. 2002; Minibaeva and Gordon 2003; Parra-Lobato et al. 2009; Podgórska et al. 2017).

Ascorbate-Glutathione Pathway Enzymes and Other Antioxidants The Ascorbate-Glutathione (AsA-GSH) pathway involves four enzymatic antioxidants viz., APX, GR, MDHAR, and DHAR, and two non-enzymatic antioxidants, ascorbate and glutathione. All the enzymes partake in reduction of H_2O_2 into H_2O and use ascorbate and glutathione as an electron donor in this cycle (Podgórska et al. 2017; Farvardin et al. 2020). In certain plants including oat, barley, and pea leaves, the activity of AsA-GSH pathway enzymes show lower activity in apoplast than symplast (Vanacker et al. 1998a, b; Farvardin et al. 2020). The apoplast registers only 0.2–2.8% of the total symplastic activity of these enzymes (Vanacker et al. 1998a, b; Hernandez et al. 2001; Podgórska et al. 2017). In other plants, all the AsA-GSH pathway enzymes do not remain active in apoplast, thus only few of these enzymes show their activities (de Pinto and De Gara 2004; Podgórska et al. 2017; Farvardin et al. 2020). APX (class I APX) involved in this pathway, quenches H_2O_2 by reducing it into H_2O, using ascorbate as electron source (Pandey et al. 2017; Farvardin et al. 2020). Under drought stress, an increased activity of APX occurs in the apoplast of *Ctenenthe setosa* leaf (Saruhan et al. 2010). Similarly, inoculation of Plum Pox Virus (PPV) increased class I APX activity in leaves of very susceptible peach cultivar GF305 and resistant apricot cultivar Stark Early Orange (Díaz-Vivancos et al. 2006). The two enzymes of AsA-GSH pathways, MDHAR and

DHAR function in maintaining the pool of AsA, which is an important non-enzymatic antioxidant, facilitating plant growth and tolerance to stress (Farvardin et al. 2020). The enzyme MDHAR regenerates AsA from monodehydroascorbate (MDHA) by utilizing NADPH as a reductant. DHAR reduces dehydroascorbate into AsA by using reduced GSH as an electron source (Ding et al. 2020; Farvardin et al. 2020). The evidence for occurrence of MDHAR in plant apoplasts are very little, however, in *Arabidopsis*, one of the six isoforms of MDHAR has been reported from apoplastic space (Bindschedler et al. 2008). Similarly, DHAR has been found localized in apoplasts of pea plants (Hernandez et al. 2001). Moreover, activity of apoplastic DHAR has been reported from leaves of barley and oats (Vanacker et al. 1998a, b). Another important enzyme of the AsA-GSH cycle is GR, which reduces glutathione disulfide (GSSG) into GSH. To catalyze the reaction, GR uses NADPH as a reducing agent (Farvardin et al. 2020). GR is an important enzyme of the pathway as it maintains cellular redox equilibrium of GSH/GSSH (Farvardin et al. 2020). Saruhan et al. (2010) have reported the increased activity of GR in the apoplast of *Ctenanthe setosa* leaf under drought stress. The GSH generated by GR is a non-enzymatic scavenger of ROS and assists in stress tolerance. In apoplast many other non-enzymatic compounds such as phenolic substances and proline are found under stress conditions. For instance, apoplast as well as symplast of common bean leaves accumulates the elevated levels of proline under drought situations (Guler et al. 2012).

8.6 Extracellular ROS Role in Signaling

The H_2O_2 is the major ROS which triggers signaling in plants in response to an array of biotic and abiotic stress, and developmental processes (Arnaud et al. 2021). The generation of ROS in the apoplast on subsequent attack of pathogens induces pathogen-associated molecular patterns (PAMPs)-triggered immunity (PTI). The plasma membrane-localized NOX and cell wall-associated POX incite transient oxidative burst, leading to PTI, as an early response toward pathogen attack (Arnaud et al. 2021). Plasma membrane receptor kinases (RKs), on the perception of PAMPs, activate the co-receptor BAK1 and the cytosolic kinase BIK1, which in turn causes phosphorylation and activation of NOX (Li et al. 2014). This triggers production of $O_2^{\bullet-}$ in extracellular space (Arnaud et al. 2021). The $O_2^{\bullet-}$ eventually dismutates and generates H_2O_2, which leads to apoplastic oxidative burst. Besides this, apoplastic ROS-induced PTI activates plasma membrane Ca^{2+} channels, increasing movement of Ca^{2+} into cytosol, which elevates its concentration (Mittler et al. 2011; Iqbal et al. 2021). Cytosolic Ca^{2+} triggers activity of NOX by binding with the EF-hand motif, creating a feedback loop (Podgórska et al. 2017). The phosphorylation and Ca^{2+} binding, two major processes required for the activation of NOX, are often reported to occur synergistically. Nevertheless, some studies also suggest that phosphorylation precedes Ca^{2+} binding (Janku et al. 2019). The POX associated with the cell wall also generates H_2O_2 under stress. The reaction requires an alkaline environment in apoplastic space to catalyze the ROS genesis (Bolwell et al. 2002). The oxidative burst in extracellular region causes diffusion of ROS, particularly, H_2O_2 into the cytosol through specific aquaporins present in plasma membrane (Bienert et al. 2007), which induces downstream signaling pathway, eventually resulting in defense response like callose deposition and/or stomatal closure (Arnaud et al. 2021).

8.7 Conclusion

The plasma membrane and the apoplast are primary sites, which perceive external stimuli and facilitate plant response accordingly. Apoplastic ROS play an instrumental role in plant growth, plant-pathogen interaction and others. The low antioxidant condition favors the accumulation of ROS in the apoplast, which either serve as signaling molecules or trigger oxidative stress. The function of ROS as signaling molecules, inducing defense responses and elevating tolerance against adverse stimuli is of great importance for crop protection under perturbed environmental situations. Thus, the extensive knowledge on mechanism switching production of extracellular ROS and their role may facilitate the development of new strategies for improving crop productivity worldwide.

References

Angelini R, Cona A, Federico R, Fincato P, Tavladoraki P, Tisi A (2010) Plant amine oxidases "on the move": an update. Plant Physiol Biochem 48(7):560–564

Arnaud D, Deeks MJ, Smirnoff N (2021) Differences between apoplastic and cytosolic reactive oxygen species production in *Arabidopsis* during pattern-triggered immunity. bioRxiv. https://doi.org/10.1101/2021.04.20.440614

Arthikala MK, Montiel J, Sánchez-López R, Nava N, Cárdenas L, Quinto C (2017) Respiratory burst oxidase homolog gene a is crucial for rhizobium infection and nodule maturation and function in common bean. Front Plant Sci 8:2003

Avramova V, AbdElgawad H, Zhang Z, Fotschki B, Casadevall R, Vergauwen L, Knapen D, Taleisnik E, Guisez Y, Asard H, Beemster GT (2015) Drought induces distinct growth response, protection, and recovery mechanisms in the maize leaf growth zone. Plant Physiol 169(2): 1382–1396

Bhaduri AM, Fulekar MH (2012) Antioxidant enzyme responses of plants to heavy metal stress. Rev Environ Sci Biotechnol 11(1):55–69

Bienert GP, Møller AL, Kristiansen KA, Schulz A, Møller IM, Schjoerring JK, Jahn TP (2007) Specific aquaporins facilitate the diffusion of hydrogen peroxide across membranes. J Biol Chem 282(2):1183–1192

Bindschedler LV, Palmblad M, Cramer R (2008) Hydroponic isotope labelling of entire plants (HILEP) for quantitative plant proteomics; an oxidative stress case study. Phytochemistry 69(10):1962–1972

Bolwell PP, Page A, Piślewska M, Wojtaszek P (2001) Pathogenic infection and the oxidative defences in plant apoplast. Protoplasma 217(1):20–32

Bolwell GP, Bindschedler LV, Blee KA, Butt VS, Davies DR, Gardner SL, Gerrish C, Minibayeva F (2002) The apoplastic oxidative burst in response to biotic stress in plants: a three-component system. J Exp Bot 53(372):1367–1376

Bridge A, Barr R, Morré DJ (2000) The plasma membrane NADH oxidase of soybean has vitamin K1 hydroquinone oxidase activity. Biochim Biophys Acta (BBA) Biomembr 1463(2):448–458

Chang Y, Li B, Shi Q, Geng R, Geng S, Liu J, Zhang Y, Cai Y (2020) Comprehensive analysis of respiratory burst oxidase homologs (Rboh) gene family and function of GbRboh5/18 on *Verticillium wilt* resistance in *Gossypium barbadense*. Front Genet 11:788

Chapman JM, Muhlemann JK, Gayomba SR, Muday GK (2019) RBOH-dependent ROS synthesis and ROS scavenging by plant specialized metabolites to modulate plant development and stress responses. Chem Res Toxicol 32(3):370–396

Chen SX, Schopfer P (1999) Hydroxyl-radical production in physiological reactions: a novel function of peroxidase. Eur J Biochem 260(3):726–735

Chen D, Cao Y, Li H, Kim D, Ahsan N, Thelen J, Stacey G (2017) Extracellular ATP elicits DORN1-mediated RBOHD phosphorylation to regulate stomatal aperture. Nat Commun 8(1): 1–13

Cona A, Rea G, Angelini R, Federico R, Tavladoraki P (2006) Functions of amine oxidases in plant development and defence. Trends Plant Sci 11(2):80–88

Cosgrove DJ (2016) Catalysts of plant cell wall loosening. F1000Res 5:F1000. https://doi.org/10.12688/f1000research.7180.1

Cuypers A, Karen S, Jos R, Kelly O, Els K, Tony R et al (2011) The cellular redox state as a modulator in cadmium and copper responses in *Arabidopsis thaliana* seedlings. J Plant Physiol 168(4):309–316

Das K, Roychoudhury A (2014) Reactive oxygen species (ROS) and response of antioxidants as ROS-scavengers during environmental stress in plants. Front Environ Sci 2:53

de Pinto MC, De Gara L (2004) Changes in the ascorbate metabolism of apoplastic and symplastic spaces are associated with cell differentiation. J Exp Bot 55(408):2559–2569

Díaz-Vivancos P, Rubio M, Mesonero V, Periago PM, Ros Barceló A, Martínez-Gómez P, Hernández JA (2006) The apoplastic antioxidant system in Prunus: response to long-term plum pox virus infection. J Exp Bot 57(14):3813–3824

Ding H, Wang B, Han Y, Li S (2020) The pivotal function of dehydroascorbate reductase in glutathione homeostasis in plants. J Exp Bot 71(12):3405–3416

Farvardin A, González-Hernández AI, Llorens E, García-Agustín P, Scalschi L, Vicedo B (2020) The apoplast: a key player in plant survival. Antioxidants 9(7):604

Furt F, Simon-Plas F, Mongrand S (2011) Lipids of the plant plasma membrane. In: The plant plasma membrane. Springer, Berlin, pp 3–30

Guler NS, Sağlam A, Demiralay M, KADIOĞLU, A. (2012) Apoplastic and symplastic solute concentrations contribute to osmotic adjustment in bean genotypes during drought stress. Turk J Biol 36(2):151–160

Hao F, Wang X, Chen J (2006) Involvement of plasma-membrane NADPH oxidase in nickel-induced oxidative stress in roots of wheat seedlings. Plant Sci 170(1):151–158

Hasanuzzaman M, Bhuyan MHM, Zulfiqar F, Raza A, Mohsin SM, Mahmud JA, Fujita M, Fotopoulos V (2020) Reactive oxygen species and antioxidant defense in plants under abiotic stress: revisiting the crucial role of a universal defense regulator. Antioxidants 9(8):681

Hernandez JA, Ferrer MA, Jiménez A, Barceló AR, Sevilla F (2001) Antioxidant systems and O2.−/H2O2 production in the apoplast of pea leaves. Its relation with salt-induced necrotic lesions in minor veins. Plant Physiol 127(3):817–831

Horemans N, Foyer CH, Asard H (2000) Transport and action of ascorbate at the plant plasma membrane. Trends Plant Sci 5(6):263–267

Hu X, Bidney DL, Yalpani N, Duvick JP, Crasta O, Folkerts O, Lu G (2003) Overexpression of a gene encoding hydrogen peroxide-generating oxalate oxidase evokes defense responses in sunflower. Plant Physiol 133(1):170–181

Idänheimo N, Gauthier A, Salojärvi J, Siligato R, Brosché M, Kollist H, Mahonen AP, Kangasjarvi J, Wrzaczek M (2014) The *Arabidopsis thaliana* cysteine-rich receptor-like kinases CRK6 and CRK7 protect against apoplastic oxidative stress. Biochem Biophys Res Commun 445(2):457–462

Iqbal Z, Iqbal MS, Hashem A, Abd Allah EF, Ansari MI (2021) Plant defense responses to biotic stress and its interplay with fluctuating dark/light conditions. Front Plant Sci 12:297
Janku M, Luhová L, Petřivalský M (2019) On the origin and fate of reactive oxygen species in plant cell compartments. Antioxidants 8(4):105
Kadota Y, Shirasu K, Zipfel C (2015) Regulation of the NADPH oxidase RBOHD during plant immunity. Plant Cell Physiol 56(8):1472–1480
Kärkönen A, Kuchitsu K (2015) Reactive oxygen species in cell wall metabolism and development in plants. Phytochemistry 112:22–32
Karpinska B, Karlsson M, Schinkel H, Streller S, Suss KH, Melzer M, Wingsle G (2001) A novel superoxide dismutase with a high isoelectric point in higher plants. Expression, regulation, and protein localization. Plant Physiol 126(4):1668–1677
Kaur G, Sharma A, Guruprasad K, Pati PK (2014) Versatile roles of plant NADPH oxidases and emerging concepts. Biotechnol Adv 32(3):551–563
Kawarazaki T, Kimura S, Iizuka A, Hanamata S, Nibori H, Michikawa M, Imai A, Abe M, Kaya H, Kuchitsu K (2013) A low temperature-inducible protein AtSRC2 enhances the ROS-producing activity of NADPH oxidase AtRbohF. Biochim Biophys Acta (BBA) Mol Cell Res 1833(12): 2775–2780
Kawasaki H, Kretsinger RH (2017) Structural and functional diversity of EF-hand proteins: evolutionary perspectives. Protein Sci 26(10):1898–1920
Kaya H, Nakajima R, Iwano M, Kanaoka MM, Kimura S, Takeda S, Kawarazaki T, Sezaki E, Hamamura Y, Higashiyama T, Takayama S, Abe M, Kuchitsu K (2014) Ca2+-activated reactive oxygen species production by *Arabidopsis* RbohH and RbohJ is essential for proper pollen tube tip growth. Plant Cell 26(3):1069–1080
Kimura S, Kaya H, Kawarazaki T, Hiraoka G, Senzaki E, Michikawa M, Kuchitsu K (2012) Protein phosphorylation is a prerequisite for the Ca2+-dependent activation of *Arabidopsis* NADPH oxidases and may function as a trigger for the positive feedback regulation of Ca2+ and reactive oxygen species. Biochim Biophys Acta (BBA) Mol Cell Res 1823(2):398–405
Kimura M, Umemoto Y, Kawano T (2014) Hydrogen peroxide-independent generation of superoxide by plant peroxidase: hypotheses and supportive data employing ferrous ion as a model stimulus. Front Plant Sci 5:285
Kravchik M, Bernstein N (2013) Effects of salinity on the transcriptome of growing maize leaf cells point at cell-age specificity in the involvement of the antioxidative response in cell growth restriction. BMC Genomics 14(1):1–13
Kukavica B, Vučinić Ž, Vuletić M (2005) Superoxide dismutase, peroxidase, and germin-like protein activity in plasma membranes and apoplast of maize roots. Protoplasma 226(3):191–197
Langebartels C, Wohlgemuth H, Kschieschan S, Grün S, Sandermann H (2002) Oxidative burst and cell death in ozone-exposed plants. Plant Physiol Biochem 40(6-8):567–575
Lee BR, Kim KY, Jung WJ, Avice JC, Ourry A, Kim TH (2007) Peroxidases and lignification in relation to the intensity of water-deficit stress in white clover (*Trifolium repens* L.). J Exp Bot 58(6):1271–1279
Li L, Li M, Yu L, Zhou Z, Liang X, Liu Z, Cai G, Gao L, Zhang X, Wang Y, Chen S, Zhou JM (2014) The FLS2-associated kinase BIK1 directly phosphorylates the NADPH oxidase RbohD to control plant immunity. Cell Host Microbe 15(3):329–338
Li X, Li Y, Ahammed GJ, Zhang XN, Ying L, Zhang L, Yan P, Zhang LP, Li QY, Han WY (2019) RBOH1-dependent apoplastic H2O2 mediates epigallocatechin-3-gallate-induced abiotic stress tolerance in *Solanum lycopersicum* L. Environ Exp Bot 161:357–366
Lin CC, Kao CH (2001) Cell wall peroxidase activity, hydrogen peroxide level and NaCl-inhibited root growth of rice seedlings. Plant Soil 230(1):135–143
Liszkay A, Kenk B, Schopfer P (2003) Evidence for the involvement of cell wall peroxidase in the generation of hydroxyl radicals mediating extension growth. Planta 217(4):658–667
Llorente F, López-Cobollo RM, Catalá R, Martínez-Zapater JM, Salinas J (2002) A novel cold-inducible gene from *Arabidopsis*, RCI3, encodes a peroxidase that constitutes a component for stress tolerance. Plant J 32(1):13–24

Łukawska-Kuźma K, Podgórska A, Rychter AM (2012) Plasma membrane-generated ROS and their possible contribution to leaf cell growth of cucumber (*Cucumis sativus*) MSC16 mitochondrial mutant. Acta Physiol Plant 34(2):721–730

Maksimović JD, Maksimović V, Živanović B, Šukalović VHT, Vuletić M (2008) Peroxidase activity and phenolic compounds content in maize root and leaf apoplast, and their association with growth. Plant Sci 175(5):656–662

Mansour MMF (2013) Plasma membrane permeability as an indicator of salt tolerance in plants. Biol Plant 57(1):1–10

Marino D, Dunand C, Puppo A, Pauly N (2012) A burst of plant NADPH oxidases. Trends Plant Sci 17(1):9–15

Marmagne A, Ferro M, Meinnel T, Bruley C, Kuhn L, Garin J et al (2007) A high content in lipid-modified peripheral proteins and integral receptor kinases features in the *Arabidopsis* plasma membrane proteome. Mol Cell Proteomics 6(11):1980–1996

Martinez C, Montillet JL, Bresson E, Agnel JP, Dai GH, Daniel JF, Geiger J, P., & Nicole, M. (1998) Apoplastic peroxidase generates superoxide anions in cells of cotton cotyledons undergoing the hypersensitive reaction to *Xanthomonas campestris* pv. malvacearum race 18. Mol Plant-Microbe Interact 11(11):1038–1047

Minibaeva FV, Gordon LK (2003) Superoxide production and the activity of extracellular peroxidase in plant tissues under stress conditions. Russ J Plant Physiol 50(3):411–416

Minibayeva F, Kolesnikov O, Chasov A, Beckett RP, Lüthje S, Vylegzhanina N, Buck F, Böttger M (2009) Wound-induced apoplastic peroxidase activities: their roles in the production and detoxification of reactive oxygen species. Plant Cell Environ 32(5):497–508

Mittler R, Vanderauwera S, Suzuki N, Miller GAD, Tognetti VB, Vandepoele K, Gollery M, Shulaev V, Van Breusegem F (2011) ROS signaling: the new wave? Trends Plant Sci 16(6): 300–309

Moschou PN, Paschalidis KA, Delis ID, Andriopoulou AH, Lagiotis GD, Yakoumakis DI, Roubelakis-Angelakis KA (2008) Spermidine exodus and oxidation in the apoplast induced by abiotic stress is responsible for H2O2 signatures that direct tolerance responses in tobacco. Plant Cell 20(6):1708–1724

Müller K, Carstens AC, Linkies A, Torres MA, Leubner-Metzger G (2009) The NADPH-oxidase AtrbohB plays a role in *Arabidopsis* seed after-ripening. New Phytol 184(4):885–897

Nagawa S, Xu T, Yang Z (2010) RHO GTPase in plants: conservation and invention of regulators and effectors. Small GTPases 1(2):78–88

O'Brien JA, Daudi A, Finch P, Butt VS, Whitelegge JP, Souda P, Ausubel FM, Bolwell GP (2012) A peroxidase-dependent apoplastic oxidative burst in cultured *Arabidopsis* cells functions in MAMP-elicited defense. Plant Physiol 158(4):2013–2027

Oda T, Hashimoto H, Kuwabara N, Akashi S, Hayashi K, Kojima C, Wong HL, Kawasaki T, Shimamoto K, Sato M, Shimizu T (2010) Structure of the N-terminal regulatory domain of a plant NADPH oxidase and its functional implications. J Biol Chem 285(2):1435–1445

Ogawa KI, Kanematsu S, Asada K (1996) Intra-and extra-cellular localization of "cytosolic" CuZn-superoxide dismutase in spinach leaf and hypocotyl. Plant Cell Physiol 37(6):790–799

Ogawa KI, Kanematsu S, Asada K (1997) Generation of superoxide anion and localization of CuZn-superoxide dismutase in the vascular tissue of spinach hypocotyls: their association with lignification. Plant Cell Physiol 38(10):1118–1126

Ogboo BC, Grabovyy UV, Maini A, Scouten S, van der Vliet A, Mattevi A, Heppner DE (2022) Architecture of the NADPH oxidase family of enzymes. Redox Biol 52:102298

Pandey S, Fartyal D, Agarwal A, Shukla T, James D, Kaul T, Negi YK, Arora S, Reddy MK (2017) Abiotic stress tolerance in plants: myriad roles of ascorbate peroxidase. Front Plant Sci 8:581

Parra-Lobato MC, Fernandez-Garcia N, Olmos E, Alvarez-Tinaut MC, Gómez-Jiménez MC (2009) Methyl jasmonate-induced antioxidant defence in root apoplast from sunflower seedlings. Environ Exp Bot 66(1):9–17

Planas-Portell J, Gallart M, Tiburcio AF, Altabella T (2013) Copper-containing amine oxidases contribute to terminal polyamine oxidation in peroxisomes and apoplast of Arabidopsis thaliana. BMC Plant Biol 13(1):1–13

Podgórska A, Burian M, Szal B (2017) Extra-cellular but extra-ordinarily important for cells: apoplastic reactive oxygen species metabolism. Front Plant Sci 8:1353

Pottosin I, Velarde-Buendía AM, Bose J, Zepeda-Jazo I, Shabala S, Dobrovinskaya O (2014) Cross-talk between reactive oxygen species and polyamines in regulation of ion transport across the plasma membrane: implications for plant adaptive responses. J Exp Bot 65(5):1271–1283

Pourrut B, Perchet G, Silvestre J, Cecchi M, Guiresse M, Pinelli E (2008) Potential role of NADPH-oxidase in early steps of lead-induced oxidative burst in *Vicia faba* roots. J Plant Physiol 165(6): 571–579

Qi J, Wang J, Gong Z, Zhou JM (2017) Apoplastic ROS signaling in plant immunity. Curr Opin Plant Biol 38:92–100

Raggi S, Ferrarini A, Delledonne M, Dunand C, Ranocha P, De Lorenzo G, Cervone F, Ferrari F, Ferrari S (2015) The *Arabidopsis* class III peroxidase AtPRX71 negatively regulates growth under physiological conditions and in response to cell wall damage. Plant Physiol 169(4): 2513–2525

Ranieri A, Castagna A, Pacini J, Baldan B, Mensuali Sodi A, Soldatini GF (2003) Early production and scavenging of hydrogen peroxide in the apoplast of sunflower plants exposed to ozone. J Exp Bot 54(392):2529–2540

Rashid M, Hampton JG, Shaw ML, Rolston MP, Khan KM, Saville DJ (2020) Oxidative damage in forage rape (*Brassica napus* L.) seeds following heat stress during seed development. J Agron Crop Sci 206(1):101–117

Remans T, Opdenakker K, Smeets K, Mathijsen D, Vangronsveld J, Cuypers A (2010) Metal-specific and NADPH oxidase dependent changes in lipoxygenase and NADPH oxidase gene expression in *Arabidopsis thaliana* exposed to cadmium or excess copper. Funct Plant Biol 37(6):532–544

Remans T, Opdenakker K, Guisez Y, Carleer R, Schat H, Vangronsveld J, Cuypers A (2012) Exposure of *Arabidopsis thaliana* to excess Zn reveals a Zn-specific oxidative stress signature. Environ Exp Bot 84:61–71

Richards SL, Wilkins KA, Swarbreck SM, Anderson AA, Habib N, Smith AG, McAinsh M, Davies JM (2015) The hydroxyl radical in plants: from seed to seed. J Exp Bot 66(1):37–46

Rodríguez AA, Córdoba AR, Ortega L, Taleisnik E (2004) Decreased reactive oxygen species concentration in the elongation zone contributes to the reduction in maize leaf growth under salinity. J Exp Bot 55(401):1383–1390

Rodríguez AA, Maiale SJ, Menéndez AB, Ruiz OA (2009) Polyamine oxidase activity contributes to sustain maize leaf elongation under saline stress. J Exp Bot 60(15):4249–4262

Sachdev S, Ansari SA, Ansari MI, Fujita M, Hasanuzzaman M (2021) Abiotic stress and reactive oxygen species: generation, signaling, and defense mechanisms. Antioxidants 10(2):277

Saidi Y, Finka A, Muriset M, Bromberg Z, Weiss YG, Maathuis FJ, Goloubinoff P (2009) The heat shock response in moss plants is regulated by specific calcium-permeable channels in the plasma membrane. Plant Cell 21(9):2829–2843

Saidi Y, Peter M, Finka A, Cicekli C, Vigh L, Goloubinoff P (2010) Membrane lipid composition affects plant heat sensing and modulates Ca2+-dependent heat shock response. Plant Signal Behav 5(12):1530–1533

Saruhan N, Terzi R, Saĝlam A, Kadioĝlu A (2010) Scavenging of reactive oxygen species in apoplastic and symplastic areas of rolled leaves in *Ctenanthe setosa* under drought stress. Acta Biol Hung 61(3):282–298

Schmidt R, Kunkowska AB, Schippers JH (2016) Role of reactive oxygen species during cell expansion in leaves. Plant Physiol 172(4):2098–2106

Shimizu N, Hosogi N, Hyon GS, Jiang S, Inoue K, Park P (2006) Reactive oxygen species (ROS) generation and ROS-induced lipid peroxidation are associated with plasma membrane

modifications in host cells in response to AK-toxin I from *Alternaria alternata* Japanese pear pathotype. J Gen Plant Pathol 72(1):6–15
Souza-Santos P, Ramos RS, Ferreira ST, Carvalho-Alves PC (2001) Iron-induced oxidative damage of corn root plasma membrane H+-ATPase. Biochim Biophys Acta Biomembr 1512(2):357–366
Survila M, Davidsson PR, Pennanen V, Kariola T, Broberg M, Sipari N, Heino P, Palva ET (2016) Peroxidase-generated apoplastic ROS impair cuticle integrity and contribute to DAMP-elicited defenses. Front Plant Sci 7:1945
Takahashi D, Li B, Nakayama T, Kawamura Y, Uemura M (2013) Plant plasma membrane proteomics for improving cold tolerance. Front Plant Sci 4:90
Tavladoraki P, Cona A, Angelini R (2016) Copper-containing amine oxidases and FAD-dependent polyamine oxidases are key players in plant tissue differentiation and organ development. Front Plant Sci 7:824
Toumi I, Moschou PN, Paschalidis KA, Bouamama B, Salem-Fnayou AB, Ghorbel AW, Mliki A, Roubelakis-Angelakis KA (2010) Abscisic acid signals reorientation of polyamine metabolism to orchestrate stress responses via the polyamine exodus pathway in grapevine. J Plant Physiol 167(7):519–525
Vaahtera L, Brosché M, Wrzaczek M, Kangasjärvi J (2014) Specificity in ROS signaling and transcript signatures. Antioxid Redox Signal 21(9):1422–1441
van Loon LC, Rep M, Pieterse CM (2006) Significance of inducible defense-related proteins in infected plants. Annu Rev Phytopathol 44:135–162
Vanacker H, Harbinson J, Ruisch J, Carver TLW, Foyer CH (1998a) Antioxidant defences of the apoplast. Protoplasma 205(1):129–140
Vanacker H, Carver TL, Foyer CH (1998b) Pathogen-induced changes in the antioxidant status of the apoplast in barley leaves. Plant Physiol 117(3):1103–1114
Voothuluru P, Sharp RE (2013) Apoplastic hydrogen peroxide in the growth zone of the maize primary root under water stress. I. Increased levels are specific to the apical region of growth maintenance. J Exp Bot 64(5):1223–1233
Vuletić M, Hadži-Tašković Šukalović V, Vučinić Ž (2003) Superoxide synthase and dismutase activity of plasma membranes from maize roots. Protoplasma 221(1):073–077
Welinder KG, Justesen AF, Kjærsgård IV, Jensen RB, Rasmussen SK, Jespersen HM, Duroux L (2002) Structural diversity and transcription of class III peroxidases from *Arabidopsis thaliana*. Eur J Biochem 269(24):6063–6081
Wolf S, Hématy K, Höfte H (2012) Growth control and cell wall signaling in plants. Annu Rev Plant Biol 63:381–407
Wong HL, Pinontoan R, Hayashi K, Tabata R, Yaeno T, Hasegawa K, Koima C, Yoshioka H, Iba K, Kawasaki T, Shimamoto K (2007) Regulation of rice NADPH oxidase by binding of Rac GTPase to its N-terminal extension. Plant Cell 19(12):4022–4034
Yu S, Kakar KU, Yang Z, Nawaz Z, Lin S, Guo Y, Ren XL, Baloch AA, Han D (2020) Systematic study of the stress-responsive *Rboh* gene family in *Nicotiana tabacum*: genome-wide identification, evolution and role in disease resistance. Genomics 112(2):1404–1418
Zhang Y, Zhu H, Zhang Q, Li M, Yan M, Wang R, Wang L, Welti R, Zhang W, Wang X (2009) Phospholipase Dα1 and phosphatidic acid regulate NADPH oxidase activity and production of reactive oxygen species in ABA-mediated stomatal closure in *Arabidopsis*. Plant Cell 21(8): 2357–2377

Zhang XY, Liang C, Wang GP, Luo Y, Wang W (2010) The protection of wheat plasma membrane under cold stress by glycine betaine overproduction. Biol Plant 54(1):83–88

Zhou J, Wang J, Shi K, Xia XJ, Zhou YH, Yu JQ (2012) Hydrogen peroxide is involved in the cold acclimation-induced chilling tolerance of tomato plants. Plant Physiol Biochem 60:141–149

Źróbek-Sokolnik A, Asard H, Górska-Koplińska K, Górecki RJ (2009) Cadmium and zinc-mediated oxidative burst in tobacco BY-2 cell suspension cultures. Acta Physiol Plant 31(1): 43–49

Zurbriggen MD, Carrillo N, Hajirezaei MR (2010) ROS signaling in the hypersensitive response: when, where and what for? Plant Signal Behav 5(4):393–396

Antioxidant Defensive Mechanisms to Regulate Cellular Redox Homeostatic Balance

9

9.1 Introduction

The stressful environmental conditions such as heat, chilling/freezing, water deficit, waterlogging, salinity, metals, and pathogens trigger the excess production of reactive oxygen species (ROS) in plants, challenging their developmental mechanism and yield production (Jalil and Ansari 2008, 2020a, b; Anjum et al. 2015; Sharma et al. 2019a, b; Ansari et al. 2021; Sachdev et al. 2021). Plants thrive best under optimal conditions with the generation of small titers of ROS, demonstrating beneficial impacts including signaling (Ansari et al. 2021). However, adverse situations prompt the exorbitant production of ROS, which induces oxidative stress. The arsenal of ROS comprising free radicals like superoxide radical ($O_2^{\bullet -}$) and hydroxyl radical ($^{\bullet}OH$); and non-radicals such as hydrogen peroxide (H_2O_2) and singlet oxygen ($^1O_2$) hold the capacity to oxidize biomolecules, causing alteration in their structure and functions (Karuppanapandian et al. 2011; Ansari et al. 2021). To minimize the harmful oxidative impact of ROS on vital cellular biomolecules and secure their functions, plants harness the potential of antioxidants. The plant's detoxification system includes a network of enzymatic and non-enzymatic antioxidants with the power to restrict ROS accumulation below the threshold limit (Apel and Hirt 2004; Ansari et al. 2021). The enzymatic antioxidants, particularly superoxide dismutase (SOD), catalase (CAT), peroxidases (POX), glutathione reductase (GR), monodehydroascorbate reductase (MDHAR), dehydroascorbate reductase (DHAR), peroxiredoxins (Prxs) and thioredoxins (Trxs) commonly appear in plant cells (Mehla et al. 2017; Hasanuzzaman et al. 2020). The low molecular weight non-enzymatic antioxidants alleviating ROS levels in plants comprise a range of vitamins like ascorbic acid or ascorbate (AsA), tocopherols, and carotenoids; glutathione (GSH); and polyphenols like flavonoids (Pandhair and Sekhon 2006; Chong et al. 2009; Sharma et al. 2019a; Hasanuzzaman et al. 2020; Sachdev et al. 2022).

S. Sachdev et al., *Reactive Oxygen Species in Plants*,
https://doi.org/10.1007/978-981-19-9884-3_9

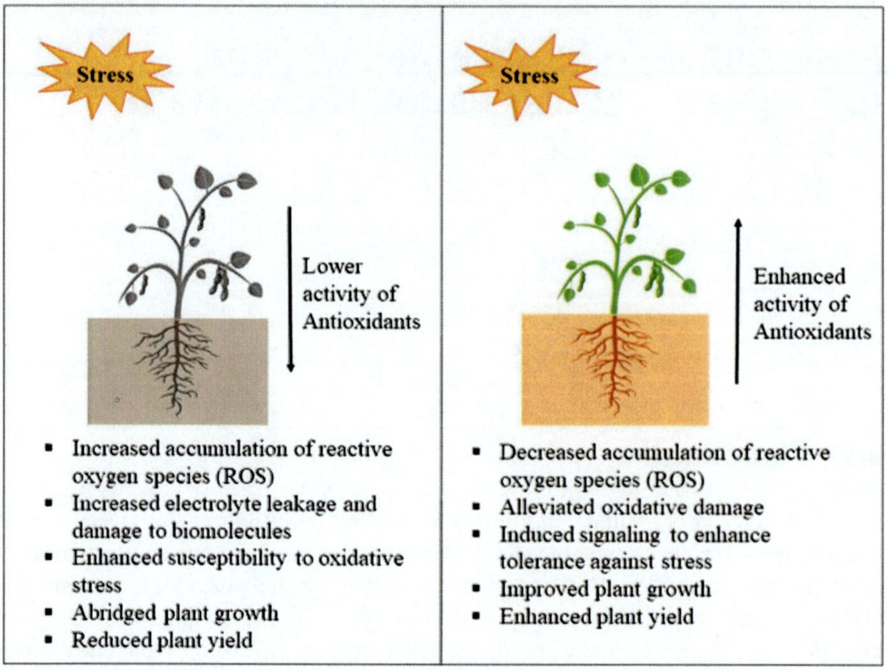

Fig. 9.1 Enhanced activity of antioxidants under stress is imperative to protect them from oxidative stress and promote their growth and productivity

The efficient working of an antioxidant-based detoxification system protects against an increased level of ROS and helps plants to withstand unfavorable circumstances (Fig. 9.1). The antioxidants neutralize the impact of ROS by scavenging them, inhibiting the oxidation process by forming chelate complexes with metals, quenching photosensitizers, inducing activation or deactivation of enzymes, and repairing and eliminating damages induced by ROS (Olszowy 2019). Plants in stress recruit multiple mechanisms, which directly or indirectly partake in quenching excess ROS, and maintain cellular redox homeostasis. Among these mechanisms, antioxidants play a major role in defense against ROS. Thus, transgenic plants with augmented plant antioxidant potentials invariably enhance their resilience under stress conditions. Moreover, understanding the regulatory aspects of ROS quenching mechanisms could be effective in promoting stress tolerance in plants. This chapter aims at highlighting the function and the coordinative mechanism of various ROS scavenging or regulating molecules, improving the defensive status of plants under stress conditions.

9.2 Enzymatic Antioxidants as ROS Scavengers

Plants cells operate an array of enzymatic antioxidants, particularly SOD, CAT, POX, GR, APX, GPX, MDHAR, DHAR, Prx, and TrxR, which effectively quench ROS from the intracellular and extracellular compartments of the cell (Tables 9.1 and 9.2). The enzymatic antioxidants are the cluster of proteins commonly occurring in plants that are involved either in the prevention of the Haber-Weiss reaction; and/or the Asada-Halliwell reaction (Stephenie et al. 2020; Rajput et al. 2021). The enzymatic antioxidants not only prevent cells from oxidative damage but also facilitate growth and development in plants by altering physiological activities like cell elongation, senescence, and cell death as well (Rajput et al. 2021).

Table 9.1 Plants' enzymatic antioxidants showing reactive oxygen species (ROS) scavenging potential

Enzymatic antioxidants	Target ROS/action	Sub cellular localization
Superoxide dismutase (SOD)	Singlet oxygen ($O_2^{\bullet -}$) $O_2^{\bullet -} + O_2^{\bullet -} + 2H^+ \rightarrow 2H_2O_2 + O_2$	Chloroplast, mitochondria, peroxisomes, apoplast, nucleus, cytosol, and endoplasmic reticulum
Catalase (CAT)	Hydrogen peroxide (H_2O_2) $H_2O_2 \rightarrow H_2O + O_2$	Chloroplast, mitochondria, peroxisomes, cytosol
Peroxidases (POX)	Hydrogen peroxide (H_2O_2) $H_2O_2 \rightarrow H_2O + O_2$	Cell wall, endoplasmic reticulum, vacuoles, apoplast, and cytoplasm
Ascorbate peroxidase (APX)	Hydrogen peroxide (H_2O_2) H_2O_2 + Ascorbate (AsA) $\rightarrow$ H_2O + Dehydroascorbate (DHA)	Chloroplast, mitochondria, peroxisomes, glyoxysomes
Glutathione peroxidase (GPX)	Hydrogen peroxide (H_2O_2) H_2O_2 + Glutathione (GSH) $\rightarrow$ H_2O + oxidized glutathione (GSSG)	Chloroplast, mitochondria, cytosol, and endoplasmic reticulum
Glutathione reductase (GR)	Participate in Ascorbate-Glutathione (AsA-GSH) cycle and catalyze and oxidize GSSG to GSH GSSG + NADPH $\rightarrow$ GSH + $NADP^+$	Chloroplast, mitochondria, cytosol, and peroxisomes
Monodehydroascorbate reductase (MDHAR)	Participate in Ascorbate-Glutathione (AsA-GSH) cycle and catalyze reduction of monodehydroascorbate (MDHA) MDHA + NADH $\rightarrow$ AsA+ NAD^+	Chloroplast, mitochondria, cytosol, and peroxisomes
Dehydroascorbate reductase (DHAR)	Participate in Ascorbate-Glutathione (AsA-GSH) cycle and catalyze reduction of dehydroascorbate (DHA) DHA + 2GSH $\rightarrow$ AsA+ GSSG	Chloroplast, mitochondria, cytosol, peroxisomes, apoplast, and vacuole

Table 9.2 Alteration in antioxidant content in plants exposed to heavy metal stress

Stress	Target plant (s)	Antioxidant activity	References
Copper (Cu)	*Zea mays*	Plant exposed to 1.5 mM concentration of Cu elevates activity of SOD in the leaves	Tanyolac et al. (2007)
Zinc (Zn)	*Brassica napus*	The activity of NADPH oxidase, POX increases whereas that of SOD, CAT, and APX activities decreases in leaves and roots at high Zn concentration	Wang et al. (2009)
Iron (Fe)	*Phaseolus vulgaris*	The activity of APX, AsA, CAT, and GR increases	Shainberg et al. (2000)
Cadmium (Cd)	*Brassica juncea*	Exposure to Cd increases glutathione content	Szőllősi et al. (2009)
Manganese (Mn)	*Oryza sativa* seedling	The content of AsA, and GSH decreases. Mn-SOD, Cu/Zn-SOD and Fe-SOD, and GPX elevate. The CAT content increases in 10 days old seedlings, but by 20 days the activity decreases. Moreover, the activity of AsA-GSH-related enzymes APX, MDHAR, DHAR, and GR also increases	Srivastava and Dubey (2011)
Heavy metal, salt and osmotic stress	*Brassica juncea*	Transgenric *B. juncea* enriched with α-tocopherol shows better tolerance against stress along with improved enzymatic antioxidants (SOD, APX, and GR) content than wild-type plant	Kumar et al. (2013)
Nickel (Ni)	*Zea mays* L. W 777	Activity of CAT gets inhibited and that of POX, APX and SOD increases	Kumar et al. (2007)
Nickel (Ni)	*Triticum aestivum* L. cv 'Zyta'	SOD activity, particularly in roots, decreases. CAT activity reduces by 30% in shoot and remains static in roots. APX, POD, GST activity gains several folds in shoots with no significant alteration in roots. Above all, proline accumulation increases	Gajewska and Skłodowska (2008)

9.2.1 Superoxide Dismutase (EC: 1.15.1.1)

Superoxide dismutase (SOD) is the major metalloenzyme antioxidant with the ability to detoxify stress-inducing $O_2^{\bullet-}$ by converting them into H_2O_2 (Kukavica et al. 2005; Hasanuzzaman et al. 2020; Rajput et al. 2021; Sachdev et al. 2022). It provides first-line protection against toxic ROS (Sachdev et al. 2022). SOD catalyzes the disproportionation or dismutation of $O_2^{\bullet-}$, involving both oxidation and reduction reactions liberating lower oxidation state H_2O_2 and higher oxidation state O_2 (Rajput et al. 2021). H_2O_2 generated so acts as a signaling molecule, hence the activity of SOD facilitates the signaling process under stressful situations (Sachdev et al. 2022). However, excess accumulation of H_2O_2 can result in the generation of destructive $^{\bullet}OH$ via the Haber-Weiss reaction hence the efficient removal of H_2O_2 from cellular compartments is important (Stephenie et al. 2020). To abridge H_2O_2

concentration build-up, the activity of SOD induces the expression of other enzymatic antioxidants, including CAT, POX, APX, and GPX, which scavenge excess H_2O_2 (Zhang et al. 2016).

The enzyme SOD is located in both intracellular and extracellular regions of a cell (Kukavica et al. 2005), where free radicals and substrates of SOD are formed (Río et al. 2018; Sachdev et al. 2022). The SODs exist in various isoforms depending on the metallic cofactors, particularly, Cu, Zn, Fe, and Mn linked with them at their active sites (Rajput et al. 2021; Sachdev et al. 2022). Metallic prosthetic groups are important for SOD enzymes to maximize their catalytic activity to metabolize reactive intermediates (Stephenie et al. 2020). The Cu/Zn-SOD, Fe-SOD, and Mn-SOD are three ubiquitous isoforms found in plants (Hasanuzzaman et al. 2020), however, their subcellular site of occurrence, structure, and functions vary greatly (Saibi and Brini 2018). For instance, Cu/Zn-SOD is generally located in the cytosol, chloroplast, peroxisomes, and apoplastic region of the plant cell, Fe-SOD occurs in the chloroplast, and Mn-SOD is found in peroxisomes and in the matrix of mitochondria (Jamdhade et al. 2017; Saibi and Brini 2018; Zafra et al. 2018; Stephenie et al. 2020; Rajput et al. 2021). SOD comprises a two-domain structure, α-helices and β-sheets, and the side chains, which include aspartate, histidine, and histamine (Stephenie et al. 2020; Rajput et al. 2021). The metal-binding site is located between these two domains and the side chains (Stephenie et al. 2020). Cu/Zn-SOD mainly comprises β-strands whereas Fe-SOD and Mn-SOD chiefly include α-helices followed by β-strands (Rajput et al. 2021). In Cu/Zn-SOD isoform, Cu and Zn are linked by a disulfide bond, forming covalent interaction (Saibi and Brini 2018). In Cu/Zn-SOD, Cu is most important as the role of Zn is to provide structural stability rather than participating in catalytic function (Rajput et al. 2021), thus elimination or substitution of Cu with another metal can cause enzyme inactivation (Saibi and Brini 2018). The Cu/Zn-SOD is present in its native homodimer form in the cytosol, the nucleus, and intermembrane space of mitochondria, whereas in chloroplast and apoplast, it is present in the homotetramer form (Stephenie et al. 2020; Rajput et al. 2021). Fe-SOD has homodimer and homotetramer forms (Saibi and Brini 2018). Analogously, Mn-SOD exists in the homodimer or homotetramer form in the matrix and inner membrane of mitochondria and peroxisomes (Palma et al. 1998; Stephenie et al. 2020; Rajput et al. 2021). The sensitivity of the three isoforms is also different from each other. For instance, Cu/Zn-SOD shows sensitivity toward KCN and H_2O_2. On the other hand, Fe-SOD is only influenced by H_2O_2; Mn-SOD is neither affected by KCN nor by H_2O_2 (Saibi and Brini 2018).

The plants subjected to stress stimulate the activity of the SOD enzyme, eliminating oxidative stress-inducing free radicals. Durum wheat plants exposed to drought stress harbor nine SOD isoenzymes including three Mn-SOD isoforms, one Fe-SOD, and five Cu/Zn-SOD isoforms in leaves (Huseynova et al. 2014). A transcription factor (TF) CsPIF8 in citrus upregulates the expression of the SOD enzyme under cold stress, enhancing the detoxification of $O_2^{\bullet -}$ (He et al. 2020). The increased activity of the SOD reportedly imparts tolerance against water-deficit stress in several plants including *Phaseolus vulgaris*, *Hordeum vulgare*, *Oryza sativa*, *Olea europaea*, and *Trifolium repens* (Sharma and Dubey 2005; Zlatev et al. 2006; Chang-Quan and Rui-Chang 2008; Ahmed et al. 2013; Doupis et al. 2013). Similarly, under salinity stress, SOD activity gains elevation in *Solanum lycopersicum*, *Cicer arietinum*, *Pisum sativum*, seedlings of *O. sativa*, and transgenic *Arabidopsis thaliana* (Gomez et al. 1999; Kukreja et al. 2005; Eyidogan and Öz 2007; Gapińska et al. 2008; Mishra et al. 2013). In *Jatropha curcas*, salinity stress in conjugation with heat stress stimulates the activity of the SOD (Silva et al. 2013).

9.2.2 Catalase (EC: 1.11.1.6)

Catalase (CAT) is a heme group containing enzymatic antioxidants, contributing to detoxification of H_2O_2 (Zamocky et al. 2008; Mhamdi et al. 2010; Anjum et al. 2016; Hasanuzzaman et al. 2020; Palma et al. 2020; Sachdev et al. 2022). The enzyme CAT is chiefly found in all differentiated forms of plant peroxisomes including leaves, root, cotyledons, glyoxysomes, as well as unspecialized peroxisomes (Su et al. 2014; Anjum et al. 2016). Moreover, CAT is reportedly localized in plant mitochondria, e.g., CAT3 from mitochondria of *Zea mays* (Roupakias et al. 1980). CAT catalyzes an energy-efficient reaction in which two H_2O_2 molecules are converted into H_2O and O_2 (Mhamdi et al. 2010; Sharma and Ahmad 2014; Anjum et al. 2016; Leung 2018). CAT performs catalytic action in two steps and does not depend on any reductant (Anjum et al. 2016; Rajput et al. 2021). The first step involves the oxidation of heme iron (Fe) of CAT by the H_2O_2 molecule, forming an iron peroxide intermediate (also known as compound I) (Anjum et al. 2016; Sachdev et al. 2022). In the second step, the intermediate form is reduced back to the original state by the second molecule of H_2O_2, which acts as a reductant and eventually releases H_2O and O_2 (Zamocky et al. 2008; Rajput et al. 2021). A high amount of H_2O_2 is needed for initiating the second step of the catalytic reaction, otherwise, at a low H_2O_2 concentration the enzyme remains in the intermediate resting state (Rajput et al. 2021). The CAT enzyme has the potential to detoxify 26 million H_2O_2 molecules in one minute by converting them into water molecules (Hasanuzzaman et al. 2020).

The antioxidants CAT and APX maintain the delicate balance of H_2O_2 in plants. Compared to APX, the K_m value of CAT for H_2O_2 is high and therefore requires a high concentration of H_2O_2 for an effective scavenging (Rajput et al. 2021). A native CAT found in plants is a homotetrameric enzyme, which consists of four identical subunits of size ranging between 54 and 59 kDa (Anjum et al. 2016; Palma et al. 2020), making the total mass of the enzyme nearly 240 kDa (Anjum et al. 2016; Rajput et al. 2021). Each monomer of the CAT consists of heme as a prosthetic group (Mhamdi et al. 2010). Some exceptions in the catalase size have also been reported. For instance, the homodimeric CAT of *Capsicum annuum* consists of a 55 kDa monomer subunit, which makes the total mass 125–135 kDa (Palma et al. 2020; Rajput et al. 2021). The azide, cyanide, and hydroxylamine inhibit CAT activity due to the presence of the Fe molecule as a cofactor (Rajput et al. 2021).

In plants, a multigene family is responsible for encoding several isozymes of CAT, whose number differs depending on the species, e.g., 14 in *Brassica napus*, seven in *Gossypium* spp., three in *Zea mays*, *Triticum aestivum*, and *O. sativa*, and two in *Hordeum vulgare* (Skadsen et al. 1995; Mhamdi et al. 2010; Wutipraditkul et al. 2011; Sytykiewicz et al. 2018; Wang et al. 2019; Palma et al. 2020; Rajput et al. 2021; Raza et al. 2021). The genomic sequencing of *A. thaliana* reveals three catalase genes, *CAT1*, *CAT2*, and *CAT3* (Frugoli et al. 1996; Du et al. 2008). Two genes (*CAT1* and *CAT3*) occur on Chromosome 1 and the third gene (*CAT2*) on Chromosome 4 (Mhamdi et al. 2010). Stress conditions modulate CAT activity. *Cicer arietinum* growing under salinity stress expresses CAT enzyme both in leaves and roots (Kukreja et al. 2005). Similarly, CAT activity in the *Faba* bean increases under salinity stress, when the accumulation of H_2O_2 reaches above 90% (Alzahrani et al. 2019). On the contrary, CAT activity in *A. thaliana* declines on exposure to photoperiod stress (Abuelsoud et al. 2020). Interestingly, the application of melatonin elevates CAT activity in roots of drought-sensitive (Qinyou8) as well as drought-tolerant (Q2) cultivars of *Brassica napus* subjected to drought stress (Dai et al. 2020).

9.2.3 Peroxidases (EC 1.11.1.7)

The plant peroxidases (POX) are the ubiquitous heme-containing glycoproteins, which partake in multiple cellular functions including plant development and alleviation of stress by scavenging H_2O_2 (Jouili et al. 2011; Meisrimler et al. 2014; Veljović Jovanović et al. 2018; Rajput et al. 2021). The glycoprotein structure of POX includes a single polypeptide chain of 300–350 amino acid residues, having a molecular weight ranging from 33 to 55 kDa (Veljović Jovanović et al. 2018; Rajput et al. 2021). Peroxidases are generally monomeric structures. However, some plants also harbor POX homotetramers. POX in coconut represents a homotetramer structure with 55 kDa molecular weight of each subunit. Similarly, Brussels sprouts maintain a homodimer form of POX with a molecular weight of 45 kDa for each submits (Regalado et al. 1999). The family of plant peroxidases (Fig. 9.2) constitutes three classes, viz., class I, class II, and class III (Jouili et al. 2011; Rajput et al. 2021). Class I comprises intracellular enzymes such as CAT-POX (Yamada et al. 2002), cytochrome c POX, and APX; class II are extracellular fungal enzymes that include Mn-depended POX and ligninase; class III POX enzymes perform a function in the extracellular region (Jouili et al. 2011; Rajput et al. 2021). The class III POXs are the classical POX of higher plants, which are also known as secretory enzymes (Lüthje and Martinez-Cortes 2018; Veljović Jovanović et al. 2018). These class III POX are oxidoreductase enzyme, which catalyzes the oxidation of various substrates, particularly secondary metabolites like phenols, indoles, and others, which donate electrons to H_2O_2, generating H_2O (Veljović Jovanović et al. 2018; Hasanuzzaman et al. 2020). The form of POX, which utilizes guaiacol as a substrate, is called guaiacol peroxidase (GP) (Rajput et al. 2021). Peroxidases are the only enzymatic antioxidants that can alleviate H_2O_2 levels from the extracellular space (Veljović

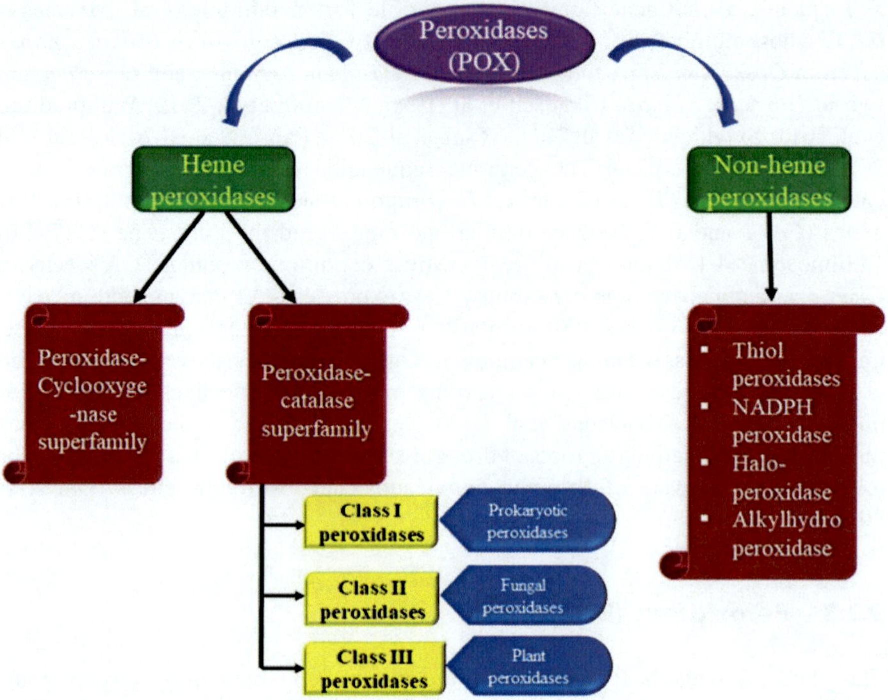

Fig. 9.2 Classification of peroxidases

Jovanović et al. 2018). Peroxidases not only act as a sink of H_2O_2 but also partake in the generation of H_2O_2 via one-electron reduction of O_2 under adverse conditions like pathogen attack, wounding ozone, and others (Gill and Tuteja 2010; Meisrimler et al. 2014; Veljović Jovanović et al. 2018). The H_2O_2 produced via POX in the extracellular region facilitates the formation of a secondary cell wall and stiffening of the cell wall (Rajput et al. 2021). Moreover, POX has the ability to generate $^{\bullet}OH$ radical (Veljović Jovanović et al. 2018). The $^{\bullet}OH$ produced via POX in apoplastic space assists in cell wall loosening, which in turn is required for the expansion of a cell or organ for growth and under conditions like drought, salinity, or osmotic stress (Kukavica et al. 2009; Tenhaken 2015; Veljović Jovanović et al. 2018).

In plants, the endoplasmic reticulum is the site for synthesis of POX, which the gets transported to extracellular space and vacuoles via the Golgi bodies, where it scavenges excess H_2O_2 (Veljović Jovanović et al. 2018). The H_2O_2 over-production in chloroplast under stress events surpasses the eliminating capacity of the AsA-GSH cycle, resulting in the evasion of H_2O_2 to the vacuole, where it is efficiently quenched by the POX enzyme (Karuppanapandian et al. 2011; Veljović Jovanović et al. 2018). *Catharanthus roseus* leaf vacuolar POX isozymes scavenge H_2O_2 at the rate up to 9 mM s^{-1}, by utilizing phenolic substrates (Ferreres et al. 2011). The activity of POX is quite sensitive to heavy metal stress and atmospheric pollution that prompts to monitor POX activity in urban pollution as well as industrial settings (Veljović Jovanović et al. 2018). Other stress conditions such as cold, heat, salinity, flooding, and water deficit also modulate the POX activity (Veljović Jovanović et al. 2018). The onset of cold stress gradually upregulates the POX activity in various *Medicago* accessions and with prolonged duration of exposure the activity gets reduced (Nourredine et al. 2015). A similar result has been obtained with naked oat plants (Liu et al. 2013). The maize plant subjected to flooding results in downregulation of neutral POX within 4 h of exposure, whereas 28 or 52 h of exposure upregulates the expression of alkaline POXs (Meisrimler et al. 2014).

9.2.4 Ascorbate Peroxidase (1.11.1.11)

Ascorbate peroxidase (APX) is a class I heme-peroxidase enzyme with protoporphyrin as a prosthetic group (Pang and Wang 2010; Yadav et al. 2014; Anjum et al. 2016; Ozyigit et al. 2016; Pandey et al. 2017; Sachdev et al. 2022). It is also known as ascorbate (AsA)-dependent peroxidase (Rajput et al. 2021). Like POX, APXs are active scavengers of H_2O_2 molecules (Sachdev et al. 2022). In higher plants, multigene families of APX occur; for instance, nine APX genes in *Arabidopsis* and in tomato encoding different isozymes (Chew et al. 2003; Najami et al. 2008). Rice harbors eight isozymes located in different organelles (Teixeira et al. 2004). The APX enzymes in plant cells are localized in chloroplast (chAPX), mitochondria (mitAPX), cytosol (cAPX), and peroxisomes or glyoxysomes (mAPX) (Anjum et al. 2016; Pandey et al. 2017). In plants, the APX isoforms are classified based on subcellular location, and these isoforms display different structural as well as kinetic properties such as molecular weight, stability, substrate specificity, catalytic rate, and optimum pH (Anjum et al. 2016; Pandey et al. 2017). APX enzyme requires reducing substrate AsA for active functioning and stability (Shigeoka et al. 2002; Pandey et al. 2017). Low concentrations (<20 μM) of AsA result in enzyme inactivation and stability loss (Pang and Wang 2010). The enzymatic action of APX involves the oxidation of heme groups by H_2O_2, resulting in the genesis of oxyferryl, which eventually gets converted into a water molecule by two subsequent one-electron reactions (Anjum et al. 2016). Besides, APX is also an important enzymatic constituent of the AsA-GSH cycle, which detoxifies H_2O_2 (Pang and Wang 2010; Saxena et al. 2020). The activity of APX is found to be stimulated in the presence of other enzymatic antioxidants such as SOD and GR, demonstrating a cross-talk between various enzymatic antioxidants (Pandey et al. 2017). Being effective quenchers of H_2O_2, APX lacks capacity to quench lipid hydroperoxides (LOOHs) and can be inhibited by cyanide, thiol-modifiers, and azide (Pandey et al. 2017).

Ascorbate peroxidase and its isozymes show their activity under both biotic and abiotic stress. *Populus tomentosa* overexpressing thylakoid APX (tAPX) displays an increased resistance toward oxidative stress on treatment with methyl viologen (Li et al. 2022). The transgenic *Brassica juncea* overexpressing cytosolic APX gene (*Apx1*) gains an enhanced enzymatic antioxidants activity, reducing the accumulation of H_2O_2 and malondialdehyde (MDA) and improving tolerance toward salinity stress (Saxena et al. 2020). *Sorghum bicolor* plants subjected to drought stress display APX expression to be upregulated in the aboveground parts and downregulated in roots (Akbudak et al. 2018). The extent of APX expression depends on the duration and intensity of the persisting stress and/or the combination of stresses (Pandey et al. 2017). Moreover, APX isoforms exhibit differential expression in different cellular compartments for the similar type of stress (Shigeoka et al. 2002).

9.2.5 Glutathione Peroxidase (EC: 1.11.1.9)

Glutathione peroxidase (GPX) is a non-heme peroxidase enzyme, which can reduce H_2O_2 to H_2O and O_2 (Rajput et al. 2021). GPX also catalyzes the reduction of peroxide radicals ($ROO^{\bullet}$) into lipid alcohols and O_2 (Rajput et al. 2021). Plant GPX is a thiol-enzyme, which utilizes thioredoxin as a reductant to mitigate H_2O_2-induced damaging effects (Rajput et al. 2021). Plant GPXs are different from their animal counterpart, firstly due to the presence of cysteine (Cys) in their active site instead of seleno-cysteine (Navrot et al. 2006; Bela et al. 2015; Passaia and Margis-Pinheiro 2015; Ozyigit et al. 2016), and secondly, plants use thioredoxin to regenerate oxidized GPX, instead of reduced glutathione (GSH), which has been noticed in animals (Rajput et al. 2021). However, Se-containing GPX in *Aloe vera* consists of four subunits containing one Se atom per subunit with the molecular weight of 16 kDa, each corresponding to the animal GPX (Sabeh et al. 1993; Rajput et al. 2021). GPX has multiple isozymes, which occur in various organelles such as chloroplast, mitochondria, cytosol, and endoplasmic reticulum (Bela et al. 2015; Ozyigit et al. 2016).

Under stressful events like salinity, oxidative stress, pathogen attack, cold, and metals, the level of GPX mRNA increases significantly (Navrot et al. 2006; Ozyigit et al. 2016). In case of rice plants, an upregulated expression of GPX genes occurs in the presence of excess H_2O_2 and cold stress (Passaia et al. 2013). Both the salt-sensitive and salt-tolerant cultivar of *O. sativa* exposed to Cd stress enhance the activity of GPX and APX. However, the salt-tolerant cultivar expresses the higher activity than the salt-sensitive cultivar (Roychoudhury et al. 2012). A transgenic tomato plant overexpressing GPX alleviates the abiotic stress, but counteract the plant's defense response under biotic stress eventually promoting necrotic lesions (Herbette et al. 2011). Thus, GPX playa an efficient role in abridging abiotic stress by attenuating ROS levels. However, excess scavenging of H_2O_2 under biotic stress interferes with the ROS-mediated signaling process. Besides the management of stress, GPX enzyme regulates cellular redox homeostasis by adjusting the balance between thiol-disulfide (Bela et al. 2015; Ozyigit et al. 2016).

9.2.6 Glutathione Reductase (EC: 1.6.4.2)

Glutathione reductase (GR) is a flavoprotein oxidoreductase enzyme that uses NADPH (nicotinamide adenine dinucleotide phosphate) as an electron donor (Yousuf et al. 2012; Cohen et al. 2020; Rajput et al. 2021). Glutathione, an important cellular tripeptide, primarily performs its activity in the reduced form (Marty et al. 2009). This reduced glutathione (GSH), however, while performing function under normal and stressed conditions gets converted into the oxidized glutathione (GSSG) by forming a disulfide bond between two glutathione molecules (Cohen et al. 2020). To maintain the content of GSH, GSSG is recycled back by the enzyme GR in the presence of a reductant NADPH (Cohen et al. 2020), which is essential for effective functioning of a cell (Noctor et al. 2012). Due to the ability to maintain high GSH/GSSG ratio, GR is a crucial part of AsA-GSH pathway (Rajput et al. 2021), where it catalyzes the reduction of disulfide GSSG (dimeric form) into the GSH (monomeric form) (Yousuf et al. 2012; Rajput et al. 2021). GR isoforms occur in chloroplast, mitochondria and cytosol, however, its maximum activity, more than 80% having been demonstrated in photosynthetic tissues (Gill et al. 2013). In *Arabidopsis thaliana*, two GR genes namely, GR1 and GR2 encode the cytosolic-peroxisomal and chloroplastic-mitochondrial form, respectively (Garnik et al. 2016; Cohen et al. 2020; Marty et al. 2019). GR1 contributes to 65% and GR2, 35% of the total measured activity of GR in both wild-type and mutant *Arabidopsis* leaves (Marty et al. 2009). GR in plants occurs in homodimer form with molecular weight ranging 100–150 kDa (Rajput et al. 2021). GR contains one flavin adenine dinucleotide (FAD) per monomer and two Cys residues at the active site (Rajput et al. 2021). The reaction catalyzed by GR takes place in two stages. The first stage involves the reduction of FAD, facilitated by NADPH, followed by the relocation of the reducing power to the thiol group of Cys at the active site. In the next stage, the thiol-disulfide interchange reaction takes place on the subsequent attack of a nucleophile on the GSSG disulfide bond, resulting in the reduction of dimeric GSSG to GSH (Gill et al. 2013).

GR, being an essential component of AsA-GSH cycle, partakes in detoxification of ROS (Gill et al. 2013). Thus, its enhanced activity imparts tolerance against stress in plants. For example, *Nicotiana tabacum* and *Oryza sativa* exposed to chilling and salt stress, respectively, exhibit a reduced GR activity along with an increased sensitivity toward oxidative stress (Ding et al. 2012; Wu et al. 2015). The increased activity of GR along with other enzymatic antioxidants and AsA-GSH cycle in mitochondria of soybean alleviates levels of H_2O_2 and MDA and induces tolerance to chilling stress (Sun et al. 2011). Similarly, the increased activity of GR in muskmelon reduces chilling injury in sensitive cultivar (Fogelman et al. 2011). The chickpea cultivar with high GR activity on exposure to chilling temperatures also displays an alleviated level of MDA, protecting seedlings from the oxidative damage (Turan and Ekmekçi 2011). Under waterlogging stress, an increased content of GR in addition to other antioxidants promotes tolerance to oxidative stress in pigeonpea genotype ICP 301 (Kumutha et al. 2009). The increased content of GR, other antioxidant enzymes and endogenous resveratrol counteract oxidative stress in peanut seedlings exposed to UV-C radiation (Tang et al. 2010).

9.2.7 Monodehydroascorbate Reductase (EC: 1.6.5.4)

Monodehydroascorbate reductase (MDHAR) is an enzyme which takes part in AsA-GSH cycle and catalyzes reduction of monodehydroascorbate (MDHA) into AsA, thus maintaining the pool of reduce AsA in plants (Eltayeb et al. 2006; Rajput et al. 2021). MDHAR utilizes a reductant (NAD(P)H), orchestrating reduction of MDHA to AsA (Sudan et al. 2015). MDHAR is a ubiquitous enzyme with its isoforms being located in chloroplast, mitochondria, cytosol, and peroxisomes (Sudan et al. 2015; Rajput et al. 2021). The structure of MDHAR found in *O. sativa* resembles with other iron (Fe)-sulphur (S) protein reductases, comprising a 60–80 residues long loop as a part of the active site (Rajput et al. 2021). The arginine and tyrosine residues localized at the 320 and 349 position, respectively, play a crucial role in substrate binding, and transfer electrons to the bound substrate from NAD(P)H through FAD (Park et al. 2016; Rajput et al. 2021). MDHAR is an important enzymatic antioxidant that induces tolerance to stress. The unfavorable conditions such as salinity, excess light, extreme temperature, drought, and UV radiations spikes MDHAR activity. The overexpression of MDHAR gene (*AtMDAR1*) in *A. thaliana* induces tolerance in transgenic tobacco against salinity, ozone and polyethylene glycol-triggered stresses (Eltayeb et al. 2007). Similarly, *A. thaliana* overexpressing chloroplastic MDHAR from tomato triggers tolerance against oxidative stress arising from extreme temperature as well as methyl viologen (Li et al. 2010). MDHAR in *A. thaliana* catalyzes a reaction generating AsA, which removes excess ROS and protects photosynthetic apparatus. The transgenic rice plant overexpressing MDHAR (AeMDHAR from *Acanthus ebracteatus*, a mangrove plant) gains tolerance to salt stress and improves yield attributes (Sultana et al. 2012).

9.2.8 Dehydroascorbate Reductase (EC: 1.8.5.1)

Dehydroascorbate reductase (DHAR) also participates in the AsA-GSH cycle and catalyzes the reduction of dehydroascorbate (DHA) in the presence of reduced glutathione (GSH), generating AsA and oxidized glutathione (GSSG) (Rajput et al. 2021). DHAR belongs to the superfamily of glutathione-*S*-transferase (GST) enzymes. Unlike most of the GST, DHAR has active site-Cys instead of serine and in place of ability to stabilize thiolate anion of GSH, it forms reversible disulphide bond with the GSH (Ding et al. 2020). The crystal structure of DHAR has been investigated in different plants including *O. sativa*, *Pennisetum glaucum*, *Populus trichocarpa*, and *A. thaliana* (Ding et al. 2020). In plants two to four DHAR genes exist, however, in some cases few additional pseudogenes have also been reported (Zhang et al. 2015; Rajput et al. 2021). In plants, DHAR is localized in chloroplast, cytosol, mitochondria, peroxisomes, apoplast, and vacuole (Hernández et al. 2001; Zhang et al. 2015; Noshi et al. 2017; Ding et al. 2020).

The reduction of DHA in plant cells occurs chemically, however, this reduction is accelerated by the DHAR (Ding et al. 2020). DHA, if not reduced rapidly due to limited reductive capacity of a cell, shows spontaneous as well as irreversible hydrolysis yielding 2,3-diketogluconic acid (Deutsch 2000; Parsons and Fry 2012). This reaction eventually decreases net AsA content. Thus, DHAR activity maintains an effective AsA pool in the plant cell (Ding et al. 2020; Rajput et al. 2021), which facilitates quenching of excess ROS. Several studies have confirmed the protecting role of DHAR against oxidative damage (Zhang et al. 2015). Tropical fig (*Ficus macrocarpa*) leaves lacking heat-stable DHAR activity exhibits high light sensitivity that indicates the critical role of DHAR in inducing tolerance in plants against photoinhibition (Yamasaki et al. 1999). Similarly, *A. thaliana* mutant deficient in cytosolic DHAR decreases apoplastic AsA content and enhances sensitivity to ozone stress (Yoshida et al. 2006). The rice plant overexpressing cytosolic DHAR gene obtains salt stress tolerance by regulating AsA pool, and ion and redox homeostasis (Kim et al. 2014). Analogously, transgenic tobacco overexpressing cytosolic DHAR from *A. thaliana* increases tolerance toward ozone and drought stress (Eltayeb et al. 2006).

9.2.9 Peroxiredoxin (EC: 1.11.1.15)

Peroxiredoxin (Prx) is a thiol-based peroxidase which functions as an antioxidant, acts as molecular chaperons and participates in the redox signaling network (Sevilla et al. 2015; Liebthal et al. 2018; Ratajczak et al. 2019; Sachdev et al. 2021). Prx detoxifies peroxidase substrates including H_2O_2 and alkyl hydroperoxide (ROOH), thereby alleviating oxidative stress (Sachdev et al. 2021). Prxs are localized in chloroplast, cytosol, mitochondria, and nucleus (Liebthal et al. 2018). Prxs are different from other peroxidases due to the deficiency of the prosthetic group containing metal ions in their structure, essentially required for the catalytic activity. Hence, while delivering their catalytic action, Prxs undergo oxidation, resulting in the generation of an inactive form of Prx (Bhatt and Tripathi 2011). To regenerate the active form, Prx requires external electron donors like thioredoxin, glutaredoxin, NADPH-dependent thioredoxin reductase C (NTRC) and others. Most of the Prx utilizes thioredoxin (Trx) as electron donor for catalyzing reduction of H_2O_2 (Lee et al. 2018). In plants, five subclasses of Prx, i.e., 1-Cys Prx, 2-Cys Prx, Prx II, Prx Q and homologue of animal Se-dependent GPX, have been identified on the basis of subunit composition and the Cys residue present (Bhatt and Tripathi 2011). 1-Cys Prx is a monomer, which contains single conserved catalytic Cys (Bhatt and Tripathi 2011). 2-Cys Prx is a homodimer with two subunits linked covalently through a disulphide bridge in the oxidized form (Bhatt and Tripathi 2011). Prx II has two Cys residues, which are catalytically active (Bhatt and Tripathi 2011). Prx Q is a monomer showing intramolecular disulphide bond between two catalytic Cys residues (Bhatt and Tripathi 2011). The fifth subclass of Prx belonging to the family of plant GPX contains three conserved Cys residues, and only two of them partake in catalytic reactions (Navrot et al. 2006; Bhatt and Tripathi 2011).

9.2.10 Thioredoxin and Thioredoxin Reductase (EC: 1.8.1.9)

Thioredoxin (Trx) is a small thiol-disulfide regulatory protein, which reduces the disulfide bond and regulates ROS level in the plant cell (Da Fonseca-Pereira et al. 2019; Sachdev et al. 2021). Trxs are ubiquitous protein with two active Cys residues, which remain separated by two amino acids (Cys-Gly-Pro-Cys), forming a disulphide bridge that is involved in the catalytic activity (Bertini et al. 1999; Da Fonseca-Pereira et al. 2019; Mata-Pérez and Spoel 2019; Sachdev et al. 2021). To undergo a new catalytic cycle, Trxs are reduced by the enzyme thioredoxin reductase (TrxR) (intracellular oxidoreductase enzyme), which utilize either NADPH or reduced ferredoxin as a reductant or electron donor (Geigenberger et al. 2017). Plant Trxs are encoded by the large gene family (Geigenberger et al. 2017). The isoforms of Trxs found in the plant cell differ from each other on the basis of amino acid sequence and subcellular location (Da Fonseca-Pereira et al. 2019). In *A. thaliana* around 20 isoforms of Trx, grouped into seven subfamilies have been identified (Belin et al. 2015). The plant Trxs isoforms are localized in chloroplast, mitochondria, nucleus, endoplasmic reticulum, peroxisomes, apoplast, and cytosol (Zhang et al. 2011; Du et al. 2015; Geigenberger et al. 2017). Trxs regulate the redox status of the plant cell, thereby alleviating impact of biotic and abiotic stress. The *O. sativa* plant overexpressing Trx (OsTrxh1) produces less H_2O_2 under salt stress and maintains the apoplastic redox state (Zhang et al. 2011). Under long-term salt stress, an increase in the level of Trx (PsTrx01) in the mitochondria of pea leaves prevents oxidative stress (Martí et al. 2011).

9.3 Non-enzymatic Antioxidants Alleviating ROS-Induced Stress

9.3.1 Ascorbate or Ascorbic Acid

Ascorbate (AsA), a resonance stabilized anionic form of ascorbic acid (vitamin C), is a water soluble, low molecular weight antioxidant, which partake in scavenging of free radicals, maintaining cellular redox balance (Gutteridge and Halliwell 2000; Eltayeb et al. 2006; Akram et al. 2017; Hasanuzzaman et al. 2019a; Xiao et al. 2021). It is the most powerful substrate which serves as an electron donor for detoxification of H_2O_2 through APX (Noctor et al. 1998; Eltayeb et al. 2006; Xiao et al. 2021). Moreover, it can directly scavenge $^1O_2$ (Xiao et al. 2021). AsA is localized in cell chloroplast, cytoplasm, peroxisomes, mitochondria, as well as in apoplast (Latowski et al. 2010; Yin et al. 2010). The substantial amount of AsA is present in cytosol (12–30%) and chloroplast/apoplast (5–10%) (Eltayeb et al. 2006; Akram et al. 2017). AsA is synthesized in mitochondria and then transported to other parts of the cell (Horemans et al. 2000; Eltayeb et al. 2006). The elevated level of AsA induces tolerance to various environmental stress including drought, salinity, extreme temperature, high light intensity, metals, and ozone (Eltayeb et al. 2006; Yin et al. 2010; Akram et al. 2017; Xiao et al. 2021). Under Al stress, the tolerant tobacco line (ALT301) possesses higher levels of AsA and GSH with lower content of H_2O_2 and lipid peroxides than sensitive cell line (Devi et al. 2003). Application of AsA alleviates the effect of salinity in *Hordeum vulgare* seedlings by improving antioxidant content, plant growth and photosynthetic pigments, and reducing electrolyte leakage (Agami 2014). The *Arabidopsis* mutant with only 30% AsA content of the wild type becomes prone to photoinhibition and photooxidation under sulphur dioxide and ultraviolet B irradiation stress (Conklin et al. 1996). AsA also acts as a cofactor for several enzymes and involves in plant metabolism, regulates the reduced state of the prosthetic metal ions, maintains activity of antioxidant enzymes, leading to regulated cell division, differentiation, and growth as a signal transduction molecule (Eltayeb et al. 2006; Yin et al. 2010; Hasanuzzaman et al. 2012, 2019b; Akram et al. 2017; Fenech et al. 2019; Foyer et al. 2020; Xiao et al. 2021). AsA acts as a cofactor of certain oxidase like 2-oxoglutarate-dependent dioxygenases (2-ODDs) and aminocyclopropane-1-carboxylic acid oxidases (ACOs), which partake in various phytohormones biosynthesis such as ethylene (Bilska et al. 2019; Houben and Van de Poel 2019; Xiao et al. 2021). AsA as a cofactor of violaxanthin de-epoxidase (VDE) mediates biosynthesis of zeaxanthin, facilitating dissipation of excess excitation energy in the thylakoid membrane, preventing oxidative stress (Saga et al. 2010; Yang et al. 2017; Xiao et al. 2021). Moreover, the role of AsA as a cofactor assists in epigenetic modification, which facilitates management of abiotic stress through epigenetic pathways (Xiao et al. 2021).

AsA, being a crucial antioxidant, its effective level is required to be managed through synthesis, recycling, degradation and transportation (Yin et al. 2010). The recycling pathway of AsA is important for elimination of ROS and management of stress (Broad et al. 2020; Xiao et al. 2021). In particular, for managing drought stress, recycling of AsA is more crucial than its synthesis, maintaining high content of AsA (Bartoli et al. 2005; Yin et al. 2010). In plants, AsA-GSH cycle performs dismutation of H_2O_2 into H_2O. In the reaction, two molecules of AsA participate, generating two molecules of MDHA. Further, AsA is recycled back from MDHA on action of MDHAR (Hossain et al. 1984; Yin et al. 2010; Xiao et al. 2021). Being an unstable radical with short life-time, MDHA spontaneously dismutates into AsA and dehydroascorbate (DHA) (Smirnoff and Wheeler 2000; Eltayeb et al. 2006; Xiao et al. 2021). Further, DHAR enzyme regenerates AsA from DHA (Xiao et al. 2021). Under drought stress, enhanced expression of AsA recycling genes confers tolerance to stress (Broad et al. 2020; Xiao et al. 2021).

9.3.2 Glutathione

Glutathione (GSH) is a ubiquitous low molecular weight tripeptide comprising glutamate, Cys, and glycine (γ-glutamyl-cysteinyl-glycine) (Gill et al. 2013; Cohen et al. 2020). The synthesis of GSH is catalyzed by two enzymes namely Glu-Cys (Glutamate-Cysteine) ligase and GSH synthetase in two steps, dependent on ATP (Gill et al. 2013). Cysteine, the first organic compound formed in the sulfur assimilation pathway, is the precursor for the synthesis of GSH (Takahashi et al. 2011; Cohen et al. 2020). GSH in cells occurs in millimolar concentration and assumes two forms, viz., reduced glutathione (GSH) with free thiol group, and oxidized glutathione (GSSG) (Gill et al. 2013; Cohen et al. 2020). The reactive Cys residues facilitate GSH to maintain its protein with thiol group in their native form under stress conditions (Gill et al. 2013). GSH partakes in several cellular functions such as detoxification of xenobiotic compounds; long distance transportation of reduced sulfur; direct post-translational modifications (PTM) of proteins by triggering reversible glutathionylation of thiol residues, which adjust stability and activity of these proteins; regulation of genes and signals; and participates in development of an embryo as well as post-embryonic root meristematic activities (Rouhier et al. 2008; Noctor et al. 2012; Seth et al. 2012; Hacham et al. 2006; Cohen et al. 2020). Moreover, GSH is crucial for detoxification of ROS via AsA-GSH cycle (Foyer and Noctor 2011), providing defense against oxidative stress, high temperature, low temperature, salinity, drought and others (Noctor et al. 2012). The AsA-GSH pathway comprising GSH and its redox couple (GSH/GSSG) can directly or indirectly quench ROS (Gill et al. 2013). The increased content of GSH along with other enzymatic and non-enzymatic antioxidants occurs in *Carrizo citrange* under high-temperature stress (Zandalinas et al. 2017). In addition to AsA content, GSH content in transgenic apple seedlings gains due to increased H_2O_2 level under low-temperature stress (Wang et al. 2016). Rice seedlings accumulating higher content of H_2O_2 and $O_2^{\bullet -}$ under low temperature display enhanced GSH/GSSG ratio along with elevated action of SOD and CAT (Han et al. 2017). Increased content of GSH and GSSG exists in seedlings of *O. sativa* exposed to Ni stress. Moreover, application of silicon further elevates the GSH level but reduced GSSG level (Hasanuzzaman et al. 2019a). GSH also acts as a cytosolic precursor of phytochelatins (PC), which show ability to bind metals and aid their movement into the vacuole, reducing incidence of toxicity (Hasanuzzaman et al. 2020).

9.3.3 Tocopherol

Tocopherol (vitamin E) is a lipophilic antioxidant found in the green plant tissues and seeds (Munné-Bosch and Falk 2004; Szymańska and Kruk 2008; Ahmad et al. 2009). Tocopherol inhibits peroxidation of membrane lipids and scavenges ROS (Szymańska and Kruk 2008). Additionally, tocopherols are instrumental in signaling and gene expression (Kumar et al. 2013). Tocopherol consists of four isomers including α-tocopherol, β-tocopherol, γ-tocopherol and δ-tocopherol (Szymańska and Kruk 2008). With few exceptions, the γ-tocopherol occurs in plant seeds and α-tocopherol in leaves constituting more than 90% of the total tocopherols (Szymańska and Kruk 2008). The α-tocopherol is an effective low molecular weight antioxidant, consisting of three methyl groups, which facilitate efficient scavenging of ROS (Ahmad et al. 2009). It prevents lipid autoxidation by eclipsing chain propagation step and therefore acts as a functional trap for free radical (Ahmad et al. 2008, 2009). With resonance energy transfer ability a single molecule of α-tocopherol can scavenge 120 $^1O_2$ molecules (Munné-Bosch 2005; Ahmad et al. 2009). The α-tocopherol reduces membrane fluidity and thus stabilizes membrane structure (Spicher et al. 2017). Perception of unfavorable conditions, gene expression responsible for tocopherol synthesis increases and elevates its level in higher plants (Ahmad et al. 2009). The enhanced tocopherol content along with AsA occurs as an early response on exposure to water-deficit stress in *O. sativa* (Boo and Jung 1999). The increased content of tocopherol prompts tolerance in plants against multiple stresses. The transgenic *Brassica juncea* plants with higher content of α-tocopherol obtain tolerance against heavy metals, salt, and osmotic stresses (Kumar et al. 2013). Similarly, the foliar application of α-tocopherol in salt-sensitive cultivars of onion ensures tolerance against salt stress as has been evidenced by low H_2O_2 accumulation and lipid peroxidation along with increased activity of enzymatic and non-enzymatic antioxidants (Semida et al. 2016).

9.3.4 Flavonoids

Flavonoids, ubiquitous secondary metabolites of low molecular weight compounds, impart color, fragrance as well as flavor to the different parts of plants including flower, fruit, bark, roots, leaves, stems, seeds, and grains (Rodríguez De Luna et al. 2020; Dias et al. 2021). Flavonoids belong to the large family of polyphenols or phenolic compounds and include more than 6000 distinct structures (Panche et al. 2016; Dias et al. 2021; Šamec et al. 2021). Flavonoids perform several functions such as regulation of cell growth, attracting pollinators and induction of tolerance against stresses in plants (Rodríguez De Luna et al. 2020; Dias et al. 2021); scavenge excess ROS, acting as a signaling molecule and UV filters (Dias et al. 2021). Flavonoids have 15-carbon flavone skeleton arranged in a pattern C6-C3-C6, where two C6 represents two rings of benzene (A and B), connected with a C3 (three carbons) pyran ring (C) (Dias et al. 2021). The antioxidants potential of flavonoids is attributed by the presence of catechol in the B ring in the structure, degree of hydroxylation and polymerization, and conjugation and other substitutions (Kumar et al. 2013; D'Amelia et al. 2018; Rodríguez De Luna et al. 2020; Dias et al. 2021). The hydroxyl group of flavonoids mediates stabilization of free radicals by donating electrons through resonance (Dias et al. 2021; Šamec et al. 2021). On the basis of the structure, the flavonoids comprise six classes, i.e., flavones, flavonols, flavan-3-ols, flavanones, isoflavones, and anthocyanins (Dias et al. 2021; Šamec et al. 2021). The flavonoids in plants can be present in free form or as derivatives, which are synthesized by the processes like polymerization, methylation, glycosidation, acetylation, and prenylation (Šamec et al. 2021).

Flavonoids display multiple mechanisms as an attribute of antioxidant potential. These mechanisms include (1) scavenging of ROS; (2) inhibition of ROS generation by forming chelates with metal ions or inhibiting enzymes participating in ROS production like NADH (nicotinamide adenine dinucleotide) oxidase and xanthine oxidase; (3) upregulation of antioxidants activities (Kumar et al. 2013; Dias et al. 2021). The occurrence of transition metals in plant cell mediate the generation of highly toxic $^{\bullet}OH$ radical by reducing H_2O_2 via Fenton reaction. Due to the ability of flavonoids to chelate metal ions, production of ROS in the presence of transition metal ions is inhibited. For instance, quercetin, a type of flavanol, can form a chelate with reduce iron, eliminating the chance for $^{\bullet}OH$ formation. Moreover, flavonoids have lower redox potential which thermodynamically facilitate reduction of oxidizing free radicals including $O_2^{\bullet-}$, $ROO^{\bullet}$, alkoxyl ($RO^{\bullet}$), and $^{\bullet}OH$ radical via hydrogen atom donation (Kumar et al. 2013). Flavonoids can prevent peroxidation of lipids under stress, thereby alleviating oxidative damage (Kumar and Pandey 2013; Kumar et al. 2013). The flavonoids are be localized in endoplasmic reticulum (the location for their biosynthesis), vacuoles and chloroplast (Ferdinando et al. 2012). Due to localization of flavonoids in chloroplast, they quench $^1O_2$ effectively which high light irradiance exposure generates (Agati et al. 2007). Moreover, the presence of flavonoids in vacuoles could be crucial for scavenging H_2O_2, which might have escaped from chloroplast under stress conditions (Ferdinando et al. 2012). Flavonoids act as a secondary antioxidant defense system under biotic and abiotic stress conditions. Moreover, the genes regulating biosynthesis of flavonoids are upregulated under stress circumstances. The concentration of hesperidin, a flavanone increases significantly in *Cyclopia subternata* callus and peppermint on exposure to UV-C and UV-B radiation, respectively (Dolzhenko et al. 2010; Kokotkiewicz et al. 2014). Similarly, an increase in flavonoid occurs in *Matricharia camomilla* leaves on exposure to the high Ni^{2+} (Kováčik et al. 2009). The chalcone synthase (CHS) and chalcone isomerase enzymes responsible for biosynthesis of flavonoids increase their activities in leaves of bean plants exposed to ozone stress (Paolacci et al. 2001).

9.3.5 Carotenoids

Carotenoids are tetraterpene pigments (red, yellow, orange, and purple color), which belong to isoprenoid group (Strzałka et al. 2003; Maoka 2020). Nearly 850 carotenoids have been reported in living organisms from green and non-green tissues spanning plants, bacteria, fungi, and animals (Maoka 2020). Carotenoid molecule is made up of eight isoprene units containing 40-carbon atoms (Maoka 2020). A general carotene structure includes a polyene chain with nine conjugated double bonds and two end groups, each present at the ends of the chain on both sides (Maoka 2020). The conjugated double bonds facilitate carotenoids to absorb light in the range of 350–500 nm (Strzałka et al. 2003). Carotenes are classified into two groups, viz., carotenes and xanthophylls (Maoka 2020). Carotenes are the hydrocarbons, whereas xanthophyll is the oxygen-containing carotenoids (Maoka 2020). Lycopene, α-carotene, and β-carotene are some examples of naturally occurring carotenes (Maoka 2020). The β-cryptoxanthin, zeaxanthin, fucoxanthin, and lutein are few naturally found xanthophylls (Maoka 2020).

Carotenoids are important compounds, which has photoprotective effects on plant photosynthetic apparatus (Strzałka et al. 2003). Under low light stress, carotenoids may act as antennae, which harvest light at the wavelength that cannot be harvested by the chlorophyll molecules and transfer electron excitation energy toward photochemical reaction centers, increasing the spectrum of light used for photosynthesis (Strzałka et al. 2003). On the other hand, during high light stress, carotenoid act as scavengers of triplet state of chlorophyll, generated due to over-excitation of photosystems and dissipate excess energy in heat form, thereby protecting chloroplast from photooxidative stress (Strzałka et al. 2003). Carotenoids can also detoxify $ROO^{\bullet}$, preventing oxidative damage to pigments as well as lipid unsaturated fatty acids (Strzałka et al. 2003). The reaction with the free radicals takes place via electron transfer, transfer of hydrogen atom or addition of hydrogen atoms (Strzałka et al. 2003). The α-tocopherol generates α-tocopherol radical after scavenging ROS, which itself acts as a pro-oxidant and induces oxidative damage. However, carotenoids like violaxanthin quench these α-tocopherol radical by forming adduct product, thereby alleviating stress imposed by phenoxy radical accumulation in cells (Maoka 2020). Moreover, the carotenoids protect plants form oxidative stress by altering physical properties of thylakoid membranes by involving xanthophyll cycle responsible for interconversion to two carotenoids, i.e., violaxanthin and zeaxanthin (Gruszecki and Strzal·ka 1991; Strzałka et al. 2003). Further, carotenoids also partake in signaling mechanism during unfavorable conditions. Carotenoids quench excess $^1O_2$ under stress, eventually producing breakdown product, which facilitates stress-induced signaling (Sachdev et al. 2022).

9.4 Ascorbate-Glutathione Cycle

Ascorbate-Glutathione (AsA-GSH) cycle, known as Asada-Halliwell cycle (Fig. 9.3), is an effective process for the elimination of ROS from cells (Hasanuzzaman et al. 2019a, b). This cycle involves two low molecular weight antioxidants, AsA and GSH along with four enzymatic antioxidants including APX, GR, MDHAR, and DHAR (Latowski et al. 2010). AsA-GSH cycle facilitates scavenging of H_2O_2 in chloroplast, mitochondria, peroxisome, cytoplasm, and apoplast (Latowski et al. 2010; Bartoli et al. 2017). In this cycle, initially H_2O_2 is scavenged and dismutated into H_2O molecule in a reaction where AsA is oxidized by APX, generating monodehydroascorbate (MDHA). MDHA is rapidly disproportionate into AsA on action of enzyme MDHAR, using NAD(P)H, whereas its non-enzymatic disproportionation generates dehydroascorbate (DHA) (Bartoli et al. 2017). Further, DHA is finally transformed into AsA in a reaction catalyzed

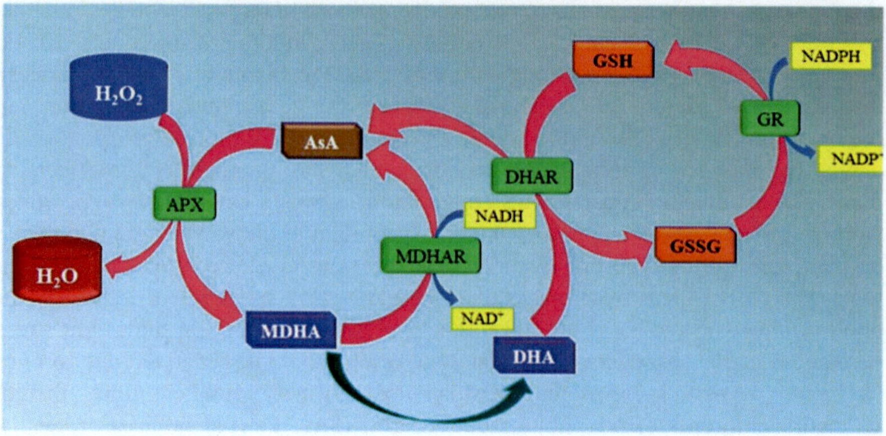

Fig. 9.3 Ascorbate-Glutathione (AsA-GSH) cycle dismutating hydrogen peroxide (H_2O_2) into water (H_2O) on action of enzymatic antioxidant APX. *APX* ascorbate peroxidase, *MDHAR* monodehydroascorbate reductase, *DHAR* dehydroascorbate reductase, *GR* glutathione reductase, *MDHA* monodehydroascorbate, *DHA* dehydroascorbate, *AsA* ascorbate, *GSH* reduced glutathione, *GSSG* oxidized glutathione

by enzyme DHAR, using GSH. This step result in oxidation of GSH to glutathione disulfide (GSSG: oxidized glutathione), which is regenerated by the enzyme GR in the presence NAD(P)H (Latowski et al. 2010; Bartoli et al. 2017).

AsA-GSH cycle is a major antioxidant defensive system involved in alleviation of ROS, generated under biotic and abiotic stress. In a study, wheat plant subjected to heat stress (40 °C) augments AsA-GSH metabolism on application of nitric oxide and hydrogen sulfide, which reduce H_2O_2 content and promote photosynthetic activity in leaves, thereby alleviating oxidative stress (Iqbal et al. 2021). Moreover, efficient functioning of AsA-GSH cycle in wheat plants exposed to Al stress decreases level of ROS and aggravates root elongation (Liu et al. 2021). Similarly, under drought stress, inoculation of fungal endophytes enhances AsA-GSH cycle enzymes in *Citrus reticulate*, which reduces the accumulation of ROS and confers tolerance against the stress (Sadeghi et al. 2020).

9.5 Conclusion

The generation of reactive oxygen species (ROS) is an inevitable by-product of aerobic activities. Over-production of ROS under adverse conditions, however, induces oxidation of biomolecules, impinging plant cellular structure and cause dysfunction. To avoid the oxidative damage, plants recruit enzymatic and non-enzymatic antioxidants, which eliminate excess ROS and induce tolerance against stress. Nevertheless, several times under stress conditions, the activities of antioxidants decline, elevating stress-triggered notorious impact on plant growth and productivity. Studies have shown that transgenic plants overexpressing particular or group of antioxidants can enhance plants' intrinsic defense mechanism, promoting stress tolerance. Moreover, exogenous treatment of plants with certain molecules fortifies plant's defense system. Therefore, pursuing future research for finding suitable compounds or developing transgenic cultivars could be an effective approach to improve plants antioxidants-based defense mechanism under varying situation.

References

Abuelsoud W, Cortleven A, Schmülling T (2020) Photoperiod stress alters the cellular redox status and is associated with an increased peroxidase and decreased catalase activity. bioRxiv. https://doi.org/10.1101/2020.03.05.978270

Agami RA (2014) Applications of ascorbic acid or proline increase resistance to salt stress in barley seedlings. Biol Plant 58(2):341–347

Agati G, Matteini P, Goti A, Tattini M (2007) Chloroplast-located flavonoids can scavenge singlet oxygen. New Phytol 174(1):77–89

Ahmad P, Sarwat M, Sharma S (2008) Reactive oxygen species, antioxidants and signaling in plants. J Plant Biol 51(3):167–173

Ahmad P, Jaleel CA, Azooz MM, Nabi G (2009) Generation of ROS and non-enzymatic antioxidants during abiotic stress in plants. Bot Res Int 2(1):11–20

Ahmed IM, Cao F, Zhang M, Chen X, Zhang G, Wu F (2013) Difference in yield and physiological features in response to drought and salinity combined stress during anthesis in Tibetan wild and cultivated barleys. PLoS One 8(10):e77869

Akbudak MA, Filiz E, Vatansever R, Kontbay K (2018) Genome-wide identification and expression profiling of ascorbate peroxidase (APX) and glutathione peroxidase (GPX) genes under drought stress in sorghum (*Sorghum bicolor* L.). J Plant Growth Regul 37(3):925–936

Akram NA, Shafiq F, Ashraf M (2017) Ascorbic acid-a potential oxidant scavenger and its role in plant development and abiotic stress tolerance. Front Plant Sci 8:613

Alzahrani SM, Alaraidh IA, Migdadi H, Alghamdi S, Khan MA, Ahmad P (2019) Physiological, biochemical, and antioxidant properties of two genotypes of *Vicia faba* grown under salinity stress. Pak J Bot 51(3):786–798

Anjum SA, Tanveer M, Hussain S, Bao M, Wang L, Khan I, Ullah E, Tung SA, Samad RA, Shahzad B (2015) Cadmium toxicity in Maize (*Zea mays* L.): consequences on antioxidative systems, reactive oxygen species and cadmium accumulation. Environ Sci Pollut Res 22(21): 17022–17030

Anjum NA, Sharma P, Gill SS, Hasanuzzaman M, Khan EA, Kachhap K, Mohamed AA, Thangavel P, Devi GD, Vasudhevan P, Sofo A, Khan NA, Misra AN, Lukatin AS, Singh HP, Pereira E, Tuteja N (2016) Catalase and ascorbate peroxidase—representative H2O2-detoxifying heme enzymes in plants. Environ Sci Pollut Res 23(19):19002–19029

Ansari MI, Jalil SU, Ansari SA, Hasanuzzaman M (2021) GABA shunt: a key-player in mitigation of ROS during stress. Plant Growth Regul 94(2):131–149

Apel K, Hirt H (2004) Reactive oxygen species: metabolism, oxidative stress, and signal transduction. Annu Rev Plant Biol 55:373–399

Bartoli CG, Guiamet JJ, Kiddle GUY, Pastori GM, Di Cagno R, Theodoulou FL, Foyer CH (2005) Ascorbate content of wheat leaves is not determined by maximal l-galactono-1, 4-lactone dehydrogenase (GalLDH) activity under drought stress. Plant Cell Environ 28(9):1073–1081

Bartoli CG, Buet A, Gergoff Grozeff G, Galatro A, Simontacchi M (2017) Ascorbate-glutathione cycle and abiotic stress tolerance in plants. In: Ascorbic acid in plant growth, development and stress tolerance. Springer, Cham, pp 177–200

Bela K, Horváth E, Gallé Á, Szabados L, Tari I, Csiszár J (2015) Plant glutathione peroxidases: emerging role of the antioxidant enzymes in plant development and stress responses. J Plant Physiol 176:192–201

Belin C, Bashandy T, Cela J, Delorme-Hinoux V, Riondet C, Reichheld JP (2015) A comprehensive study of thiol reduction gene expression under stress conditions in Arabidopsis thaliana. Plant Cell Environ 38(2):299–314

Bertini R, Zack Howard OM, Dong HF, Oppenheim JJ, Bizzarri C, Sergi R, Caselli G, Pagliei S, Romines B, Wilshire JA, Mengozzi M, Nakamura H, Yodoi J, Pekkari K, Gurunath R, Holmgren A, Herzenberg LA, Herzenberg LA, Ghezzi P (1999) Thioredoxin, a redox enzyme released in infection and inflammation, is a unique chemoattractant for neutrophils, monocytes, and T cells. J Exp Med 189(11):1783–1789

Bhatt I, Tripathi BN (2011) Plant peroxiredoxins: catalytic mechanisms, functional significance and future perspectives. Biotechnol Adv 29(6):850–859

Bilska K, Wojciechowska N, Alipour S, Kalemba EM (2019) Ascorbic acid—The little-known antioxidant in woody plants. Antioxidants 8(12):645

Boo YC, Jung J (1999) Water deficit—induced oxidative stress and antioxidative defenses in rice plants. J Plant Physiol 155(2):255–261

Broad RC, Bonneau JP, Hellens RP, Johnson AA (2020) Manipulation of ascorbate biosynthetic, recycling, and regulatory pathways for improved abiotic stress tolerance in plants. Int J Mol Sci 21(5):1790

Chang-Quan W, Rui-Chang L (2008) Enhancement of superoxide dismutase activity in the leaves of white clover (*Trifolium repens* L.) in response to polyethylene glycol-induced water stress. Acta Physiol Plant 30(6):841–847

Chew O, Whelan J, Millar AH (2003) Molecular definition of the ascorbate-glutathione cycle in *Arabidopsis* mitochondria reveals dual targeting of antioxidant defenses in plants. J Biol Chem 278(47):46869–46877

Chong J, Poutaraud A, Hugueney P (2009) Metabolism and roles of stilbenes in plants. Plant Sci 177(3):143–155

Cohen A, Hacham Y, Welfe Y, Khatib S, Avice JC, Amir R (2020) Evidence of a significant role of glutathione reductase in the sulfur assimilation pathway. Plant J 102(2):246–261

Conklin PL, Williams EH, Last RL (1996) Environmental stress sensitivity of an ascorbic acid-deficient *Arabidopsis* mutant. Proc Natl Acad Sci 93(18):9970–9974

D'Amelia V, Aversano R, Chiaiese P, Carputo D (2018) The antioxidant properties of plant flavonoids: their exploitation by molecular plant breeding. Phytochem Rev 17(3):611–625

Da Fonseca-Pereira P, Daloso DM, Gago J, Nunes-Nesi A, Araújo WL (2019) On the role of the plant mitochondrial thioredoxin system during abiotic stress. Plant Signal Behav 14(6):1592536

Dai L, Li J, Harmens H, Zheng X, Zhang C (2020) Melatonin enhances drought resistance by regulating leaf stomatal behaviour, root growth and catalase activity in two contrasting rapeseed (*Brassica napus* L.) genotypes. Plant Physiol Biochem 149:86–95

Deutsch JC (2000) Dehydroascorbic acid. J Chromatogr A 881(1-2):299–307

Devi SR, Yamamoto Y, Matsumoto H (2003) An intracellular mechanism of aluminum tolerance associated with high antioxidant status in cultured tobacco cells. J Inorg Biochem 97(1):59–68

Dias MC, Pinto DC, Silva A (2021) Plant flavonoids: chemical characteristics and biological activity. Molecules 26(17):5377

Ding S, Lei M, Lu Q, Zhang A, Yin Y, Wen X, Zhang L, Lu C (2012) Enhanced sensitivity and characterization of photosystem II in transgenic tobacco plants with decreased chloroplast glutathione reductase under chilling stress. Biochim Biophys Acta Bioener 1817(11): 1979–1991

Ding H, Wang B, Han Y, Li S (2020) The pivotal function of dehydroascorbate reductase in glutathione homeostasis in plants. J Exp Bot 71(12):3405–3416

Dolzhenko Y, Bertea CM, Occhipinti A, Bossi S, Maffei ME (2010) UV-B modulates the interplay between terpenoids and flavonoids in peppermint (*Mentha× piperita* L.). J Photochem Photobiol B Biol 100(2):67–75

Doupis G, Bertaki M, Psarras G, Kasapakis I, Chartzoulakis K (2013) Water relations, physiological behavior and antioxidant defence mechanism of olive plants subjected to different irrigation regimes. Sci Hortic 153:150–156

Du YY, Wang PC, Chen J, Song CP (2008) Comprehensive functional analysis of the catalase gene family in *Arabidopsis thaliana*. J Integr Plant Biol 50(10):1318–1326

Du H, Kim S, Hur YS, Lee MS, Lee SH, Cheon CI (2015) A cytosolic thioredoxin acts as a molecular chaperone for peroxisome matrix proteins as well as antioxidant in peroxisome. Mol Cells 38(2):187

Eltayeb AE, Kawano N, Badawi GH, Kaminaka H, Sanekata T, Morishima I, Shibahara T, Inanaga S, Tanaka K (2006) Enhanced tolerance to ozone and drought stresses in transgenic tobacco overexpressing dehydroascorbate reductase in cytosol. Physiol Plant 127(1):57–65

Eltayeb AE, Kawano N, Badawi GH, Kaminaka H, Sanekata T, Shibahara T, Inanaga S, Tanaka K (2007) Overexpression of monodehydroascorbate reductase in transgenic tobacco confers enhanced tolerance to ozone, salt and polyethylene glycol stresses. Planta 225(5):1255–1264

Eyidogan F, Öz MT (2007) Effect of salinity on antioxidant responses of chickpea seedlings. Acta Physiol Plant 29(5):485–493

Fenech M, Amaya I, Valpuesta V, Botella MA (2019) Vitamin C content in fruits: biosynthesis and regulation. Front Plant Sci 9:2006

Ferdinando MD, Brunetti C, Fini A, Tattini M (2012) Flavonoids as antioxidants in plants under abiotic stresses. In: Ahmad P, Prasad M (eds) Abiotic stress responses in plants. Springer, New York, NY, pp 159–179

Ferreres F, Figueiredo R, Bettencourt S, Carqueijeiro I, Oliveira J, Gil-Izquierdo A, Pereira DM, Valentao P, Andrade PB, Duarte P, Barcelo AR, Sottomayor M (2011) Identification of phenolic compounds in isolated vacuoles of the medicinal plant Catharanthus roseus and their interaction with vacuolar class III peroxidase: an H2O2 affair? J Exp Bot 62(8):2841–2854

Fogelman E, Kaplan A, Tanami Z, Ginzberg I (2011) Antioxidative activity associated with chilling injury tolerance of muskmelon (*Cucumis melo* L.) rind. Sci Hortic 128(3):267–273

Foyer CH, Noctor G (2011) Ascorbate and glutathione: the heart of the redox hub. Plant Physiol 155(1):2–18

Foyer CH, Kyndt T, Hancock RD (2020) Vitamin C in plants: novel concepts, new perspectives, and outstanding issues. Antioxid Redox Signal 32(7):463–485

Frugoli JA, Zhong HH, Nuccio ML, McCourt P, McPeek MA, Thomas TL, McClung CR (1996) Catalase is encoded by a multigene family in *Arabidopsis thaliana* (L.) Heynh. Plant Physiol 112(1):327–336

Gajewska E, Skłodowska M (2008) Differential biochemical responses of wheat shoots and roots to nickel stress: antioxidative reactions and proline accumulation. Plant Growth Regul 54(2): 179–188

Gapińska M, Skłodowska M, Gabara B (2008) Effect of short-and long-term salinity on the activities of antioxidative enzymes and lipid peroxidation in tomato roots. Acta Physiol Plant 30(1):11–18

Garnik EY, Belkov VI, Tarasenko VI, Korzun MA, Konstantinov YM (2016) Glutathione reductase gene expression depends on chloroplast signals in *Arabidopsis thaliana*. Biochem Mosc 81(4): 364–372

Geigenberger P, Thormählen I, Daloso DM, Fernie AR (2017) The unprecedented versatility of the plant thioredoxin system. Trends Plant Sci 22(3):249–262

Gill SS, Tuteja N (2010) Reactive oxygen species and antioxidant machinery in abiotic stress tolerance in crop plants. Plant Physiol Biochem 48(12):909–930

Gill SS, Anjum NA, Hasanuzzaman M, Gill R, Trivedi DK, Ahmad I, Pereira E, Tuteja N (2013) Glutathione and glutathione reductase: a boon in disguise for plant abiotic stress defense operations. Plant Physiol Biochem 70:204–212

Gomez JM, Hernandez JA, Jimenez A, Del Rio LA, Sevilla F (1999) Differential response of antioxidative enzymes of chloroplasts and mitochondria to long-term NaCl stress of pea plants. Free Radic Res 31(sup1):11–18

Gruszecki WI, Strzalka K (1991) Does the xanthophyll cycle take part in the regulation of fluidity of the thylakoid membrane? Biochim Biophys Acta Bioener 1060(3):310–314

Gutteridge JM, Halliwell B (2000) Free radicals and antioxidants in the year 2000: a historical look to the future. Ann N Y Acad Sci 899(1):136–147

Hacham Y, Schuster G, Amir R (2006) An in vivo internal deletion in the N-terminus region of Arabidopsis cystathionine γ-synthase results in CGS expression that is insensitive to methionine. Plant J 45(6):955–967

Han QH, Huang B, Ding CB, Zhang ZW, Chen YE, Hu C, Zhou LJ, Huang Y, Liao JQ, Yuan S, Yuan M (2017) Effects of melatonin on anti-oxidative systems and photosystem II in cold-stressed rice seedlings. Front Plant Sci 8:785

Hasanuzzaman M, Hossain MA, Silva JA, Fujita M (2012) Plant response and tolerance to abiotic oxidative stress: antioxidant defense is a key factor. In: Crop stress and its management: perspectives and strategies. Springer, Dordrecht, pp 261–315

Hasanuzzaman M, Alam M, Nahar K, Mohsin SM, Bhuyan MHM, Parvin K, Hawrylak-Nowak B, Fujita M (2019a) Silicon-induced antioxidant defense and methylglyoxal detoxification works coordinately in alleviating nickel toxicity in *Oryza sativa* L. Ecotoxicology 28(3):261–276

Hasanuzzaman M, Bhuyan MHM, Anee TI, Parvin K, Nahar K, Mahmud JA, Fujita M (2019b) Regulation of ascorbate-glutathione pathway in mitigating oxidative damage in plants under abiotic stress. Antioxidants 8(9):384

Hasanuzzaman M, Bhuyan MHM, Zulfiqar F, Raza A, Mohsin SM, Mahmud JA, Fujita M, Fotopoulos V (2020) Reactive oxygen species and antioxidant defense in plants under abiotic stress: revisiting the crucial role of a universal defense regulator. Antioxidants 9(8):681

He Z, Zhao T, Yin Z, Liu J, Cheng Y, Xu J (2020) The phytochrome-interacting transcription factor CsPIF8 contributes to cold tolerance in citrus by regulating superoxide dismutase expression. Plant Sci 298:110584

Herbette S, de Labrouhe DT, Drevet JR, Roeckel-Drevet P (2011) Transgenic tomatoes showing higher glutathione peroxydase antioxidant activity are more resistant to an abiotic stress but more susceptible to biotic stresses. Plant Sci 180(3):548–553

Hernández JA, Ferrer MA, Jiménez A, Barceló AR, Sevilla F (2001) Antioxidant systems and O2.−/H2O2 production in the apoplast of pea leaves. Its relation with salt-induced necrotic lesions in minor veins. Plant Physiol 127(3):817–831

Horemans N, Foyer CH, Potters G, Asard H (2000) Ascorbate function and associated transport systems in plants. Plant Physiol Biochem 38(7-8):531–540

Hossain MA, Nakano Y, Asada K (1984) Monodehydroascorbate reductase in spinach chloroplasts and its participation in regeneration of ascorbate for scavenging hydrogen peroxide. Plant Cell Physiol 25(3):385–395

Houben M, Van de Poel B (2019) 1-Aminocyclopropane-1-carboxylic acid oxidase (ACO): the enzyme that makes the plant hormone ethylene. Front Plant Sci 10:695

Huseynova IM, Aliyeva DR, Aliyev JA (2014) Subcellular localization and responses of superoxide dismutase isoforms in local wheat varieties subjected to continuous soil drought. Plant Physiol Biochem 81:54–60

Iqbal N, Umar S, Khan NA, Corpas FJ (2021) Nitric oxide and hydrogen sulfide coordinately reduce glucose sensitivity and decrease oxidative stress via ascorbate-glutathione cycle in heat-stressed wheat (*Triticum aestivum* L.) plants. Antioxidants 10(1):108

Jalil SU, Ansari MI (2008) Plant microbiome and its functional mechanism in response to environmental stress. Int J Green Pharm 12:S81. https://doi.org/10.22377/ijgp.v12i01.1603

Jalil SU, Ansari MI (2020a) Stress implications and crop productivity. In: Plant ecophysiology and adaptation under climate change: mechanisms and perspectives I. Springer, Singapore, pp 73–86

Jalil SU, Ansari MI (2020b) Physiological role of Gamma-aminobutyric acid in salt stress tolerance. In: Salt and drought stress tolerance in plants. Springer, Cham, pp 337–350

Jamdhade AR, Sunkar R, Hivrale VK (2017) Zymographic method for distinguishing different classes of superoxide dismutases in plants. In: Plant stress tolerance. Humana Press, New York, NY, pp 221–227

Jouili H, Bouazizi H, El Ferjani E (2011) Plant peroxidases: biomarkers of metallic stress. Acta Physiol Plant 33(6):2075–2082

Karuppanapandian T, Moon JC, Kim C, Manoharan K, Kim W (2011) Reactive oxygen species in plants: their generation, signal transduction, and scavenging mechanisms. Aust J Crop Sci 5(6): 709–725

Kim YS, Kim IS, Shin SY, Park TH, Park HM, Kim YH, Lee GS, Kang HG, Lee SH, Yoon HS (2014) Overexpression of dehydroascorbate reductase confers enhanced tolerance to salt stress in rice plants (*Oryza sativa* L. japonica). J Agron Crop Sci 200(6):444–456

Kokotkiewicz A, Bucinski A, Luczkiewicz M (2014) Light and temperature conditions affect bioflavonoid accumulation in callus cultures of *Cyclopia subternata* Vogel (honeybush). Plant Cell Tissue Organ Cult 118(3):589–593

Kováčik J, Klejdus B, Bačkor M (2009) Phenolic metabolism of *Matricaria chamomilla* plants exposed to nickel. J Plant Physiol 166(13):1460–1464

Kukavica B, Vučinić Ž, Vuletić M (2005) Superoxide dismutase, peroxidase, and germin-like protein activity in plasma membranes and apoplast of maize roots. Protoplasma 226(3):191–197

Kukavica B, Mojović M, Vuccinić Ž, Maksimović V, Takahama U, Jovanović SV (2009) Generation of hydroxyl radical in isolated pea root cell wall, and the role of cell wall-bound peroxidase, Mn-SOD and phenolics in their production. Plant Cell Physiol 50(2):304–317

Kukreja S, Nandwal AS, Kumar N, Sharma SK, Unvi V, Sharma PK (2005) Plant water status, H2O2 scavenging enzymes, ethylene evolution and membrane integrity of *Cicer arietinum* roots as affected by salinity. Biol Plant 49(2):305–308

Kumar S, Pandey AK (2013) Chemistry and biological activities of flavonoids: an overview. Sci World J 2013:162750

Kumar P, Kumar Tewari R, Nand Sharma P (2007) Excess nickel–induced changes in antioxidative processes in maize leaves. J Plant Nutr Soil Sci 170(6):796–802

Kumar D, Yusuf MA, Singh P, Sardar M, Sarin NB (2013) Modulation of antioxidant machinery in α-tocopherol-enriched transgenic *Brassica juncea* plants tolerant to abiotic stress conditions. Protoplasma 250(5):1079–1089

Kumutha D, Ezhilmathi K, Sairam RK, Srivastava GC, Deshmukh PS, Meena RC (2009) Waterlogging induced oxidative stress and antioxidant activity in pigeonpea genotypes. Biol Plant 53(1):75–84

Latowski D, Surówka E, Strzałka K (2010) Regulatory role of components of ascorbate–glutathione pathway in plant stress tolerance. In: Ascorbate-glutathione pathway and stress tolerance in plants. Springer, Dordrecht, pp 1–53

Lee ES, Kang CH, Park JH, Lee SY (2018) Physiological significance of plant peroxiredoxins and the structure-related and multifunctional biochemistry of peroxiredoxin 1. Antioxid Redox Signal 28(7):625–639

Leung DW (2018) Studies of catalase in plants under abiotic stress. In: Antioxidants and antioxidant enzymes in higher plants. Springer, Cham, pp 27–39

Li F, Wu QY, Sun YL, Wang LY, Yang XH, Meng QW (2010) Overexpression of chloroplastic monodehydroascorbate reductase enhanced tolerance to temperature and methyl viologen-mediated oxidative stresses. Physiol Plant 139(4):421–434

Li C, Li J, Du X, Zhang J, Zou Y, Liu Y, Li Y, Lin H, Li H, Liu D, Lu H (2022) Chloroplast thylakoidal ascorbate peroxidase, PtotAPX, has enhanced resistance to oxidative stress in *Populus tomentosa*. Int J Mol Sci 23(6):3340

Liebthal M, Maynard D, Dietz KJ (2018) Peroxiredoxins and redox signaling in plants. Antioxid Redox Signal 28(7):609–624

Liu W, Yu K, He T, Li F, Zhang D, Liu J (2013) The low temperature induced physiological responses of *Avena nuda* L., a cold-tolerant plant species. Sci World J 2013:658793

Liu W, Huang L, Liang X, Liu L, Sun C, Lin X (2021) Heat shock induces cross adaptation to aluminum stress through enhancing ascorbate-glutathione cycle in wheat seedlings. Chemosphere 278:130397

Lüthje S, Martinez-Cortes T (2018) Membrane-bound class III peroxidases: unexpected enzymes with exciting functions. Int J Mol Sci 19(10):2876

Maoka T (2020) Carotenoids as natural functional pigments. J Nat Med 74(1):1–16

Martí MC, Florez-Sarasa I, Camejo D, Ribas-Carbó M, Lázaro JJ, Sevilla F, Jiménez A (2011) Response of mitochondrial thioredoxin PsTrx o 1, antioxidant enzymes, and respiration to salinity in pea (*Pisum sativum* L.) leaves. J Exp Bot 62(11):3863–3874

Marty L, Siala W, Schwarzländer M, Fricker MD, Wirtz M, Sweetlove LJ, Meyer Y, Meyer AJ, Reiccheld JP, Hell R (2009) The NADPH-dependent thioredoxin system constitutes a functional backup for cytosolic glutathione reductase in *Arabidopsis*. Proc Natl Acad Sci 106(22): 9109–9114

Marty L, Bausewein D, Müller C, Bangash SAK, Moseler A, Schwarzländer M, Müller-Schüssele SJ, Zechmann B, Riondet C, Balk J, Wirtz M, Hell R, Reichheld JP, Meyer AJ (2019) Arabidopsis glutathione reductase 2 is indispensable in plastids, while mitochondrial glutathione is safeguarded by additional reduction and transport systems. New Phytol 224(4): 1569–1584

Mata-Pérez C, Spoel SH (2019) Thioredoxin-mediated redox signalling in plant immunity. Plant Sci 279:27–33

Mehla N, Sindhi V, Josula D, Bisht P, Wani SH (2017) An introduction to antioxidants and their roles in plant stress tolerance. In: Reactive oxygen species and antioxidant Systems in Plants: role and regulation under abiotic stress. Springer, Singapore, pp 1–23

Meisrimler CN, Buck F, Lüthje S (2014) Alterations in soluble Class III peroxidases of maize shoots by flooding stress. Proteomes 2(3):303–322

Mhamdi A, Queval G, Chaouch S, Vanderauwera S, Van Breusegem F, Noctor G (2010) Catalase function in plants: a focus on *Arabidopsis* mutants as stress-mimic models. J Exp Bot 61(15): 4197–4220

Mishra P, Bhoomika K, Dubey RS (2013) Differential responses of antioxidative defense system to prolonged salinity stress in salt-tolerant and salt-sensitive Indica rice (*Oryza sativa* L.) seedlings. Protoplasma 250(1):3–19

Munné-Bosch S (2005) The role of α-tocopherol in plant stress tolerance. J Plant Physiol 162(7): 743–748

Munné-Bosch S, Falk J (2004) New insights into the function of tocopherols in plants. Planta 218(3):323–326

Najami N, Janda T, Barriah W, Kayam G, Tal M, Guy M, Volokita M (2008) Ascorbate peroxidase gene family in tomato: its identification and characterization. Mol Gen Genomics 279(2): 171–182

Navrot N, Collin V, Gualberto J, Gelhaye E, Hirasawa M, Rey P, Knaff DB, Issakidis E, Jacquot JP, Rouhier N (2006) Plant glutathione peroxidases are functional peroxiredoxins distributed in several subcellular compartments and regulated during biotic and abiotic stresses. Plant Physiol 142(4):1364–1379

Noctor G, Arisi ACM, Jouanin L, Foyer CH (1998) Manipulation of glutathione and amino acid biosynthesis in the chloroplast. Plant Physiol 118(2):471–482

Noctor G, Mhamdi A, Chaouch S, Han YI, Neukermans J, Marquez-Garcia B, Queval G, Foyer CH (2012) Glutathione in plants: an integrated overview. Plant Cell Environ 35(2):454–484

Noshi M, Yamada H, Hatanaka R, Tanabe N, Tamoi M, Shigeoka S (2017) Arabidopsis dehydroascorbate reductase 1 and 2 modulate redox states of ascorbate-glutathione cycle in the cytosol in response to photooxidative stress. Biosci Biotechnol Biochem 81(3):523–533

Nourredine Y, Naima A, Dalila H, Habib S, Karim S (2015) Changes of peroxidase activities under cold stress in annuals populations of Medicago. Mol Plant Breed 6:1–9

Olszowy M (2019) What is responsible for antioxidant properties of polyphenolic compounds from plants? Plant Physiol Biochem 144:135–143

Ozyigit II, Filiz E, Vatansever R, Kurtoglu KY, Koc I, Öztürk MX, Anjum NA (2016) Identification and comparative analysis of H2O2-scavenging enzymes (ascorbate peroxidase and glutathione peroxidase) in selected plants employing bioinformatics approaches. Front Plant Sci 7:301

Palma JM, López-Huertas E, Corpas FJ, Sandalio LM, Gómez M, Del Río LA (1998) Peroxisomal manganese superoxide dismutase: purification and properties of the isozyme from pea leaves. Physiol Plant 104(4):720–726

Palma JM, Mateos RM, López-Jaramillo J, Rodríguez-Ruiz M, González-Gordo S, Lechuga-Sancho AM, Corpas FJ (2020) Plant catalases as NO and H2S targets. Redox Biol 34:101525

Panche AN, Diwan AD, Chandra SR (2016) Flavonoids: an overview. J Nutr Sci 5:e47

Pandey S, Fartyal D, Agarwal A, Shukla T, James D, Kaul T, Negi YK, Arora S, Reddy MK (2017) Abiotic stress tolerance in plants: myriad roles of ascorbate peroxidase. Front Plant Sci 8:581

Pandhair V, Sekhon BS (2006) Reactive oxygen species and antioxidants in plants: an overview. J Plant Biochem Biotechnol 15(2):71–78

Pang CH, Wang BS (2010) Role of ascorbate peroxidase and glutathione reductase in ascorbate–glutathione cycle and stress tolerance in plants. In: Ascorbate-glutathione pathway and stress tolerance in plants. Springer, Dordrecht, pp 91–113

Paolacci AR, D'Ovidio R, Marabottini R, Nali C, Lorenzini G, Abenavoli MR, Badiani M (2001) Research note: ozone induces a differential accumulation of phenyalanine ammonia-lyase, chalcone synthase and chalcone isomerase RNA transcripts in sensitive and resistant bean cultivars. Funct Plant Biol 28(5):425–428

Park AK, Kim IS, Do H, Jeon BW, Lee CW, Roh SJ, Shin SC, Park H, Kim YS, Kim YH, Yoon HS, Lee JH, Kim HW (2016) Structure and catalytic mechanism of monodehydroascorbate reductase, MDHAR, from *Oryza sativa* L japonica. Sci Rep 6(1):1–10

Parsons HT, Fry SC (2012) Oxidation of dehydroascorbic acid and 2, 3-diketogulonate under plant apoplastic conditions. Phytochemistry 75:41–49

Passaia G, Margis-Pinheiro M (2015) Glutathione peroxidases as redox sensor proteins in plant cells. Plant Sci 234:22–26

Passaia G, Fonini LS, Caverzan A, Jardim-Messeder D, Christoff AP, Gaeta ML, Mariath JEDA, Margis R, Margis-Pinheiro M (2013) The mitochondrial glutathione peroxidase GPX3 is essential for H2O2 homeostasis and root and shoot development in rice. Plant Sci 208:93–101

Rajput VD, Singh RK, Verma KK, Sharma L, Quiroz-Figueroa FR, Meena M, Gour VS, Minkina T, Sushkova S, Mandzhieva S (2021) Recent developments in enzymatic antioxidant defence mechanism in plants with special reference to abiotic stress. Biology 10(4):267

Ratajczak E, Dietz KJ, Kalemba EM (2019) The occurrence of peroxiredoxins and changes in redox state in *Acer platanoides* and Acer pseudoplatanus during seed development. J Plant Growth Regul 38(1):298–314

Raza A, Su W, Gao A, Mehmood SS, Hussain MA, Nie W, Lv Y, Zou X, Zhang X (2021) Catalase (CAT) gene family in rapeseed (*Brassica napus* L.): genome-wide analysis, identification, and expression pattern in response to multiple hormones and abiotic stress conditions. Int J Mol Sci 22(8):4281

Regalado C, Arvizu OP, Garcia-Almendarez BE, Whitaker JR (1999) Purification and properties of two acid peroxidases from brussels sprouts (*Brassica oleraceae* L.). J Food Biochem 23(4): 435–450

Río LAD, Corpas FJ, López-Huertas E, Palma JM (2018) Plant superoxide dismutases: function under abiotic stress conditions. In: Antioxidants and antioxidant enzymes in higher plants. Springer, Cham, pp 1–26

Rodríguez De Luna SL, Ramírez-Garza RE, Serna Saldívar SO (2020) Environmentally friendly methods for flavonoid extraction from plant material: impact of their operating conditions on yield and antioxidant properties. Sci World J 2020:6792069

Rouhier N, Lemaire SD, Jacquot JP (2008) The role of glutathione in photosynthetic organisms: emerging functions for glutaredoxins and glutathionylation. Annu Rev Plant Biol 59:143–166

Roupakias DG, McMillin DE, Scandalios JG (1980) Chromosomal location of the catalase structural genes in *Zea mays* using BA translocations. Theor Appl Genet 58(5):211–218

Roychoudhury A, Basu S, Sengupta DN (2012) Antioxidants and stress-related metabolites in the seedlings of two indica rice varieties exposed to cadmium chloride toxicity. Acta Physiol Plant 34(3):835–847

Sabeh F, Wright T, Norton SJ (1993) Purification and characterization of a glutathione peroxidase from the *Aloe vera* plant. Enzyme Protein 47:92–98

Sachdev S, Ansari SA, Ansari MI, Fujita M, Hasanuzzaman M (2021) Abiotic stress and reactive oxygen species: generation, signaling, and defense mechanisms. Antioxidants 10(2):277

Sachdev S, Jaiswal P, Ansari MI (2022) Coordinated of reactive oxygen functions species metabolism and defense systems in abiotic stress tolerance. In: Advancements in developing abiotic stress-resilient plants: basic mechanisms to trait improvements. CRC Press, Boca Raton, p 23

Sadeghi F, Samsampour D, Seyahooei MA, Bagheri A, Soltani J (2020) Fungal endophytes alleviate drought-induced oxidative stress in mandarin (*Citrus reticulata* L.): toward regulating the ascorbate–glutathione cycle. Sci Hortic 261:108991

Saga G, Giorgetti A, Fufezan C, Giacometti GM, Bassi R, Morosinotto T (2010) Mutation analysis of violaxanthin de-epoxidase identifies substrate-binding sites and residues involved in catalysis. J Biol Chem 285(31):23763–23770

Saibi W, Brini F (2018) Superoxide dismutase (SOD) and abiotic stress tolerance in plants: an overview. In: Magliozzi S (ed) Superoxide dismutase: structure, synthesis and applications. Nova Science Publishers, New York, pp 101–142

Šamec D, Karalija E, Šola I, Vujčić Bok V, Salopek-Sondi B (2021) The role of polyphenols in abiotic stress response: the influence of molecular structure. Plants 10(1):118

Saxena SC, Salvi P, Kamble NU, Joshi PK, Majee M, Arora S (2020) Ectopic overexpression of cytosolic ascorbate peroxidase gene (Apx1) improves salinity stress tolerance in *Brassica juncea* by strengthening antioxidative defense mechanism. Acta Physiol Plant 42(4):1–14

Semida WM, Abd El-Mageed TA, Howladar SM, Rady MM (2016) Foliar-applied alpha-tocopherol enhances salt-tolerance in onion plants by improving antioxidant defence system. Aust J Crop Sci 10(7):1030–1039

Seth CS, Remans T, Keunen E, Jozefczak M, Gielen H, Opdenakker K, Vangronsveld WJ, Cuypers A (2012) Phytoextraction of toxic metals: a central role for glutathione. Plant Cell Environ 35(2):334–346

Sevilla F, Camejo D, Ortiz-Espín A, Calderón A, Lázaro JJ, Jiménez A (2015) The thioredoxin/peroxiredoxin/sulfiredoxin system: current overview on its redox function in plants and regulation by reactive oxygen and nitrogen species. J Exp Bot 66(10):2945–2955

Shainberg O, Rubin B, Rabinowitch HD, Libal Y, Tel-Or E (2000) Acclimation of beans to oxidative stress by treatment with sublethal iron levels. J Plant Physiol 157(1):93–99

Sharma I, Ahmad P (2014) Catalase: a versatile antioxidant in plants. In: Oxidative damage to plants. Academic Press, Amsterdam, pp 131–148

Sharma P, Dubey RS (2005) Drought induces oxidative stress and enhances the activities of antioxidant enzymes in growing rice seedlings. Plant Growth Regul 46(3):209–221

Sharma A, Shahzad B, Rehman A, Bhardwaj R, Landi M, Zheng B (2019a) Response of phenylpropanoid pathway and the role of polyphenols in plants under abiotic stress. Molecules 24(13):2452

Sharma A, Shahzad B, Kumar V, Kohli SK, Sidhu GPS, Bali AS, Handa N, Kapoor D, Bhardwaj R, Zheng B (2019b) Phytohormones regulate accumulation of osmolytes under abiotic stress. Biomolecules 9(7):285

Shigeoka S, Ishikawa T, Tamoi M, Miyagawa Y, Takeda T, Yabuta Y, Yoshimura K (2002) Regulation and function of ascorbate peroxidase isoenzymes. J Exp Bot 53(372):1305–1319

Silva EN, Vieira SA, Ribeiro RV, Ponte LF, Ferreira-Silva SL, Silveira JA (2013) Contrasting physiological responses of *Jatropha curcas* plants to single and combined stresses of salinity and heat. J Plant Growth Regul 32(1):159–169

Skadsen RW, Schulze-Lefert P, Herbst JM (1995) Molecular cloning, characterization and expression analysis of two catalase isozyme genes in barley. Plant Mol Biol 29(5):1005–1014

Smirnoff N, Wheeler GL (2000) Ascorbic acid in plants: biosyn (Doctoral dissertation, thesis and function). Crit Rev Biochem Mol Biol 35:291–314

Spicher L, Almeida J, Gutbrod K, Pipitone R, Dörmann P, Glauser G, Rossi M, Kessler F (2017) Essential role for phytol kinase and tocopherol in tolerance to combined light and temperature stress in tomato. J Exp Bot 68(21-22):5845–5856

Srivastava S, Dubey RS (2011) Manganese-excess induces oxidative stress, lowers the pool of antioxidants and elevates activities of key antioxidative enzymes in rice seedlings. Plant Growth Regul 64(1):1–16

Stephenie S, Chang YP, Gnanasekaran A, Esa NM, Gnanaraj C (2020) An insight on superoxide dismutase (SOD) from plants for mammalian health enhancement. J Funct Foods 68:103917

Strzałka K, Kostecka-Gugała A, Latowski D (2003) Carotenoids and environmental stress in plants: significance of carotenoid-mediated modulation of membrane physical properties. Russ J Plant Physiol 50(2):168–173

Su Y, Guo J, Ling H, Chen S, Wang S, Xu L, Allan AC, Que Y (2014) Isolation of a novel peroxisomal catalase gene from sugarcane, which is responsive to biotic and abiotic stresses. PLoS One 9(1):e84426

Sudan J, Negi B, Arora S (2015) Oxidative stress induced expression of monodehydroascorbate reductase gene in *Eleusine coracana*. Physiol Mol Biol Plants 21(4):551–558

Sultana S, Khew CY, Morshed MM, Namasivayam P, Napis S, Ho CL (2012) Overexpression of monodehydroascorbate reductase from a mangrove plant (AeMDHAR) confers salt tolerance on rice. J Plant Physiol 169(3):311–318

Sun H, Li L, Wang X, Wu S, Wang X (2011) Ascorbate–glutathione cycle of mitochondria in osmoprimed soybean cotyledons in response to imbibitional chilling injury. J Plant Physiol 168(3):226–232

Sytykiewicz H, Kozak A, Leszczyński B, Sempruch C, Łukasik I, Sprawka I, Kmiec K, Kurowska M, Kopczynska A, Czerniewicz P (2018) Transcriptional profiling of catalase genes in juglone-treated seeds of maize (*Zea mays* L.) and wheat (*Triticum aestivum* L.). Acta Biol Hung 69(4):449–463

Szőllősi R, Varga IS, Erdei L, Mihalik E (2009) Cadmium-induced oxidative stress and antioxidative mechanisms in germinating Indian mustard (*Brassica juncea* L.) seeds. Ecotoxicol Environ Saf 72(5):1337–1342

Szymańska R, Kruk J (2008) Tocopherol content and isomers' composition in selected plant species. Plant Physiol Biochem 46(1):29–33

Takahashi H, Kopriva S, Giordano M, Saito K, Hell R (2011) Sulfur assimilation in photosynthetic organisms: molecular functions and regulations of transporters and assimilatory enzymes. Annu Rev Plant Biol 62:157–184

Tang K, Zhan JC, Yang HR, Huang WD (2010) Changes of resveratrol and antioxidant enzymes during UV-induced plant defense response in peanut seedlings. J Plant Physiol 167(2):95–102

Tanyolac D, Ekmekçi Y, Ünalan Ş (2007) Changes in photochemical and antioxidant enzyme activities in maize (*Zea mays* L.) leaves exposed to excess copper. Chemosphere 67(1):89–98

Teixeira FK, Menezes-Benavente L, Margis R, Margis-Pinheiro M (2004) Analysis of the molecular evolutionary history of the ascorbate peroxidase gene family: inferences from the rice genome. J Mol Evol 59(6):761–770

Tenhaken R (2015) Cell wall remodeling under abiotic stress. Front Plant Sci 5:771

Turan Ö, Ekmekçi Y (2011) Activities of photosystem II and antioxidant enzymes in chickpea (*Cicer arietinum* L.) cultivars exposed to chilling temperatures. Acta Physiol Plant 33(1):67–78

Veljović Jovanović S, Kukavica B, Vidović M, Morina F, Menckhoff L (2018) Class III peroxidases: functions, localization and redox regulation of isoenzymes. In: Antioxidants and antioxidant enzymes in higher plants. Springer, Cham, pp 269–300

Wang C, Zhang SH, Wang PF, Hou J, Zhang WJ, Li W, Lin ZP (2009) The effect of excess Zn on mineral nutrition and antioxidative response in rapeseed seedlings. Chemosphere 75(11): 1468–1476

Wang QJ, Sun H, Dong QL, Sun TY, Jin ZX, Hao YJ, Yao YX (2016) The enhancement of tolerance to salt and cold stresses by modifying the redox state and salicylic acid content via the cytosolic malate dehydrogenase gene in transgenic apple plants. Plant Biotechnol J 14(10): 1986–1997

Wang W, Cheng Y, Chen D, Liu D, Hu M, Dong J, Zhang X, Song L, Shen F (2019) The catalase gene family in cotton: genome-wide characterization and bioinformatics analysis. Cells 8(2):86

Wu TM, Lin WR, Kao CH, Hong CY (2015) Gene knockout of glutathione reductase 3 results in increased sensitivity to salt stress in rice. Plant Mol Biol 87(6):555–564

Wutipraditkul N, Boonkomrat S, Buaboocha T (2011) Cloning and characterization of catalases from rice, *Oryza sativa* L. Biosci Biotechnol Biochem 75(10):1900–1906

Xiao M, Li Z, Zhu L, Wang J, Zhang B, Zheng F, Zhao B, Zhang H, Wang Y, Zhang Z (2021) The multiple roles of ascorbate in the abiotic stress response of plants: antioxidant, cofactor, and regulator. Front Plant Sci 12:592

Yadav P, Yadav T, Kumar S, Rani B, Jain V, Malhotra SP (2014) Partial purification and characterization of ascorbate peroxidase from ripening ber (*Ziziphus mauritiana* L) fruits. Afr J Biotechnol 13(32). https://doi.org/10.5897/AJB2013.12193

Yamada Y, Fujiwara T, Sato T, Igarashi N, Tanaka N (2002) The 2.0 Å crystal structure of catalase-peroxidase from *Haloarcula marismortui*. Nat Struct Biol 9(9):691–695

Yamasaki H, Takahashi S, Heshiki R (1999) The tropical fig *Ficus microcarpa* L. f. cv. golden leaves lacks heat-stable dehydroascorbate reductase activity. Plant Cell Physiol 40(6):640–646

Yang DY, Ma NN, Zhuang KY, Zhu SB, Liu ZM, Yang XH (2017) Overexpression of tomato SlGGP-LIKE gene improves tobacco tolerance to methyl viologen-mediated oxidative stress. J Plant Physiol 209:31–41

Yin L, Wang S, Eltayeb AE, Uddin M, Yamamoto Y, Tsuji W, Takeuchi Y, Tanaka K (2010) Overexpression of dehydroascorbate reductase, but not monodehydroascorbate reductase, confers tolerance to aluminum stress in transgenic tobacco. Planta 231(3):609–621

Yoshida S, Tamaoki M, Shikano T, Nakajima N, Ogawa D, Ioki M, Aono M, Kubo A, Kamada H, Inoue Y, Saji H (2006) Cytosolic dehydroascorbate reductase is important for ozone tolerance in *Arabidopsis thaliana*. Plant Cell Physiol 47(2):304–308

Yousuf PY, Hakeem KUR, Chandna R, Ahmad P (2012) Role of glutathione reductase in plant abiotic stress. In: Abiotic stress responses in plants. Springer, New York, NY, pp 149–158

Zafra A, Castro AJ, Alché JDD (2018) Identification of novel superoxide dismutase isoenzymes in the olive (*Olea europaea* L.) pollen. BMC Plant Biol 18(1):1–16

Zamocky M, Furtmüller PG, Obinger C (2008) Evolution of catalases from bacteria to humans. Antioxid Redox Signal 10(9):1527–1548

Zandalinas SI, Balfagón D, Arbona V, Gómez-Cadenas A (2017) Modulation of antioxidant defense system is associated with combined drought and heat stress tolerance in citrus. Front Plant Sci 8:953

Zhang CJ, Zhao BC, Ge WN, Zhang YF, Song Y, Sun DY, Guo Y (2011) An apoplastic h-type thioredoxin is involved in the stress response through regulation of the apoplastic reactive oxygen species in rice. Plant Physiol 157(4):1884–1899

Zhang YJ, Wang W, Yang HL, Li Y, Kang XY, Wang XR, Yang ZL (2015) Molecular properties and functional divergence of the dehydroascorbate reductase gene family in lower and higher plants. PLoS One 10(12):e0145038

Zhang M, Smith JAC, Harberd NP, Jiang C (2016) The regulatory roles of ethylene and reactive oxygen species (ROS) in plant salt stress responses. Plant Mol Biol 91(6):651–659

Zlatev ZS, Lidon FC, Ramalho JC, Yordanov IT (2006) Comparison of resistance to drought of three bean cultivars. Biol Plant 50(3):389–394

Role of Osmolytes in Alleviation of Oxidative Stress

10

10.1 Introduction

The plants exposed to adverse environmental cues, exorbitantly generates reactive oxygen species (ROS), which induces detrimental alteration to cellular biomolecules, eventually leading to oxidative stress. Besides antioxidants, to eliminate the harmful free radicals and prevent cells from stress-induced injuries, plants recruit another detoxification channel comprising osmolytes. The de-novo synthesis of osmolytes, the low molecular weight compatible solutes, regulates the cell osmotic potential by managing the water content without altering the normal cellular mechanism (Hossain et al. 2014a; Sharma et al. 2019). The osmolytes in plants generally include amino acids (proline, gamma-aminobutyric acid (GABA), sugars (sucrose, trehalose, fructans), quaternary amines (glycine betaine, polyamines), and sugar alcohols or polyols (mannitol, sorbitol) (Montesinos-Pereira et al. 2014; Puniran-Hartley et al. 2014; Annunziata et al. 2019; Hasanuzzaman et al. 2019; Sharma et al. 2019; Ejaz et al. 2020; Savchenko and Tikhonov 2021). These osmolytes, in addition to osmotic potential regulation, prevent oxidative damage by inhibiting the generation of ROS and/or scavenging free radicals, mediating stress tolerance (Smirnoff and Cumbes 1989; Montesinos-Pereira et al. 2014; Hasanuzzaman et al. 2019; Rajasheker et al. 2019; Sharma et al. 2019; Ejaz et al. 2020). Moreover, osmolytes maintain stability of protein molecules under stress, protect lipid membrane affected by dehydration or desiccation, suppress the process of photorespiration, bind toxic metals, regulate potassium homeostasis within cytoplasm, act as antifreeze agents and activate stress-responsive transcription factors (Hincha et al. 2003; Yancey 2005; Sharma and Dietz 2006; Cuin and Shabala 2007; Puniran-Hartley et al. 2014; Rajasheker et al. 2019). Several studies have evidently demonstrated the accumulation of osmolytes such as sugar and proline in a variety of plants facilitating tolerance against stress (Parvanova et al. 2004; Gupta et al. 2012; Sharma et al. 2019; Datir et al. 2020).

S. Sachdev et al., *Reactive Oxygen Species in Plants*,
https://doi.org/10.1007/978-981-19-9884-3_10

Phytohormones, the plant growth regulators have also been found efficient in imparting tolerance toward stress by regulating the synthesis and accumulation of osmolytes and/or promoting the activity of antioxidants in plants under stress (Rajasheker et al. 2019; Sharma et al. 2019; Shemi et al. 2021; Zhang et al. 2021). In particular, treatment of *Sorghum bicolor* with jasmonic acid (JA) in combination with humic acid increases the activity of ascorbate peroxidase (APX), imparting tolerance to salinity stress (Ali et al. 2020). Similarly, exogenous treatment of JA and abscisic acid (ABA) improves the activity of antioxidants enzymes in pearl millet under drought stress and promoted seedling growth profile (Awan et al. 2021). The micronutrients such as selenium (Se), silicon (Si), copper (Cu), and others are essential constituents of enzymes or required for plant growth. The treatment of plants with a low concentration of micronutrients reportedly upregulates the activity of antioxidants, imparting tolerance toward stress (Alyemeni et al. 2018; Hasanuzzaman et al. 2020; Shemi et al. 2021; Tavanti et al. 2021). Supplementation of Si improves antioxidant activities, abridges ROS and malondialdehyde (MDA) accumulation, and prevents electrolyte leakage in *Vigna radiata* exposed to salinity stress (Ahmad et al. 2019). The defense mechanism restricting stress-induced oxidative damages engages osmolytes, phytohormones, and other compounds like micronutrients which mediate stress tolerance and improve plant growth. Enhanced understanding of intricate mechanisms displayed by such compounds and their expression under variable situations is imperative to explore new opportunities to promote plant survivability in changing environmental scenarios.

10.2 Amino Acids Attenuating Oxidative Stress

10.2.1 Proline

Proline, an amino acid, is synthesized in plants and serves multiple functions under both unstressed and stressed conditions. It acts as an osmolyte, non-enzymatic antioxidants, organic nitrogen reserve, and cellular redox homeostasis regulator (Kishor et al. 2005; Hayat et al. 2012; Gupta and Huang 2014; Hossain et al. 2014a, b). Proline reduces osmotic stress by abridging uptake of toxic ions and balancing cell turgor pressure (Hayat et al. 2012; Liang et al. 2013; Datir et al. 2020; Ejaz et al. 2020). Plants accumulate proline abundantly under a wide range of stressful events like drought, salinity, heavy metals, UV radiation, extreme temperatures, excess light, and pathogen infection, and mediates stress tolerance (Chen et al. 2001; Choudhary et al. 2005; Munns 2005; Hayat et al. 2012; Liang et al. 2013; Hossain et al. 2014a, b; Jahantigh et al. 2016; Liu et al. 2020). The accumulation of proline in large number of plant varieties elevates up to 100 times the normal level on exposure to stressful situations, constituting nearly 80% of the total amino acid pool (Matysik et al. 2002). The biosynthesis of proline in plants occurs via glutamate pathway and/or ornithine pathway (Hayat et al. 2012; Hossain et al. 2014a, b; Singh et al. 2015). An increased production of glutamate under stress accelerates glutamate pathway for proline synthesis in plants (Hossain et al.

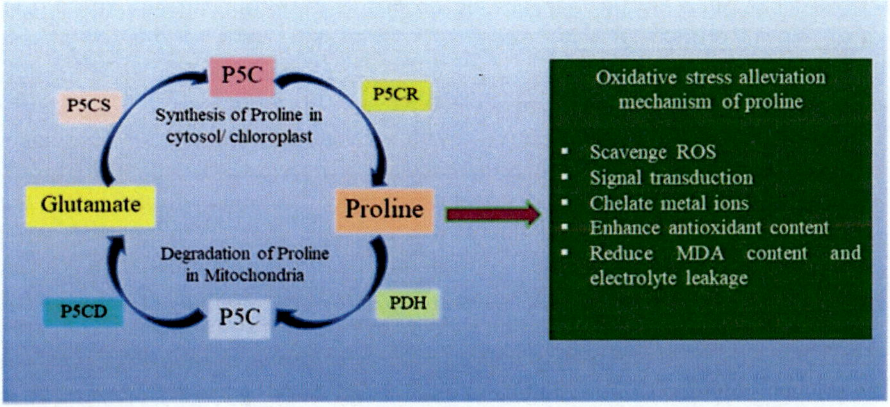

Fig. 10.1 Proline synthesized in plants alleviates oxidative stress via multiple mechanisms. The synthesis of proline takes place in cytosol and chloroplast from glutamate which is converted into P5C (△1-pyrroline-5-carboxylate) by enzyme P5CS (△1-pyrroline-5-carboxylate synthase) and then to proline by enzyme P5CR (△1-pyrroline-5-carboxylate reductase). The degradation of proline occurs in mitochondria where it gets converted into P5C by PDH (proline dehydrogenase) and then to glutamate on action of P5CD (△1-pyrroline-5-carboxylate dehydrogenase)

2014a, b). This pathway involves conversion of glutamate into proline via production of an intermediate compound △1-pyrroline-5-carboxylate (P5C) in the presence of enzymes △1-pyrroline-5-carboxylate synthase (P5CS) and △1-pyrroline-5-carboxylate reductase (P5CR) (Hayat et al. 2012; Liang et al. 2013; Hossain et al. 2014a, b) (Fig. 10.1). Alternatively, the enzyme (ornithine-δ-aminotransferase) catalyzes proline synthesis from ornithine (Hayat et al. 2012). In plant cells, cytosol is the major site for synthesis of proline, however, its production is also introduced in chloroplast under stress conditions (Hossain et al. 2014a, b). The degradation of proline in plant cell takes place in mitochondria, involving reactions sequentially catalyzed by proline dehydrogenase (PDH) or proline oxidase and P5C dehydrogenase (P5CD) (Hossain et al. 2014a, b; Rejeb et al. 2014). The accumulation of proline in plants under stressful circumstances depends cumulatively on increased proline production rates and inhibition of degradation pathway (Hossain et al. 2014a, b; Rejeb et al. 2014; Zouari et al. 2019).

The enhanced accumulation of proline alleviates oxidative stress by acting as a signaling molecule, scavenging excess amount of ROS, stimulating activity of enzymatic antioxidants, and chelating metal ions (Hayat et al. 2012; Rejeb et al. 2014; Ejaz et al. 2020) (Fig. 10.1). Proline facilitates stabilization of cellular membranes, DNA, and protein molecules, thus preventing oxidative damage (Rejeb et al. 2014). Under salt stress, accumulation of proline in *Pancratium maritimum* promotes antioxidant activities of catalase (CAT) and peroxidases (POX) (Khedr et al. 2003). In the thylakoids of *Brassica juncea* cotyledons, proline abridges the generation of singlet oxygen ($^1O_2$) under excess illumination (Saradhi and Mohanty 1997). Proline also regulates the transcription of salt stress-responsive proteins, improving stress tolerance ability (Khedr et al. 2003). Besides the accumulation of proline in plant cells, its exogenous application effectively mediates stress tolerance (Roy et al. 1993; Wang et al. 2009; Hayat et al. 2012; Merwad et al. 2018). Pre-treatment with proline of the mung bean exposed to salt stress displays the increased content of ascorbate (AsA) and glutathione (GSH), elevated GSH/glutathione disulfide (GSSG) ratio, enhanced activity of enzymatic antioxidants such as APX, glutathione reductase (GR) and CAT and reduced H_2O_2 and MDA levels (Hossain et al. 2011). Similarly, *Solanum nigrum* callus and regenerated shoots receiving proline as pre-treatment under cadmium (Cd) stress detoxifies ROS by enhancing activity of superoxide dismutase (SOD) and CAT, and intracellular content of total GSH (Xu et al. 2009). The exogenous proline application at low concentration facilitates mitigation of oxidative stress (Wang et al. 2009; Hayat et al. 2012). At high concentration, whether applied exogenously or excessive intracellular accumulation, it induces toxic effects (Rodriguez and Heyser 1988; Hayat et al. 2012). Exogenous treatment of rice seedlings with low concentration of proline (30 mM) alleviates negative effects of salinity; however, application of high concentration (40–50 mM) of proline produces toxic effects, by altering K^+/Na^+ ratio and stunted growth of the seedlings (Roy et al. 1993).

Proline, by and large, targets and scavenges the singlet oxygen ($^1O_2$), hydrogen peroxide (H_2O_2), and hydroxyl radicals ($^{\bullet}OH$) (Banu et al. 2010; Rejeb et al. 2014). A proline-proline cycle as a mechanism for scavenging of $^{\bullet}OH$ has been proposed by Signorelli et al. (2014). This cycle involves capturing of $^{\bullet}OH$ by proline molecule via two subsequent abstraction of H^+ from its structure by two $^{\bullet}OH$, resulting in production of P5C, which later recycled back to proline on the action of the P5CR/NADPH (reduced nicotinamide adenine dinucleotide phosphate) enzymatic system. However, controversy shrouds with regard to the role of proline in quenching, scavenging, or interaction with the $^1O_2$ (Signorelli et al. 2013, 2016). Recently, Rehman and his coworkers (Rehman et al. 2021) have, however, demonstrated $^1O_2$ quenching ability of proline in photosynthetically active isolated thylakoid membranes under in vitro conditions. The study concludes that the proline scavenges $^1O_2$ via a physical mechanism in which excited $^1O_2$ is converted back to the ground state molecular oxygen (O_2) through energy transfer. Further, proline also eliminates superoxide radicals ($O_2^{\bullet -}$) under in vitro condition via electron transfer reaction, without accumulating an intermediate product H_2O_2 (Rehman et al. 2021).

10.2.2 Gamma-Aminobutyric Acid

Gamma-aminobutyric acid (GABA) is a four-carbon biomolecule which ubiquitously occurs in most of the organisms (Ramos-Ruiz et al. 2019; Ejaz et al. 2020; Jalil and Ansari 2020; Ansari et al. 2021). It is a non-proteinogenic amino acid firstly reported from potato tuber, more than seven decades ago (Steward 1949; Li et al. 2021). In plants, GABA caters to multiple functions like plant growth and development, pH regulation, carbon/nitrogen metabolism, signaling, and management of biotic and abiotic stress (Ansari et al. 2005, 2014, 2021; Ramesh et al. 2017;

Ramos-Ruiz et al. 2019; Li et al. 2021; Sachdev et al. 2021). The biosynthesis of GABA in plants takes place via GABA shunt pathway, which involves decarboxylation of glutamate by enzyme glutamate decarboxylase (GAD) (Ramos-Ruiz et al. 2019; Ansari et al. 2014, 2021; Jalil and Ansari 2020). Additionally, the synthesis of GABA can also occur through polyamine metabolic pathway (Li et al. 2021). The production and accumulation of endogenous GABA in plants are relatively low, however, it occurs in a significantly high concentration under adverse conditions such as water-deficit, metal toxicity, hypoxia, γ-radiation, extreme temperature, low light, salinity, and biotic stress, resulting in alleviation of negative impacts (Ramos-Ruiz et al. 2019; Ejaz et al. 2020; Ansari et al. 2021; Li et al. 2021; Hasan et al. 2021). For instance, plants accumulate an increased amount of GABA in plants during cold stress (Mazzucotelli et al. 2006; Toubiana et al. 2020). Accumulation of GABA in higher concentration (Table 10.1) mediates tolerance toward stress by regulating stomatal opening during water deficiency, enhancing photosynthesis, inhibiting ROS production, and activating antioxidant enzymes (Jalil et al. 2019a, b; Li et al. 2021).

The accumulation of GABA in higher quantities remarkably performs a positive role in stress alleviation. The exogenous treatment of plants with GABA increases the content of endogenous GABA (Ramos-Ruiz et al. 2019). Treatment with GABA to stress-subjected plants promotes GABA shunt activity, photosynthetic performance, elevation in endogenous level of GABA and antioxidants and decreases the content of ROS and MDA, which help in maintaining membrane integrity (Ramos-Ruiz et al. 2019; Sita and Kumar 2020). Moreover, the elevated concentration of endogenous GABA confers osmo-protection against both osmotic and oxidative stress (Jalil et al. 2017; Ansari et al. 2021). The exogenous treatment of GABA in a variety of plants including banana peel, tomato seedlings, tea plants and anthurium cut-flowers confers tolerance against low-temperature-induced oxidative stress, invoking various mechanisms such as increased endogenous GABA content, fortification of antioxidant system, reduction in ROS content, reduced electrolyte leakage and MDA level, and/or increased accumulation of proline-induced osmoregulation (Malekzadeh et al. 2014; Wang et al. 2014; Aghdam et al. 2016; Zhu et al. 2019). GABA efficiently and significantly offsets production of $O_2^{\bullet -}$ and H_2O_2 in tobacco leaves under water stress (Liu et al. 2011) and promotes activity of enzymatic antioxidants, reducing level of MDA and $O_2^{\bullet -}$ coupled with protection against oxidative damage in the maize seedling exposed to salt stress (Wang et al. 2017). GABA also enhances the activities of enzymes participating in ascorbate-glutathione (AsA-GSH) cycle viz., APX, dehydroascorbate reductase (DHAR), and monodehydroascorbate reductase (MDHAR). GABA treatment to peaches experiencing chilling stress upregulates the activities of APX, DHAR, and MDHAR along with glutathione peroxidase (GPX), glutathione-*S*-transferase (GST), CAT and SOD (Yang et al. 2011). Similarly, the exogenous GABA application to sunflowers under drought and heat stress improves the osmolyte metabolism, expression of genes such as heat shock protein, and activities of antioxidants including SOD, POX, GR, APX, and MDHAR and counteracts the negative effects of both stresses (Abdel Razik et al. 2021).

Table 10.1 Role of GABA in alleviation of oxidative damage and increased tolerance to stress in various crops

Plant	Stress	Mechanism	References
Triticum aestivum	Salt stress	Application of GABA increases tolerance toward stress by reducing oxidative damage	Al-Quraan et al. (2013)
Banana peel	Chilling injury	Enhanced antioxidant capacity triggers protection against chilling stress	Wang et al. (2014)
Rice	Heat stress	Improved turgor, upregulation of antioxidants and osmoprotectants confer protection toward stress	Nayyar et al. (2014)
Tomato seedling	Chilling stress	GABA protects tomato seedling from stress by enhancing some antioxidant enzyme activity, reducing MDA content, maintaining membrane integrity, and significantly increasing sugar and proline content	Malekzadeh et al. (2014)
Cassia italic	Salt stress	GABA enhances activity of antioxidant enzymes and reduces toxic effects of stress	Alqarawi et al. (2016)
Nigella sativa L.	Water-deficit stress	GABA improves growth and productivity by regulating osmotic stress and antioxidant defense system	Rezaei-Chiyaneh et al. (2018)
Rice seedling	Arsenite toxicity	GABA at 0.5 mM reduces H_2O_2 significantly and TBARs level, increases biosynthesis of unsaturated fatty acids, amino acid and polyamines and induces tolerance against arsenite	Kumar et al. (2019)
Rice	Osmotic and salinity stress	Priming with GABA at 0.5 mM concentration reduces detrimental stress effects on seed germination and seedling growth. The treatment increases activity of antioxidants, phenolic metabolites and thiol transcription level along with reduction of free radicals and MDA content	Sheteiwy et al. (2019)
Tomato seedling	Salt stress	Exogenous GABA alleviates salt stress damage index, reduces Na+ uptake and accumulation, increases ascorbic acid content and antioxidants, altering ROS metabolism	Wu et al. (2020)
Arabidopsis thaliana	Hypoxia stress	Elevated GABA controls ROS-Ca^{2+} hub, preventing excess H_2O_2 accumulation and disturbance to ion homeostasis and improving membrane potential	Wu et al. (2021)

10.3 Quaternary Ammonium Compounds

10.3.1 Glycine Betaine

Glycine betaine (*N,N,N,* trimethylglycine) is an amphoteric quaternary polyamine which is electrically-neutral over a large range of pH and is highly soluble in water (Gupta and Huang 2014; Annunziata et al. 2019; Acosta-Motos et al. 2020; Gupta et al. 2021). It accumulates in plants as an osmolyte and maintains osmotic pressure under stress (Chen and Murata 2008; Gupta and Huang 2014; Annunziata et al. 2019; Gupta et al. 2021). It alleviates the potassium efflux by regulating the ion channels under salt stress (Wei et al. 2017; Annunziata et al. 2019). Moreover, glycine betaine promotes uptake of phosphate and regulates phosphate homeostasis by elevating enzymatic activity of plasma membrane H^+-ATPase (Li et al. 2019; Annunziata et al. 2019). Glycine betaine also protects the stability and integrity of cellular membranes, enzymes and protein complexes such as ribulose bisphosphate carboxylase/oxygenase (Rubisco) (Mäkelä et al. 2000; Hasanuzzaman et al. 2019), abridging cellular ROS level (Ashraf and Foolad 2007; Annunziata et al. 2019; Gupta et al. 2021; Sharma et al. 2019). The synthesis of glycine betaine in plant cells occurs either from choline or glycine. Synthesis from choline involves two oxidation reactions, which are catalyzed by enzymes ferredoxin (FAD)-dependent choline monooxygenase (CMO) and nicotinamide adenine dehydrogenase (NAD^+)-dependent betaine aldehyde dehydrogenase (BADH), respectively (Chen and Murata 2011; Gupta and Huang 2014; Annunziata et al. 2019). The glycine betaine synthesis from glycine mainly occurs in halophytes, which include three successive methylation, catalyzed by two *S*-adenosyl methionine dependent methyl transferases, i.e., glycine sarcosine *N*-methyltransferase (GSMT) and sarcosine dimethylglycine *N*-methyltransferase (SDMT) (Gupta and Huang 2014).

Plants accumulate glycine betaine in chloroplast and cytosol (Zulfiqar et al. 2022). However, it predominantly occurs in chloroplasts, protects thylakoid membrane and maintains photosynthetic efficiency (Ashraf and Foolad 2007), safeguarding from oxidative stress (Ali et al. 2020; Zulfiqar et al. 2022). The glycine betaine content in chloroplast positively relates to stress tolerance, and that in cytosol may not impart stress resilience. Thus, the increased level of glycine betaine may not necessarily mediate stress tolerance (Zulfiqar et al. 2022). Under oxidative, chilling and other abiotic stresses, the uptake and accumulation of glycine betaine in chloroplast significantly mitigate stress-induced effects than that in other cellular compartments (Park et al. 2007; Ali et al. 2020). The accumulation of glycine betaine does not take place in all plant species, even the level of glycine betaine accumulated in plants differs from species to species (Ali et al. 2020). Plants such as spinach, barley, maize, bean, sorghum, and wheat are accumulator of glycine betaine and rice, mustard, tomato, and tobacco, non-accumulators of glycine betaine (Kumar et al. 2017; Zulfiqar et al. 2022). Glycine betaine is applied exogenously. Alternatively, the transgenic plants (Table 10.2) fortified with glycine betaine biosynthesis pathway obtain tolerance acquisition against stresses (Chen and Murata 2008). The exogenous application of glycine betaine ameliorates plant growth and physiological attributes, mitigating oxidative stress under abiotic stress (Yang and Lu 2005; Hasanuzzaman et al. 2014; Tisarum et al. 2020). Glycine betaine attenuates environmental oxidative stress via scavenging ROS indirectly by upregulating activity of antioxidants such as CAT, SOD, POX, and AsA-GSH cycle enzymes, alleviating MDA level, and improving plant growth and photosynthetic attributes (Yang and Lu 2005; Hu et al. 2012; Hasanuzzaman et al. 2014; Ahmad et al. 2020; Ali et al. 2020; Ejaz et al. 2020). Application of 5 mM of glycine betaine on mung beans exposed to

Table 10.2 Glycine betaine exogenous application or overexpression in transgenic plants mediates tolerance to adverse conditions

Target plant	Stress	Action	References
Cultured tobacco bright yellow (BY-2) cells	Salinity stress	Exogenous application of glycine betaine modifies activities of ascorbate-glutathione (AsA-GSH) cycle enzymes, facilitating tolerance against salinity-induced oxidative stress	Hoque et al. (2007)
Maize	Chilling stress	Pre-treatment of hybrid maize seeds with glycine betaine regulates high water content, electrolyte leakage and activity of catalase (CAT) and superoxide dismutase (SOD), inducing tolerance against the stress	Farooq et al. (2008)
Transgenic wheat lines overexpressing glycine betaine	Salt stress	Over production of glycine betaine establishes ion homeostasis, increases osmolytes, stabilizes the activity of ROS scavenging enzymatic antioxidants, mitigating oxidative damage to cellular membranes and reducing salt stress effects	Liang et al. (2009)
Transgenic sweet potato overexpressing glycine betaine biosynthetic gene (BADH)	Salt, Chilling and methyl viologen-induced oxidative stress	The transgenic potato exhibits low content of malondialdehyde (MDA) and hydrogen peroxide (H_2O_2) and improved photosynthetic activity and obtains tolerance against abiotic (oxidative) stress due to ROS scavenging	Fan et al. (2012)
Cotton	Lead (Pb) toxicity	Exogenously applied glycine betaine at 1 mM reduces MDA content, H_2O_2 level, and electrolyte leakage and increases antioxidant enzymes activity, thereby significantly counteracting Pb toxicity	Bharwana et al. (2014)
Mustard (*Brassica juncea*)	Drought stress	Application of glycine betaine alters drought-induced oxidative stress by promoting ROS scavenging enzyme activities, eventually reducing H_2O_2 content and lipid peroxidation	Hossain et al. (2014b)

(continued)

Table 10.2 (continued)

Target plant	Stress	Action	References
Lentil (*Lens culinaris*) seedling	Drought stress	Exogenous application of 15 mM of glycine betaine reduces drought stress-induced oxidative stress by significantly improving glutathione (GSH) content, activity of Glutathione-*S*-transferase (GST) and glyoxalase (gly I) with simultaneous reduction of oxidized glutathione (GSSH) and H_2O_2	Molla et al. (2014)
Transgenic tomato overexpressing BADH gene	Heat stress	Improved plant tolerance against stress, augments photosynthetic and antioxidant activity and reduces level of H_2O_2/ superoxide ($O_2^{\bullet-}$) and lipid peroxidation	Li et al. (2014a, b)
Soybean seedlings	Salt stress	Application of 50 mM glycine betaine increases the activity of catalase (CAT) and superoxide dismutase (SOD), decreases proline, MDA, and sodium ion content, improving seedling growth	Malekzadeh (2015)
Marigold cultivar treated with GB	Heat stress	Foliar application of glycine betaine significantly alleviates H_2O_2, $O_2^{\bullet-}$, and MDA content, increases gas exchange and protects photosynthetic efficiency	Sorwong and Sakhonwasee (2015)
Safflower	Salt stress	Exogenous application with 30–60 mM glycine betaine increases the activity of CAT and SOD, improves ion homeostasis and reduces lipid peroxidation, eventually preventing oxidative damage to cellular membranes	Alasvandyari et al. (2017)
Millet (*Sorghum bicolor* L.)	Chromium (Cr) toxicity	Exogenous treatment with 100 mM glycine betaine along with 10 g arbuscular mycorrhiza fungi (AMF) improves the activity of antioxidants and reduces oxidative stress, ameliorating Cr toxicity	Kumar (2021)

Cd stress facilitates the activities of both enzymatic and non-enzymatic antioxidants such as DHAR, MDHAR, APX, GST, GPX, and glyoxalase (Gly I, II), CAT, GR, AsA, and GSH responsible for ROS scavenging and maintains high GSH/GSSH ratio on, eventually declining content of H_2O_2 and reducing lipid peroxidation (Hossain et al. 2010). Similarly, exogenous treatment with 20 mM glycine betaine of perennial Cd stressed ryegrass alleviates Cd accumulation, improves antioxidants enzyme CAT, SOD and POX, and reduces electrolyte leakage and level of MDA (Lou et al. 2015). In *Brassica oleracea* exposed to Cr stress, the exogenous application of glycine betaine relieves from oxidative stress (Ahmad et al. 2020). Glycine betaine in plants alleviates the chromium (Cr) uptake and toxicity, improves photosynthetic attributes, and promotes the activity of antioxidant enzymes. Foliar treatment with glycine betaine of turfgrass ensures security from prolonged drought or salinity stress and improves antioxidant level, reducing electrolyte leakage and lower MDA accumulation (Yang et al. 2012). Moreover, the application of 0.5 mM salicylic acid to mung bean under salinity stress helps accumulate glycine betaine, which increases activity of antioxidants (GSH, GR), and GSH/GSSH ratio, improves photosynthetic and growth parameters, and decreases accumulation of Na^+ and Cl^- ions, Na^+/K^+ ratio, lipid peroxidation, and oxidative stress (Syeed et al. 2021). The transgenic tomato expressing betaine aldehyde dehydrogenase (BADH) genes accumulates glycine betaine, which induces tolerance to heat-enhanced photoinhibition, reduces H_2O_2 and $O_2^{\bullet-}$ contents and prevents oxidative damage to membranes (Li et al. 2014a). The study further demonstrates that the increased activity of antioxidants due to glycine betaine accumulation reduces ROS content in plants under stress.

10.4 Sugars

Plant cells are storehouse various soluble sugars including monosaccharides, disaccharides, raffinose family of oligosaccharides and fructans that facilitate diverse structural and metabolic functions, and partake in signaling pathway (Couée et al. 2006; Bolouri-Moghaddam et al. 2010; Keunen et al. 2013; Afzal et al. 2021). Sugars, in particular, sucrose, trehalose, and fructans are important osmolytes, which maintain cell osmotic pressure under stress and assist plants to withstand unfavorable conditions (Ejaz et al. 2020). Moreover, these sugars have been reported to participate in both generation and quenching of toxic ROS, alleviating environmental (oxidative) stress (Couée et al. 2006). Sugars mediate the generation of NADPH via oxidative pentose phosphate (OPP) pathway, helping ultimately the synthesis of GSH that eventually scavenges ROS (Bolouri-Moghaddam et al. 2010; Ejaz et al. 2020). Besides, sugar itself acts as an antioxidant and partakes in scavenging of ROS, in particular, $^{\bullet}OH$ and $O_2^{\bullet-}$ (Ende and Peshev 2013; Matros et al. 2015). Carbohydrates have been anticipated to scavenge $^{\bullet}OH$ mainly via abstraction of H^+ atom, forming a carbon-centered radical (Hernandez-Marin and Martínez 2012; Peshev et al. 2013).

10.4.1 Sucrose

Sucrose is a disaccharide sugar consisting of glucose and fructose (Keunen et al. 2013). It is one of the most abundant sugars in plants, produced as the major product of photosynthesis and partakes in multiple cellular processes such as growth, signaling, transport, stress tolerance, and others (Ende and Peshev 2013; Keunen et al. 2013; Cao et al. 2014). Sucrose forms a raffinose family of oligosaccharides (tri-, tetra-, and penta-saccharides), when combined with galactosyl moieties and also participates in production of fructans in vacuoles (Keunen et al. 2013). Sucrose plays an instrumental role in ROS scavenging by acting as an antioxidant, facilitating biosynthesis and accumulation of antioxidants and other metabolites such as AsA, phenolics, flavonoids, and anthocyanin, and/or performing signaling function in plants (Baque et al. 2012; Ende and Peshev 2013; Ejaz et al. 2020). The exogenous sucrose application increases the accumulation of AsA in the florets of harvested broccoli (Nishikawa et al. 2005). Moreover, sucrose improves nutritional values and increases content of AsA, anthocyanins and other antioxidants in broccoli sprouts (Guo et al. 2011). Treatment of *Morinda citrifolia* root suspension culture with 1% sucrose solution augments anthraquinone, phenolics, and flavonoids contents and reduces the oxidative damage (Baque et al. 2012). Further, the exogenous sucrose supplementation upregulates gene expression involved in defense mechanism against ROS, rendering atrazine-induced oxidative stress in *Arabidopsis thaliana* (Ramel et al. 2007, 2009). Cucumber seedlings pretreated with 50mM sucrose obtain tolerance against chilling-induced oxidative stress via activation of antioxidant enzymes (SOD, GPX, APX, GR, DHAR, and MDHAR) (Cao et al. 2014). It also increases endogenous sucrose, proline, soluble acid invertase as well as neutral invertase content, alleviating the levels of $O_2^{\bullet -}$, H_2O_2 and MDA. The potato seedlings exposed to heat stress gain thermotolerance due to sucrose pre-treatment (Gong and Chen 2021). Sucrose reduces the level of electrolyte leakage in potato leaves, abridges accumulation of ROS such as H_2O_2, and $O_2^{\bullet -}$, promotes the activities of antioxidant enzymes, e.g., POX, APX, CAT, and SOD as well as enhances endogenous sucrose and proline content (Gong and Chen 2021).

10.4.2 Trehalose

Trehalose, a non-reducing disaccharide sugar, is a versatile osmo-protectant, mediating osmotic balance along with stabilization of biomolecules under stress in plants (Ejaz et al. 2020). Trehalose is naturally present in minute quantities, and elevates its endogenous level in stressful events, in higher plants (Kosar et al. 2019; Ejaz et al. 2020). For instance, *Arabidopsis thaliana* increases its endogenous trehalose level by twofold at high (40 °C) temperature stress for 4 h and at eightfold cold (4 °C) temperature stress for 4 days (Kaplan et al. 2004; Kosar et al. 2019). Similarly, winter wheat exposed to heat stress increases its trehalose level to 1250 μg g^{-1} fresh weight from 502 μg g^{-1} fresh weight under normal conditions (Luo et al. 2010). Moreover, exogenous trehalose supplementation in small quantities serves a putative role in the alleviation of stress-induced detrimental effects on plants (Kosar et al. 2019). The pre-treatment with trehalose of winter wheat seedlings increases endogenous trehalose content which induces protection to proteins in the thylakoid membrane and safeguards photosynthetic activity along with reduction in electrolyte leakage, activity of lipoxygenase, and accumulation of MDA, $O_2^{\bullet -}$, and H_2O_2 (Luo et al. 2010). Even transgenic plants expressing trehalose synthesizing genes enrich their trehalose content and display tolerance against stress (Kosar et al. 2019). In plants, two enzymes, trehalose-6-phosphate synthase and trehalose-6-phosphate phosphatase catalyze trehalose synthesis (Ejaz et al. 2020).

Trehalose is an important sugar signaling molecule and demonstrates the ability to quench ROS either directly or indirectly (Luo et al. 2008; Chang et al. 2014; Ejaz et al. 2020). The exogenous application of 50 mM trehalose neutralizes the harmful effects of high-temperature stress in wheat by direct eliminating $O_2^{\bullet -}$ and H_2O_2 in concentration-dependent manner along with slight conservation of SOD activity, inducing protection of membranes against lipid peroxidation (Luo et al. 2008). Trehalose activates antioxidant-based defense machinery of plants encompassing enzymes like SOD, POX and CAT, and non-enzymatic antioxidants such as AsA and total phenolics, indirectly abridging ROS level (Shafiq et al. 2015; Ejaz et al. 2020). In rice, pre-treatment with 25 mM trehalose improves the content of proline, total soluble sugar and trehalose, and increases the activity of antioxidant enzymes, SOD, POX, and CAT, which mediates tolerance against salt stress (Abdallah et al. 2016). Trehalose application to radish under water-stressed condition restricts ROS and MDA level by increasing activities of SOD, POX, and CAT, and enhancing accumulation of AsA, total phenolics, glycine betaine, total soluble protein, and total tocopherol (Shafiq et al. 2015). Exogenous treatment with trehalose improves the level of antioxidants including flavonoids, tocopherols, and phenolics as well as 2,2-diphenyl-1-picrylhydrazyl (DPPH) radical neutralizing activity in maize seed oil under water-deficit condition (Ali et al. 2012).

10.4.3 Fructans

Fructans, a class of oligosaccharides, scavenge $^{\bullet}OH$ and other ROS and are, therefore, considered as an antioxidant (Keunen et al. 2013; Soares et al. 2019; Rigui et al. 2019). Fructans are more efficient $^{\bullet}OH$ scavengers than to $O_2^{\bullet -}$ (Keunen et al. 2013). The degree of polymerization (DP) determines antioxidant capacity of fructans. For instance, fructans with high DP rapidly eliminate $^{\bullet}OH$ and provide protection from oxidative stress in wheat seedling subjected to drought stress (Nemati et al. 2018). Fructans are generated from sucrose and get accumulated in vacuoles (Darwen and John 1989; Keunen et al. 2013), where they are involved in ROS detoxification (Peshev et al. 2013) (Fig. 10.2). In vacuole, fructans are localized near tonoplast inner space and in vacuole lumen. Fructans strongly interact and integrate with head groups (Is it mean glycerol moiety? Clarify) of the tonoplastic membrane, where they directly scavenge ROS and maintain the stability of membrane under stress (Elaborate mechanism!!!) (Keunen et al. 2013; Peshev et al. 2013; Ejaz et al. 2020).

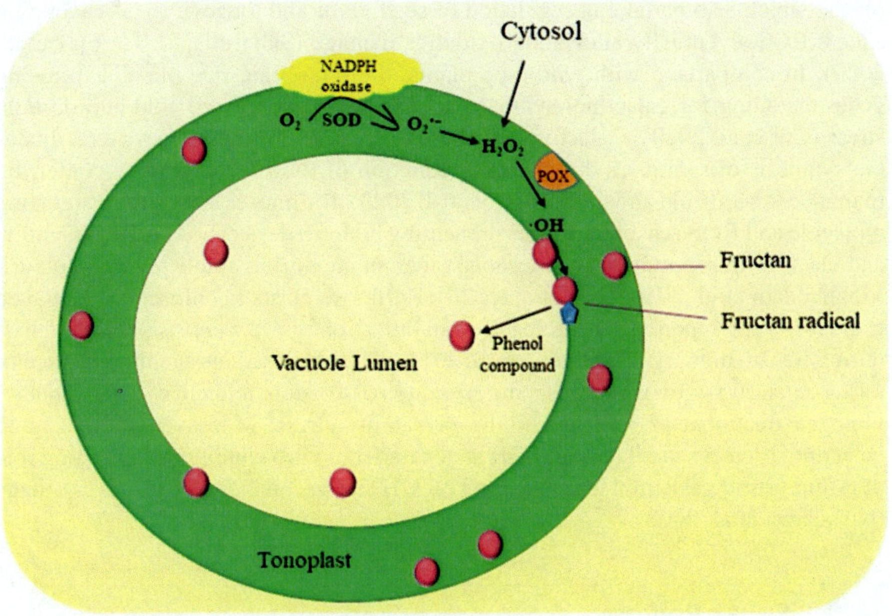

Fig. 10.2 Detoxification of ROS by fructans in vacuole

The excessively generated H_2O_2 in cytoplasm under stress diffuses to the vacuole through the tonoplast. Moreover, NADPH oxidases localized in vacuoles generate $O_2^{\bullet-}$, which on action of SOD converts to H_2O_2 (Peshev et al. 2013). Tonoplast-associated POX (type III) converts the excessive H_2O_2 in vacuole to $^{\bullet}OH$ and peroxyl ($OOH^{\bullet}$) radical via hydroxylic cycle (Dunand et al. 2007; Peshev et al. 2013; Ejaz et al. 2020). In vacuole, fructans present in tonoplast react with the harmful radicals and detoxify them. During the detoxification process, fructans are converted to fructan radicals which diffuse to vacuolar lumen where they are recycled back to fructans by secondary metabolite, particularly, phenolic compounds (Takahama 2004; Peshev et al. 2013; Ejaz et al. 2020). Fructans further neutralize ROS indirectly by promoting the accumulation and activity of antioxidants (Ejaz et al. 2020).

10.4.4 Raffinose Family Oligosaccharides

D-galactose derived from the sucrose comprises oligosaccharide family of raffinoses (Sengupta et al. 2015; Ejaz et al. 2020). Raffinoses are water soluble, low molecular weight carbohydrates ubiquitously found in higher plants (Ejaz et al. 2020; Salvi et al. 2020). They comprise non-reducing trisaccharide (raffinose), tetrasaccharide (stachyose), pentasaccharide (verbascose), and hexasaccharide (Ajugose) (Ejaz et al. 2020; Elango et al. 2022). Galactinol synthase catalyzes the synthesis of raffinoses in plants, which also partake in regulation of seed vigor and longevity via scavenging excess ROS, eventually alleviating oxidative damage (Salvi et al. 2016; Ejaz et al. 2020). In comparison with wild-type plants, the transgenic rice plants expressing genes encoding for galactinol synthase acquire tolerance toward cold and drought stress (Cui et al. 2020). Galactinol synthase expression in transgenic rice facilitates the synthesis of raffinoses that mediate reduction of ROS level, thereby conferring tolerance toward cold and drought (Cui et al. 2020). Raffinoses also act as a signaling molecule and ROS scavengers, thus preventing embryo desiccation, ATP generation and electron transportation in chloroplast membrane during abiotic stress (Bolouri-Moghaddam et al. 2010; Ejaz et al. 2020). Raffinose exists in chloroplast (ElSayed et al. 2014) and reportedly participates in stabilization of *Arabidopsis* photosystem II during freeze–thaw cycle (Knaupp et al. 2011). The increased intracellular content of raffinose and galactinol in transgenic *Arabidopsis* triggers tolerance against methyl viologen treatment as well as chilling or salinity stress. Moreover, raffinose and galactinol protect salicylate from $^{\bullet}OH$ attack under in vitro condition, suggesting role of raffinose and galactinol in scavenging of $^{\bullet}OH$ and attenuation of oxidative damage (Nishizawa et al. 2008).

10.5 Sugar Polyols

Sugar polyols, also known as sugar alcohols, polyalcohol, and polyhydric alcohol, are the reduced form of aldose or ketose sugars, forming respective hydroxyl residues (Williamson et al. 2002; Bhattacharya and Kundu 2020). These are the water-soluble compounds that are classified into cyclic and acyclic structure. Cyclic sugar polyols include myo-inositol, ononitol, pinitol, and others, whereas acyclic structures commonly include mannitol, sorbitol and inositols (Bhattacharya and Kundu 2020). These compounds occur in a wide range of plants exposed to environmental stresses like dehydration, salinity as well as osmotic stress (Bhattacharya and Kundu 2020). However, a single kind of sugar polyol is present in particular plant species, hence the specific polyol found is the characteristic of plant families or genera (Williamson et al. 2002). Polyols act as an osmolyte and molecular chaperones, and facilitate reduction of oxidative damage (Bohnert et al. 1995; Sengupta et al. 2008; Bhattacharya and Kundu 2020). The ability of mannitol, sorbitol, and myo-inositol to quench $^{\bullet}OH$ has been reported several years ago (Ejaz et al. 2020).

10.5.1 Mannitol

It is a water-soluble hexitol having structural similarity with aldohexose sugar mannose. Mannitol is found in a diverse range of plants, fungi, bacteria, and algae (Patel and Williamson 2016; Bhattacharya and Kundu 2020). Its occurrence has been reported in more than 100 plants representing 70 families of angiosperms (Bhattacharya and Kundu 2020). Mannitol is an efficient osmoregulator in plants and scavenges ROS under stress. In higher plants, NADPH-mannose-6-phosphate reductase catalyzes the synthesis of mannitol using mannose-6-phosphate as a substrate. This reaction yields mannitol-1-phosphate which is subsequently dephosphorylated by enzyme mannitol-1-phosphate phosphatase generating mannitol (Loescher et al. 1992; Bhattacharya and Kundu 2020). The catabolism of mannitol into mannose in plants is catalyzed by the enzyme mannitol dehydrogenase, which is used as an indicator of mannitol pool within the cell and activates plant defense system.

Mannitol alleviates oxidative stress by quenching ROS or by increasing activity of the antioxidants in plants (Seckin et al. 2009; Ejaz et al. 2020). More than three decades ago, an in vitro study has demonstrated that 33 mM mannitol quenches approximately 60% of $^{\bullet}OH$ (Smirnoff and Cumbes 1989). Shen et al. (1997) have further provided evidence that the in vivo transgenic tobacco line expressing bacterial mannitol biosynthetic enzyme accumulates 100 mM mannitol in chloroplast that quenches endogenous $^{\bullet}OH$ and imparts tolerance against methyl viologen triggered oxidative damage. The literature cites several instances of exogenously applied mannitol-inducing enzymatic antioxidants in plants exposed to environmental stress. The pre-treatment with 100 mM mannitol of salt-sensitive wheat genotype facilitates the activity of SOD, POX, CAT, APX, and GR antioxidants in roots under salt stress, which reduces lipid peroxidation and eventually alleviates oxidative damage (Seckin et al. 2009). The exogenous application of mannitol also alleviates Cr uptake and toxicity, improves growth parameters, photosynthetic pigments, and antioxidant activity, and abridges ROS level in maize and wheat plants subjected to Cr stress (Adrees et al. 2015; Habiba et al. 2019).

10.5.2 Sorbitol

It is a six-carbon alcohol found in a wide number of organisms including plants, animals, fungi, yeast, and bacteria. Sorbitol is a photosynthetic end product in various species belonging to the Rosaceae family (Li et al. 2012). The accumulation of sorbitol occurs in mature plant leaves and is translocated via phloem to young leaves, shoot tips, and fruits (sink tissues) (Lo Bianco et al. 2000; Williamson et al. 2002; Bhattacharya and Kundu 2020). The synthesis of sorbitol occurring in source tissue involves conversion of glucose-6-phosphate to sorbitol-6-phosphate by an enzyme, sorbitol-6-phosphate dehydrogenase, which later sorbitol-6-phosphate phosphatase transforms into sorbitol (Deguchi et al. 2006; Bhattacharya and Kundu 2020). In the sink tissues, sorbitol dehydrogenase converts sorbitol into fructose (Deguchi et al. 2006; Bhattacharya and Kundu 2020). The exogenous application of sorbitol reduces oxidative damage in plants. The maize seedling subjected to polyethylene glycol (PEG)-induced drought stress when treated with 10 mM sorbitol displays increased contents of soluble sugar, soluble protein, proline, and sorbitol with enhanced activities of antioxidants and aldose reductase, and reduced level of MDA and H_2O_2, mediating resistance to drought stress (Wang et al. 2022). Nevertheless, the application of sorbitol, particularly at high concentration develops oxidative stress and expresses symptoms like desiccation, ROS generation at the photosystem reaction center of chloroplast, etc. (Abbasi et al. 2007). The excess accumulation of sorbitol in transgenic tobacco impairs chlorophyll pigment development and develops necrotic lesions in leaves (Sheveleva et al. 1998). Besides, the treatment with sorbitol of *Artemisia annua* cell suspension culture enhances level of MDA, and H_2O_2, inducing oxidative stress (Salehi et al. 2019).

10.5.3 Myo-Inositols

Myo-inositol and its derivatives are found in all eukaryotes and mediate diverse function in plants such as signal transduction, stress management, alleviation of ROS, osmoregulation, cell wall formation and others (Yildizli et al. 2018; Jia et al. 2019). They show solubility in both water as well as in lipids (Valluru and Van den Ende 2011; Jia et al. 2019). The biosynthesis of myo-inositol starts with D-glucose-6-phosphate, which D-myo-inositol-3-phosphate synthase converts into myo-inositol-3-phosphate (Siracusa et al. 2022). Myo-inositol-3-phosphate subsequently undergoes dephosphorylation on action of specific Mg^{2+}-dependent inositol monophosphate phosphatase, eventually generating free myo-inositol (Valluru and Van den Ende 2011). Myo-inositol like many other sugar polyols, mediates neutralization of excess ROS. Yildizli et al. (2018) have reported that the exogenous treatment with myo-inositol of *Capsicum annuum* significantly reduces the content of H_2O_2 and proline, and membrane leakage, lowers the activity of APX and CAT, but improves water status, calcium level, and GR activity, which ultimately ameliorate drought stress. The exogenous application of 50 μM myo-inositol significantly alleviates salinity stress in *Malus hupehensis* Rehd. and restores photosynthetic activity by abridging ROS accumulation, enhancing activity of antioxidants, elevating the accumulation of sugar and sugar polyols, and maintaining Na^+ uptake by upregulating expression of genes involved in Na^+ uptake (Hu et al. 2018).

10.6 Plant Metabolites/Compounds Attenuating Oxidative Stress

10.6.1 Phytohormones

Phytohormones such as salicylic acid (SA), ethylene (ET), JA, ABA, gibberellic acid (GA), auxin (AUX), cytokinins (CKs), strigolactones (SLs), and brassinosteroids (BRs), are the signaling molecules which accumulate in small titres and partake in regulation of growth and developmental processes and promote stress tolerance in plants (Ahmad et al. 2010; Sharma et al. 2019; Jogawat et al. 2021). Phytohormones positively regulate the activity of enzymatic and non-enzymatic antioxidants, and the accumulation of osmolytes in plants, which facilitate protection against oxidative stress and ability to withstand under adverse situations (Li et al. 1998; Hasan et al.

2008; Ahmad et al. 2010; Sharma et al. 2019). Phytohormone brassinosteroids (BRs) are the steroidal compounds that have been isolated from the pollens of *Brassica* for the first time (Ahmad et al. 2010). BRs participate in signaling transduction during unfavorable conditions, eventually stimulating biosynthesis of key osmolytes, which induce stress tolerance. The overexpression of BRL3, a vascular BR receptor, promotes accumulation of major osmoprotectants such as proline and sugars in *Arabidopsis* under drought conditions and triggers stress tolerance (Fàbregas et al. 2018). BR application elevates activity of enzymatic antioxidants in salt-stressed rice seedling (Nunez et al. 2003) and promotes activity of antioxidants as well as photosynthesis in mustard plants subjected to Cd stress (Hayat et al. 2007). The application of 24-epibrassinolide and 28-homobrassinolide improves growth parameters, activity of antioxidant CAT, POX, and SOD, and proline content in mungbean seedlings exposed to Al stress (Ali et al. 2008). The treatment with epibrassinolide of radish seedling under Cu stress raises activity of CAT, SOD, GR, and guaiacol peroxidase, increases proline content, total phenol, AsA, and phytochelatin, and reduces level of MDA, eventually leading to attenuation of Cu-triggered oxidative stress (Choudhary et al. 2010). Application of salicylic acid (SA) amends salt-triggered oxidative stress by enhancing activity of antioxidants, chlorophyll content and photosynthetic activity as well as recovers membrane injury in *Torreya grandis* trees (Li et al. 2014b). Similarly, application of 0.1–1 $\mu mol.L^{-1}$ methyl jasmonate, a methyl ester of JA, repairs oxidative damage triggered by Cd toxicity in leaves of *Kandelia obovata* by enhancing concentration of AsA and promoting activities of CAT and APX (Chen et al. 2014). Table 10.3 enlists role of phytohormones in regulation of oxidative stress in plants.

10.6.2 Micronutrients

Micronutrients application not only improves nutritional status of plants but also imparts tolerance against stress. Micronutrients, in particular, Se, Cu, Si, boron (B), iron (Fe), nickel (Ni), manganese (Mn), molybdenum (Mo), and zinc (Zn) show capacity to up-regulate the activities of antioxidant enzymes and sugar metabolism at low concentration, which ameliorate stress-induced oxidative damage (Tavanti et al. 2021). The application of Cu has been demonstrated to improve activity of SOD, POX, and CAT in *Prunus cerasifera* (Lombardi and Sebastiani 2005). Similarly, B, Mn, Ni, and Se elicit the activity of enzymatic antioxidants in chickpeas, cotton, tomato, and rice, respectively (Ardıc et al. 2009; Kumar et al. 2015; dos Reis et al. 2017; Tavanti et al. 2020). Zn application reduces lipid peroxidation of cell membrane and alleviates Cu-induced oxidative stress in *Spirodela polyrhiza* L. (Upadhyay and Panda 2010). Further, the exogenous application of Zn also reduces ROS level, electrolyte leakage and MDA level in wheat and rice shoots subjected to Cd stress (Hassan et al. 2005; Rizwan et al. 2017). In cowpea, the treatment with Se in the form of selenate and selenite reduces the lipid peroxidation by stimulating different antioxidants (Silva et al. 2020). Selenate improves the activity of APX and GR, whereas selenite upregulates the activity of CAT. In

Table 10.3 The phytohormones regulating oxidative damage in plants under stress conditions

Plant	Stress	Phytohormone	Mechanism	References
Loquat fruit	Chilling injury	Methyl Jasmonate (MeJA)-methyl ester of jasmonic acid (JA)	The treatment reduces production of $O_2^{\bullet -}$ and H_2O_2 contents and activity of lipoxygenase with significant increase in SOD, CAT, APX for 14 days of cold storage	Cao et al. (2009)
Rice	Cadmium (Cd) toxicity	Salicylic acid (SA)	Pre-treatment with SA alleviates Cd toxicity by enhancing antioxidant defense system component such as proline level, SOD, CAT, POX, GSH and non-protein thiols, and repairs oxidative damage	Guo et al. (2009)
Mustard (*Brassica juncea* L.)	High temperature	Salicylic acid (SA)		Hayat et al. (2009)
Phragmites communis Trin	Heat stress	Abscisic acid (ABA)	Exogenous treatment increases the activity of SOD, CAT, APX, and POX, and decreases H_2O_2 and MDA content	Ding et al. (2010)
Creeping bentgrass (*Agrostis stolonifera* L.)	Drought stress	Cytokinin	The treatment increases endogenous cytokinin level, metabolites like GABA, proline alanine and glycine; carbohydrates like sucrose, fructose, maltose and ribose, and organic acid, inducing tolerance to stress	Merewitz et al. (2012)
Capsicum frutescens	Cadmium (Cd) toxicity	Methyl Jasmonate (MeJA)-methyl ester of jasmonic acid (JA)	0.1 μmol. L^{-1} MeJA reduces MDA content and increases chlorophyll content, CAT and APX activity, and elevates endogenous JA	Yan et al. (2013)
Linum usitatissimum	Cadmium (Cd) toxicity	Salicylic acid (SA)	Pre-treatment with SA increases seedling antioxidant capacities by activating H_2O_2 detoxifying enzymes	Belkadhi et al. (2014)
Triticum aestivum	Drought stress	Abscisic acid (ABA)/ Cytokinin	Combined treatment of ABA and cytokinin increases level of glycine betaine and proline, improves activity of CAT and POX, reduces H_2O_2 and MDA content	Sarafraz-Ardakani et al. (2014)

(continued)

Table 10.3 (continued)

Plant	Stress	Phytohormone	Mechanism	References
Grapevine	Low temperature	Abscisic acid (ABA)	Foliar application reduces electrolyte leakage and visible damage on leaf; improves proline, total phenol, soluble carbohydrate content and reduces relative water content (RWC)	Karimi and Ershadi (2015)
Brassica rapa subsp. *pekinensis*	Drought stress	Salicylic acid (SA)	SA facilitates activity of SOD, CAT, guaiacol peroxidase, APX, proline content and reduces $O_2^{\bullet-}$, H_2O_2, and MDA level	La et al. (2019)

some cases, however, high doses of certain micronutrients such as Cu, Ni, Zn, and Fe, produce toxic effects (Upadhyay and Panda 2010; Hassan et al. 2019; Zaid et al. 2020). The treatment with low Zn concentration (10 μM) reduces oxidative stress induced by Cd in tomato seedling by reducing level of thiobarbituric acid reacting substances (TBARS), a marker of oxidative damage to membrane lipids, and activity of enzyme lipoxygenase. On the other hand, the high Zn concentration (50 μM or above) increases the content of TBARS in tomato leaves inducing detrimental effects (Ben Ammar et al. 2015).

10.7 Conclusion

The generation of reactive oxygen species (ROS) in plants under stress induces oxidative damage to proteins and enzymes, reducing scavenging ability of enzymatic antioxidants. However, plants synthesize osmolytes under stress, which not only regulate osmotic balance but manage excess ROS. The general mechanisms involved by osmolytes that confer tolerance from oxidative stress include signal transduction, chelation of metal ions, direct scavenging of ROS, and enhanced activity of enzymatic antioxidants, which reduce level of ROS, thereby alleviating oxidative stress. The extensive literature shows that the exogenous application and/or development of transgenic plants overexpressing genes involved in synthesis of osmolytes have capacity to mitigate oxidative stress. However, there is a need to conduct more on-field research and find the signaling pathways employed by various osmolytes to elucidate the detailed mechanism involved in management of oxidative stress in different crops under climate change scenarios.

References

Abbasi AR, Hajirezaei M, Hofius D, Sonnewald U, Voll LM (2007) Specific roles of α-and γ-tocopherol in abiotic stress responses of transgenic tobacco. Plant Physiol 143(4):1720–1738

Abdallah MS, Abdelgawad ZA, El-Bassiouny HMS (2016) Alleviation of the adverse effects of salinity stress using trehalose in two rice varieties. S Afr J Bot 103:275–282

Abdel Razik ES, Alharbi BM, Pirzadah TB, Alnusairi GS, Soliman MH, Hakeem KR (2021) γ-Aminobutyric acid (GABA) mitigates drought and heat stress in sunflower (*Helianthus annuus* L.) by regulating its physiological, biochemical and molecular pathways. Physiol Plant 172(2):505–527

Acosta-Motos JR, Penella C, Hernández JA, Díaz-Vivancos P, Sánchez-Blanco MJ, Navarro JM, Gomez-Ballot MJ, Barba-Espín G (2020) Towards a sustainable agriculture: strategies involving phytoprotectants against salt stress. Agronomy 10(2):194

Adrees M, Ali S, Iqbal M, Bharwana SA, Siddiqi Z, Farid M, Ali Q, Saeed R, Rizwan M (2015) Mannitol alleviates chromium toxicity in wheat plants in relation to growth, yield, stimulation of anti-oxidative enzymes, oxidative stress and Cr uptake in sand and soil media. Ecotoxicol Environ Saf 122:1–8

Afzal S, Chaudhary N, Singh NK (2021) Role of soluble sugars in metabolism and sensing under abiotic stress. In: Plant growth regulators. Springer, Cham, pp 305–334

Aghdam MS, Naderi R, Jannatizadeh A, Babalar M, Sarcheshmeh MAA, Faradonbe MZ (2016) Impact of exogenous GABA treatments on endogenous GABA metabolism in anthurium cut flowers in response to postharvest chilling temperature. Plant Physiol Biochem 106:11–15

Ahmad P, Umar S, Sharma S (2010) Mechanism of free radical scavenging and role of phytohormones in plants under abiotic stresses. In: Plant adaptation and phytoremediation. Springer, Dordrecht, pp 99–118

Ahmad P, Ahanger MA, Alam P, Alyemeni MN, Wijaya L, Ali S, Ashraf M (2019) Silicon (Si) supplementation alleviates NaCl toxicity in mung bean [*Vigna radiata* (L.) Wilczek] through the modifications of physio-biochemical attributes and key antioxidant enzymes. J Plant Growth Regul 38(1):70–82

Ahmad R, Ali S, Abid M, Rizwan M, Ali B, Tanveer A, Ahmad I, Azam A, Ghani MA (2020) Glycinebetaine alleviates the chromium toxicity in *Brassica oleracea* L. by suppressing oxidative stress and modulating the plant morphology and photosynthetic attributes. Environ Sci Pollut Res 27(1):1101–1111

Alasvandyari F, Mahdavi B, Hosseini SM (2017) Glycine betaine affects the antioxidant system and ion accumulation and reduces salinity-induced damage in safflower seedlings. Arch Biol Sci 69(1):139–147

Ali B, Hasan SA, Hayat S, Hayat Q, Yadav S, Fariduddin Q, Ahmad A (2008) A role for brassinosteroids in the amelioration of aluminium stress through antioxidant system in mung bean (*Vigna radiata* L. Wilczek). Environ Exp Bot 62(2):153–159

Ali Q, Ashraf M, Anwar F, Al-Qurainy F (2012) Trehalose-induced changes in seed oil composition and antioxidant potential of maize grown under drought stress. J Am Oil Chem Soc 89(8): 1485–1493

Ali S, Abbas Z, Seleiman MF, Rizwan M, YavaŞ İ, Alhammad BA, Shami A, Hasanuzzaman M, Kalderis D (2020) Glycine betaine accumulation, significance and interests for heavy metal tolerance in plants. Plants 9(7):896

Alqarawi AA, Hashem A, Elsayed FA, Al-Huqail AA, Alshahrani TS, Alshalawi SAR, Egamberdieva D (2016) Protective role of gamma amminobutyric acid on *Cassia italica* Mill under salt stress. Legum Res 39(3):396–404

Al-Quraan NA, Sartawe FAB, Qaryouti MM (2013) Characterization of γ-aminobutyric acid metabolism and oxidative damage in wheat (*Triticum aestivum* L.) seedlings under salt and osmotic stress. J Plant Physiol 170(11):1003–1009

Alyemeni MN, Ahanger MA, Wijaya L, Alam P, Bhardwaj R, Ahmad P (2018) Selenium mitigates cadmium-induced oxidative stress in tomato (*Solanum lycopersicum* L.) plants by modulating chlorophyll fluorescence, osmolyte accumulation, and antioxidant system. Protoplasma 255(2): 459–469

Annunziata MG, Ciarmiello LF, Woodrow P, Dell'Aversana E, Carillo P (2019) Spatial and temporal profile of glycine betaine accumulation in plants under abiotic stresses. Front Plant Sci 10:230

Ansari MI, Lee RH, Chen SCG (2005) A novel senescence-associated gene encoding γ-aminobutyric acid (GABA): pyruvate transaminase is upregulated during rice leaf senescence. Physiol Plant 123(1):1–8

Ansari MI, Hasan S, Jalil SU (2014) Leaf senescence and GABA shunt. Bioinformation 10(12):734

Ansari MI, Jalil SU, Ansari SA, Hasanuzzaman M (2021) GABA shunt: a key-player in mitigation of ROS during stress. Plant Growth Regul 94(2):131–149

Ardıc M, Selman A, Tour S, Ozdemir OF, Turkan I (2009) Antioxidant responses of chickpea plants subjected to boron toxicity. Plant Biol 11(3):328–338. https://doi.org/10.1111/j.1438-8677.2008.00132.x

Ashraf MFMR, Foolad MR (2007) Roles of glycine betaine and proline in improving plant abiotic stress resistance. Environ Exp Bot 59(2):206–216

Awan SA, Khan I, Rizwan M, Zhang X, Brestic M, Khan A, AEl-Sheikh MA, Alyemeni MN, Ali S, Huang L (2021) Exogenous abscisic acid and jasmonic acid restrain polyethylene glycol-induced drought by improving the growth and antioxidative enzyme activities in pearl millet. Physiol Plant 172(2):809–819

Banu MNA, Hoque MA, Watanabe-Sugimoto M, Islam MM, Uraji M, Matsuoka K, Nakamura Y, Murata Y (2010) Proline and glycinebetaine ameliorated NaCl stress via scavenging of hydrogen peroxide and methylglyoxal but not superoxide or nitric oxide in tobacco cultured cells. Biosci Biotechnol Biochem 74:1010012170

Baque MD, Elgirban A, Lee EJ, Paek KY (2012) Sucrose regulated enhanced induction of anthraquinone, phenolics, flavonoids biosynthesis and activities of antioxidant enzymes in adventitious root suspension cultures of *Morinda citrifolia* (L.). Acta Physiol Plant 34(2): 405–415

Belkadhi A, De Haro A, Soengas P, Obregon S, Cartea ME, Chaibi W, Djebali W (2014) Salicylic acid increases tolerance to oxidative stress induced by hydrogen peroxide accumulation in leaves of cadmium-exposed flax (*Linum usitatissimum* L.). J Plant Interact 9(1):647–654

Ben Ammar W, Zarrouk M, Nouairi I (2015) Zinc alleviates cadmium effects on growth, membrane lipid biosynthesis and peroxidation in *Solanum lycopersicum* leaves. Biologia 70(2):198–207

Bharwana SA, Ali S, Farooq MA, Iqbal N, Hameed A, Abbas F, Ahmad MSA (2014) Glycine betaine-induced lead toxicity tolerance related to elevated photosynthesis, antioxidant enzymes suppressed lead uptake and oxidative stress in cotton. Turk J Bot 38(2):281–292

Bhattacharya S, Kundu A (2020) Sugars and sugar polyols in overcoming environmental stresses. In: Protective chemical agents in the amelioration of plant abiotic stress: biochemical and molecular perspectives. Wiley, pp 71–101

Bohnert HJ, Nelson DE, Jensen RG (1995) Adaptations to environmental stresses. Plant Cell 7(7): 1099

Bolouri-Moghaddam MR, Le Roy K, Xiang L, Rolland F, Van den Ende W (2010) Sugar signalling and antioxidant network connections in plant cells. FEBS J 277(9):2022–2037

Cao S, Zheng Y, Wang K, Jin P, Rui H (2009) Methyl jasmonate reduces chilling injury and enhances antioxidant enzyme activity in postharvest loquat fruit. Food Chem 115(4):1458–1463

Cao YY, Yang MT, Li X, Zhou ZQ, Wang XJ, Bai JG (2014) Exogenous sucrose increases chilling tolerance in cucumber seedlings by modulating antioxidant enzyme activity and regulating proline and soluble sugar contents. Sci Hortic 179:67–77

Chang B, Yang L, Cong W, Zu Y, Tang Z (2014) The improved resistance to high salinity induced by trehalose is associated with ionic regulation and osmotic adjustment in *Catharanthus roseus*. Plant Physiol Biochem 77:140–148

Chen TH, Murata N (2008) Glycinebetaine: an effective protectant against abiotic stress in plants. Trends Plant Sci 13(9):499–505

Chen TH, Murata N (2011) Glycinebetaine protects plants against abiotic stress: mechanisms and biotechnological applications. Plant Cell Environ 34(1):1–20

Chen CT, Chen LM, Lin CC, Kao CH (2001) Regulation of proline accumulation in detached rice leaves exposed to excess copper. Plant Sci 160(2):283–290

Chen J, Yan Z, Li X (2014) Effect of methyl jasmonate on cadmium uptake and antioxidative capacity in *Kandelia obovata* seedlings under cadmium stress. Ecotoxicol Environ Saf 104: 349–356

Choudhary NL, Sairam RK, Tyagi A (2005) Expression of Δ^1-pyrroline-5-carboxylate synthetase gene during drought in rice (*Oryza sativa* L.). Indian J Biochem Biophys 42(6):366–370

Choudhary SP, Bhardwaj R, Gupta BD, Dutt P, Gupta RK, Biondi S, Kanwar M (2010) Epibrassinolide induces changes in indole-3-acetic acid, abscisic acid and polyamine concentrations and enhances antioxidant potential of radish seedlings under copper stress. Physiol Plant 140(3):280–296

Couée I, Sulmon C, Gouesbet G, El Amrani A (2006) Involvement of soluble sugars in reactive oxygen species balance and responses to oxidative stress in plants. J Exp Bot 57(3):449–459

Cui LH, Byun MY, Oh HG, Kim SJ, Lee J, Park H, Lee H, Kim WT (2020) Poaceae Type II galactinol synthase 2 from antarctic flowering plant *Deschampsia antarctica* and rice improves cold and drought tolerance by accumulation of raffinose family oligosaccharides in transgenic rice plants. Plant Cell Physiol 61(1):88–104

Cuin TA, Shabala S (2007) Compatible solutes reduce ROS-induced potassium efflux in *Arabidopsis* roots. Plant Cell Environ 30(7):875–885

Darwen CW, John P (1989) Localization of the enzymes of fructan metabolism in vacuoles isolated by a mechanical method from tubers of Jerusalem artichoke (*Helianthus tuberosus* L.). Plant Physiol 89(2):658–663

Datir S, Singh N, Joshi I (2020) Effect of NaCl-induced salinity stress on growth, osmolytes and enzyme activities in wheat genotypes. Bull Environ Contam Toxicol 104(3):351–357

Deguchi M, Bennett AB, Yamaki S, Yamada K, Kanahama K, Kanayama Y (2006) An engineered sorbitol cycle alters sugar composition, not growth, in transformed tobacco. Plant Cell Environ 29(10):1980–1988

Ding W, Song L, Wang X, Bi Y (2010) Effect of abscisic acid on heat stress tolerance in the calli from two ecotypes of Phragmites communis. Biol Plant 54(4):607–613

dos Reis AR, de Queiroz Barcelos JP, de Souza Osório CRW, Santos EF, Lisboa LAM, Santini JMK, dos Santos MJD, Junior EF, Campos M, de Figueiredo PAM, Lavres J, Gratao PL (2017) A glimpse into the physiological, biochemical and nutritional status of soybean plants under Ni-stress conditions. Environ Exp Bot 144:76–87

Dunand C, Crèvecoeur M, Penel C (2007) Distribution of superoxide and hydrogen peroxide in *Arabidopsis* root and their influence on root development: possible interaction with peroxidases. New Phytol 174(2):332–341

Ejaz S, Fahad S, Anjum MA, Nawaz A, Naz S, Hussain S, Ahmad S (2020) Role of osmolytes in the mechanisms of antioxidant defense of plants. In: Sustainable agriculture reviews, vol 39. Springer, Cham, pp 95–117

Elango D, Rajendran K, Van der Laan L, Sebastiar S, Raigne J, Thaiparambil NA, El Haddad N, Raja B, Wang W, Ferela A, Chiteri KO, Thudi M, Varshney RK, Chopra S, Singh A, Singh AK (2022) Raffinose family oligosaccharides: friend or foe for human and plant health? Front Plant Sci 13. https://doi.org/10.3389/fpls.2022.829118

ElSayed AI, Rafudeen MS, Golldack D (2014) Physiological aspects of raffinose family oligosaccharides in plants: protection against abiotic stress. Plant Biol 16(1):1–8

Ende WVD, Peshev D (2013) Sugars as antioxidants in plants. In: Crop improvement under adverse conditions. Springer, New York, NY, pp 285–307

Fàbregas N, Lozano-Elena F, Blasco-Escámez D, Tohge T, Martínez-Andújar C, Albacete A, Osorio S, Bustamante M, Riechmann JL, Nomura T, Yokota T, Conesa A, Alfocea FP, Fernie AR, Caño-Delgado AI (2018) Overexpression of the vascular brassinosteroid receptor BRL3 confers drought resistance without penalizing plant growth. Nat Commun 9(1):1–13

Fan W, Zhang M, Zhang H, Zhang P (2012) Improved tolerance to various abiotic stresses in transgenic sweet potato (Ipomoea batatas) expressing spinach betaine aldehyde dehydrogenase. PLoS One 7(5):e37344

Farooq M, Aziz T, Hussain M, Rehman H, Jabran K, Khan MB (2008) Glycinebetaine improves chilling tolerance in hybrid maize. J Agron Crop Sci 194(2):152–160

Gong HL, Chen QQ (2021) Exogenous sucrose protects potato seedlings against heat stress by enhancing the antioxidant defense system. J Soil Sci Plant Nutr 21(2):1511–1519

Guo H, Li L, Ye H, Yu X, Algreen A, Yin Y (2009) Three related receptor-like kinases are required for optimal cell elongation in *Arabidopsis thaliana*. Proc Natl Acad Sci 106(18):7648–7653

Guo R, Yuan G, Wang Q (2011) Effect of sucrose and mannitol on the accumulation of health-promoting compounds and the activity of metabolic enzymes in broccoli sprouts. Sci Hortic 128(3):159–165

Gupta B, Huang B (2014) Mechanism of salinity tolerance in plants: physiological, biochemical, and molecular characterization. Int J Genomics 2014:701596

Gupta B, Gupta K, Sengupta DN (2012) Spermidine-mediated in vitro phosphorylation of transcriptional regulator OSBZ8 by SNF1-type serine/threonine protein kinase SAPK4 homolog in indica rice. Acta Physiol Plant 34(4):1321–1336

Gupta P, Rai R, Vasudev S, Yadava DK, Dash PK (2021) Ex-foliar application of glycine betaine and its impact on protein, carbohydrates and induction of ROS scavenging system during drought stress in flax (*Linum usitatissimum*). J Biotechnol 337:80–89

Habiba U, Ali S, Rizwan M, Ibrahim M, Hussain A, Shahid MR, Alamri SA, Alyemeni MN, Ahmad P (2019) Alleviative role of exogenously applied mannitol in maize cultivars differing in chromium stress tolerance. Environ Sci Pollut Res 26(5):5111–5121

Hasan SA, Hayat S, Ali B, Ahmad A (2008) 28-Homobrassinolide protects chickpea (*Cicer arietinum*) from cadmium toxicity by stimulating antioxidants. Environ Pollut 151(1):60–66

Hasan MM, Alabdallah NM, Alharbi BM, Waseem M, Yao G, Liu XD, El-Gawad HGA, El-Yazied AA, Ibrahim MFM, Jahan MS, Fang XW (2021) GABA: a key player in drought stress resistance in plants. Int J Mol Sci 22(18):10136

Hasanuzzaman M, Alam M, Rahman A, Hasanuzzaman M, Nahar K, Fujita M (2014) Exogenous proline and glycine betaine mediated upregulation of antioxidant defense and glyoxalase systems provides better protection against salt-induced oxidative stress in two rice (*Oryza sativa* L.) varieties. BioMed Res Int 2014:757219

Hasanuzzaman M, Banerjee A, Bhuyan MB, Roychoudhury A, Al Mahmud J, Fujita M (2019) Targeting glycinebetaine for abiotic stress tolerance in crop plants: physiological mechanism, molecular interaction and signaling. Phyton 88(3):185

Hasanuzzaman M, Bhuyan MB, Raza A, Hawrylak-Nowak B, Matraszek-Gawron R, Al Mahmud J, Nahar K, Fujita M (2020) Selenium in plants: boon or bane? Environ Exp Bot 178:104170

Hassan MJ, Zhang G, Wu F, Wei K, Chen Z (2005) Zinc alleviates growth inhibition and oxidative stress caused by cadmium in rice. J Plant Nutr Soil Sci 168(2):255–261

Hassan MU, Chattha MU, Khan I, Chattha MB, Aamer M, Nawaz M, Ali A, Khan MAU, Khan TA (2019) Nickel toxicity in plants: reasons, toxic effects, tolerance mechanisms, and remediation possibilities—a review. Environ Sci Pollut Res 26(13):12673–12688

Hayat S, Ali B, Hasan SA, Ahmad A (2007) Brassinosteroid enhanced the level of antioxidants under cadmium stress in *Brassica juncea*. Environ Exp Bot 60(1):33–41

Hayat S, Masood A, Yusuf M, Fariduddin Q, Ahmad A (2009) Growth of Indian mustard (*Brassica juncea* L.) in response to salicylic acid under high-temperature stress. Braz J Plant Physiol 21: 187–195

Hayat S, Hayat Q, Alyemeni MN, Wani AS, Pichtel J, Ahmad A (2012) Role of proline under changing environments: a review. Plant Signal Behav 7(11):1456–1466

Hernandez-Marin E, Martínez A (2012) Carbohydrates and their free radical scavenging capability: a theoretical study. J Phys Chem B 116(32):9668–9675

Hincha DK, Zuther E, Heyer AG (2003) The preservation of liposomes by raffinose family oligosaccharides during drying is mediated by effects on fusion and lipid phase transitions. Biochim Biophy Acta Biomembr 1612(2):172–177

Hoque MA, Banu MNA, Okuma E, Amako K, Nakamura Y, Shimoishi Y, Murata Y (2007) Exogenous proline and glycinebetaine increase NaCl-induced ascorbate–glutathione cycle enzyme activities, and proline improves salt tolerance more than glycinebetaine in tobacco Bright Yellow-2 suspension-cultured cells. J Plant Physiol 164(11):1457–1468

Hossain MA, Hasanuzzaman M, Fujita M (2010) Up-regulation of antioxidant and glyoxalase systems by exogenous glycinebetaine and proline in mung bean confer tolerance to cadmium stress. Physiol Mol Biol Plants 16(3):259–272

Hossain MA, Hasanuzzaman M, Fujita M (2011) Coordinate induction of antioxidant defense and glyoxalase system by exogenous proline and glycinebetaine is correlated with salt tolerance in mung bean. Front Agric China 5(1):1–14

Hossain MA, Hoque MA, Burritt DJ, Fujita M (2014a) Proline protects plants against abiotic oxidative stress: biochemical and molecular mechanisms. In: Oxidative damage to plants. Academic Press, Amsterdam, pp 477–522

Hossain MA, Golam Mostof M, Fujita M (2014b) Modulation of reactive oxygen species and methylglyoxal detoxification systems by exogenous glycinebetaine and proline improves drought tolerance in mustard (*Brassica juncea* L.). Int J Plant Biol Res 2:14

Hu L, Hu T, Zhang X, Pang H, Fu J (2012) Exogenous glycine betaine ameliorates the adverse effect of salt stress on perennial ryegrass. J Am Soc Hortic Sci 137(1):38–46

Hu L, Zhou K, Li Y, Chen X, Liu B, Li C, Gong X, Ma F (2018) Exogenous myo-inositol alleviates salinity-induced stress in *Malus hupehensis* Rehd. Plant Physiol Biochem 133:116–126

Jahantigh O, Najafi F, Badi HN, Khavari-Nejad RA, Sanjarian F (2016) Changes in antioxidant enzymes activities and proline, total phenol and anthocyanine contents in *Hyssopus officinalis* L. plants under salt stress. Acta Biol Hung 67(2):195–204

Jalil SU, Ansari MI (2020) Physiological role of Gamma-aminobutyric acid in salt stress tolerance. In: Salt and drought stress tolerance in plants. Springer, Cham, pp 337–350

Jalil SU, Ahmad I, Ansari MI (2017) Functional loss of GABA transaminase (GABA-T) expressed early leaf senescence under various stress conditions in *Arabidopsis thaliana*. Curr Plant Biol 9: 11–22

Jalil SU, Khan MAA, Ansari MI (2019a) Tremendous role of GABA shunt metabolic pathway in plant system. J Biol Chem Res 36(1):188–196

Jalil SU, Khan MIR, Ansari MI (2019b) Role of GABA transaminase in the regulation of development and senescence in *Arabidopsis thaliana*. Curr Plant Biol 19:100119

Jia Q, Kong D, Li Q, Sun S, Song J, Zhu Y, Liang K, Ke Q, Lin W, Huang J (2019) The function of inositol phosphatases in plant tolerance to abiotic stress. Int J Mol Sci 20(16):3999

Jogawat A, Yadav B, Lakra N, Singh AK, Narayan OP (2021) Crosstalk between phytohormones and secondary metabolites in the drought stress tolerance of crop plants: a review. Physiol Plant 172(2):1106–1132

Kaplan F, Kopka J, Haskell DW, Zhao W, Schiller KC, Gatzke N, Sung DL, Guy CL (2004) Exploring the temperature-stress metabolome of *Arabidopsis*. Plant Physiol 136(4):4159–4168

Karimi R, Ershadi A (2015) Role of exogenous abscisic acid in adapting of 'Sultana' grapevine to low-temperature stress. Acta Physiol Plant 37(8):1–11

Keunen ELS, Peshev D, Vangronsveld J, Van Den Ende WIM, Cuypers ANN (2013) Plant sugars are crucial players in the oxidative challenge during abiotic stress: extending the traditional concept. Plant Cell Environ 36(7):1242–1255

Khedr AHA, Abbas MA, Wahid AAA, Quick WP, Abogadallah GM (2003) Proline induces the expression of salt-stress-responsive proteins and may improve the adaptation of *Pancratium maritimum* L. to salt-stress. J Exp Bot 54(392):2553–2562

Kishor PK, Sangam S, Amrutha RN, Laxmi PS, Naidu KR, Rao KRSS, Rao S, Reddy KJ, Theriappan P, Sreenivasulu N (2005) Regulation of proline biosynthesis, degradation, uptake and transport in higher plants: its implications in plant growth and abiotic stress tolerance. Curr Sci 88:424–438

Knaupp M, Mishra KB, Nedbal L, Heyer AG (2011) Evidence for a role of raffinose in stabilizing photosystem II during freeze–thaw cycles. Planta 234(3):477–486

Kosar F, Akram NA, Sadiq M, Al-Qurainy F, Ashraf M (2019) Trehalose: a key organic osmolyte effectively involved in plant abiotic stress tolerance. J Plant Growth Regul 38(2):606–618

Kumar P (2021) Soil applied glycine betaine with Arbuscular mycorrhizal fungi reduces chromium uptake and ameliorates chromium toxicity by suppressing the oxidative stress in three genetically different Sorghum (*Sorghum bicolor* L.) cultivars. BMC Plant Biol 21(1):1–16

Kumar P, Rouphael Y, Cardarelli M, Colla G (2015) Effect of nickel and grafting combination on yield, fruit quality, antioxidative enzyme activities, lipid peroxidation, and mineral composition of tomato. J Plant Nutr Soil Sci 178(6):848–860

Kumar V, Shriram V, Hoque TS, Hasan M, Burritt DJ, Hossain MA (2017) Glycinebetaine-mediated abiotic oxidative-stress tolerance in plants: physiological and biochemical mechanisms. In: Stress signaling in plants: genomics and proteomics perspective, vol 2. Springer, Cham, pp 111–133

Kumar N, Gautam A, Dubey AK, Ranjan R, Pandey A, Kumari B, Sngh G, Mandotra S, Chauhan PS, Srikishna S, Dutta V, Mallick S (2019) GABA mediated reduction of arsenite toxicity in rice seedling through modulation of fatty acids, stress responsive amino acids and polyamines biosynthesis. Ecotoxicol Environ Saf 173:15–27

La VH, Lee BR, Zhang Q, Park SH, Islam M, Kim TH (2019) Salicylic acid improves drought-stress tolerance by regulating the redox status and proline metabolism in *Brassica rapa*. Hortic Environ Biotechnol 60(1):31–40

Li L, Staden JV, Jäger AK (1998) Effects of plant growth regulators on the antioxidant system in seedlings of two maize cultivars subjected to water stress. Plant Growth Regul 25(2):81–87

Li F, Lei H, Zhao X, Tian R, Li T (2012) Characterization of three sorbitol transporter genes in micropropagated apple plants grown under drought stress. Plant Mol Biol Report 30(1): 123–130

Li M, Li Z, Li S, Guo S, Meng Q, Li G, Yang X (2014a) Genetic engineering of glycine betaine biosynthesis reduces heat-enhanced photoinhibition by enhancing antioxidative defense and alleviating lipid peroxidation in tomato. Plant Mol Biol Report 32(1):42–51

Li T, Hu Y, Du X, Tang H, Shen C, Wu J (2014b) Salicylic acid alleviates the adverse effects of salt stress in *Torreya grandis* cv. Merrillii seedlings by activating photosynthesis and enhancing antioxidant systems. PLOS One 9(10):e109492

Li D, Zhang T, Wang M, Liu Y, Brestic M, Chen TH, Yang X (2019) Genetic engineering of the biosynthesis of glycine betaine modulates phosphate homeostasis by regulating phosphate acquisition in tomato. Front Plant Sci 9:1995

Li L, Dou N, Zhang H, Wu C (2021) The versatile GABA in plants. Plant Signal Behav 16(3): 1862565

Liang C, Zhang XY, Luo Y, Wang GP, Zou Q, Wang W (2009) Overaccumulation of glycine betaine alleviates the negative effects of salt stress in wheat. Russ J Plant Physiol 56(3):370–376

Liang X, Zhang L, Natarajan SK, Becker DF (2013) Proline mechanisms of stress survival. Antioxid Redox Signal 19(9):998–1011

Liu C, Zhao L, Yu G (2011) The dominant glutamic acid metabolic flux to produce γ-amino butyric acid over proline in *Nicotiana tabacum* leaves under water stress relates to its significant role in antioxidant activity. J Integr Plant Biol 53(8):608–618

Liu L, Huang L, Lin X, Sun C (2020) Hydrogen peroxide alleviates salinity-induced damage through enhancing proline accumulation in wheat seedlings. Plant Cell Rep 39(5):567–575

Lo Bianco R, Rieger M, Sung SJS (2000) Effect of drought on sorbitol and sucrose metabolism in sinks and sources of peach. Physiol Plant 108(1):71–78

Loescher WH, Tyson RH, Everard JD, Redgwell RJ, Bieleski RL (1992) Mannitol synthesis in higher plants: evidence for the role and characterization of a NADPH-dependent mannose 6-phosphate reductase. Plant Physiol 98(4):1396–1402

Lombardi L, Sebastiani L (2005) Copper toxicity in *Prunus cerasifera*: growth and antioxidant enzymes responses of in vitro grown plants. Plant Sci 168(3):797–802

Lou Y, Yang Y, Hu L, Liu H, Xu Q (2015) Exogenous glycinebetaine alleviates the detrimental effect of Cd stress on perennial ryegrass. Ecotoxicology 24(6):1330–1340

Luo Y, Li WM, Wang W (2008) Trehalose: protector of antioxidant enzymes or reactive oxygen species scavenger under heat stress? Environ Exp Bot 63(1-3):378–384

Luo Y, Li F, Wang GP, Yang XH, Wang W (2010) Exogenously-supplied trehalose protects thylakoid membranes of winter wheat from heat-induced damage. Biol Plant 54(3):495–501

Mäkelä P, Kärkkäinen J, Somersalo SJBP (2000) Effect of glycinebetaine on chloroplast ultrastructure, chlorophyll and protein content, and RuBPCO activities in tomato grown under drought or salinity. Biol Plant 43(3):471–475

Malekzadeh P (2015) Influence of exogenous application of glycinebetaine on antioxidative system and growth of salt-stressed soybean seedlings (*Glycine max* L.). Physiol Mol Biol Plants 21(2): 225–232

Malekzadeh P, Khara J, Heydari R (2014) Alleviating effects of exogenous Gamma-aminobutiric acid on tomato seedling under chilling stress. Physiol Mol Biol Plants 20(1):133–137

Matros A, Peshev D, Peukert M, Mock HP, Van den Ende W (2015) Sugars as hydroxyl radical scavengers: proof-of-concept by studying the fate of sucralose in *Arabidopsis*. Plant J 82(5): 822–839

Matysik J, Alia B, Mohanty P (2002) Molecular mechanisms of quenching of reactive oxygen species by proline under stress in plants. Curr Sci 82:525–532

Mazzucotelli E, Tartari A, Cattivelli L, Forlani G (2006) Metabolism of γ-aminobutyric acid during cold acclimation and freezing and its relationship to frost tolerance in barley and wheat. J Exp Bot 57(14):3755–3766

Merewitz EB, Du H, Yu W, Liu Y, Gianfagna T, Huang B (2012) Elevated cytokinin content in ipt transgenic creeping bentgrass promotes drought tolerance through regulating metabolite accumulation. J Exp Bot 63(3):1315–1328

Merwad ARM, Desoky ESM, Rady MM (2018) Response of water deficit-stressed *Vigna unguiculata* performances to silicon, proline or methionine foliar application. Sci Hortic 228: 132–144

Molla MR, Ali MR, Hasanuzzaman M, Al-Mamun MH, Ahmed A, Nazim-Ud-Dowla MAN, Rohman MM (2014) Exogenous proline and betaine-induced upregulation of glutathione transferase and glyoxalase I in lentil (*Lens culinaris*) under drought stress. Not Bot Horti Agrobot Cluj Napoca 42(1):73–80

Montesinos-Pereira D, Barrameda-Medina Y, Romero L, Ruiz JM, Sánchez-Rodríguez E (2014) Genotype differences in the metabolism of proline and polyamines under moderate drought in tomato plants. Plant Biol 16(6):1050–1057

Munns R (2005) Genes and salt tolerance: bringing them together. New Phytol 167(3):645–663

Nayyar H, Kaur R, Kaur S, Singh R (2014) γ-Aminobutyric acid (GABA) imparts partial protection from heat stress injury to rice seedlings by improving leaf turgor and upregulating osmoprotectants and antioxidants. J Plant Growth Regul 33(2):408–419

Nemati F, Ghanati F, Gavlighi HA, Sharifi M (2018) Fructan dynamics and antioxidant capacity of 4-day-old seedlings of wheat (*Triticum aestivum*) cultivars during drought stress and recovery. Funct Plant Biol 45(10):1000–1008

Nishikawa F, Kato M, Hyodo H, Ikoma Y, Sugiura M, Yano M (2005) Effect of sucrose on ascorbate level and expression of genes involved in the ascorbate biosynthesis and recycling pathway in harvested broccoli florets. J Exp Bot 56(409):65–72

Nishizawa A, Yabuta Y, Shigeoka S (2008) Galactinol and raffinose constitute a novel function to protect plants from oxidative damage. Plant Physiol 147(3):1251–1263

Nunez M, Mazzafera P, Mazorra LM, Siqueira WJ, Zullo MAT (2003) Influence of a brassinosteroid analogue on antioxidant enzymes in rice grown in culture medium with NaCl. Biol Plant 47(1):67–70

Park EJ, Jeknić Z, Pino MT, Murata N, Chen THH (2007) Glycinebetaine accumulation is more effective in chloroplasts than in the cytosol for protecting transgenic tomato plants against abiotic stress. Plant Cell Environ 30(8):994–1005

Parvanova D, Ivanov S, Konstantinova T, Karanov E, Atanassov A, Tsvetkov T, Alexieva V, Djilianov D (2004) Transgenic tobacco plants accumulating osmolytes show reduced oxidative damage under freezing stress. Plant Physiol Biochem 42(1):57–63

Patel TK, Williamson JD (2016) Mannitol in plants, fungi, and plant–fungal interactions. Trends Plant Sci 21(6):486–497

Peshev D, Vergauwen R, Moglia A, Hideg É, Van den Ende W (2013) Towards understanding vacuolar antioxidant mechanisms: a role for fructans? J Exp Bot 64(4):1025–1038

Puniran-Hartley N, Hartley J, Shabala L, Shabala S (2014) Salinity-induced accumulation of organic osmolytes in barley and wheat leaves correlates with increased oxidative stress tolerance: in planta evidence for cross-tolerance. Plant Physiol Biochem 83:32–39

Rajasheker G, Jawahar G, Jalaja N, Kumar SA, Kumari PH, Punita DL, Karumanchi AP, Reddy PS, Rathnagiri P, Sreenivasulu N, Kishor PBK (2019) Role and regulation of osmolytes and ABA interaction in salt and drought stress tolerance. In: Plant signaling molecules. Woodhead Publishing, Duxford, pp 417–436

Ramel F, Sulmon C, Cabello-Hurtado F, Taconnat L, Martin-Magniette ML, Renou JP, El Armani A, Couee I, Gouesbet G (2007) Genome-wide interacting effects of sucrose and herbicide-mediated stress in *Arabidopsis thaliana*: novel insights into atrazine toxicity and sucrose-induced tolerance. BMC Genomics 8(1):1–20

Ramel F, Sulmon C, Bogard M, Couée I, Gouesbet G (2009) Differential patterns of reactive oxygen species and antioxidative mechanisms during atrazine injury and sucrose-induced tolerance in *Arabidopsis thaliana* plantlets. BMC Plant Biol 9(1):1–18

Ramesh SA, Tyerman SD, Gilliham M, Xu B (2017) γ-Aminobutyric acid (GABA) signalling in plants. Cell Mol Life Sci 74(9):1577–1603

Ramos-Ruiz R, Martinez F, Knauf-Beiter G (2019) The effects of GABA in plants. Cogent Food Agric 5(1):1670553

Rehman AU, Bashir F, Ayaydin F, Kóta Z, Páli T, Vass I (2021) Proline is a quencher of singlet oxygen and superoxide both in in vitro systems and isolated thylakoids. Physiol Plant 172(1): 7–18

Rejeb KB, Abdelly C, Savouré A (2014) How reactive oxygen species and proline face stress together. Plant Physiol Biochem 80:278–284

Rezaei-Chiyaneh E, Seyyedi SM, Ebrahimian E, Moghaddam SS, Damalas CA (2018) Exogenous application of gamma-aminobutyric acid (GABA) alleviates the effect of water deficit stress in black cumin (Nigella sativa L.). Ind Crop Prod 112:741–748

Rigui AP, Carvalho V, Dos Santos ALW, Morvan-Bertrand A, Prud'Homme MP, de Carvalho MAM, Gaspar M (2019) Fructan and antioxidant metabolisms in plants of *Lolium perenne* under drought are modulated by exogenous nitric oxide. Plant Physiol Biochem 145:205–215

Rizwan M, Ali S, Hussain A, Ali Q, Shakoor MB, Zia-ur-Rehman M, Farid M, Asma M (2017) Effect of zinc-lysine on growth, yield and cadmium uptake in wheat (*Triticum aestivum* L.) and health risk assessment. Chemosphere 187:35–42

Rodriguez MM, Heyser JW (1988) Growth inhibition by exogenous proline and its metabolism in saltgrass (*Distichlis spicata*) suspension cultures. Plant Cell Rep 7(5):305–308

Roy D, Basu N, Bhunia A, Banerjee SK (1993) Counteraction of exogenous L-proline with NaCl in salt-sensitive cultivar of rice. Biol Plant 35(1):69–72

Sachdev S, Ansari SA, Ansari MI, Fujita M, Hasanuzzaman M (2021) Abiotic stress and reactive oxygen species: generation, signaling, and defense mechanisms. Antioxidants 10(2):277

Salehi M, Karimzadeh G, Naghavi MR (2019) Synergistic effect of coronatine and sorbitol on artemisinin production in cell suspension culture of Artemisia annua L. cv. Anamed. Plant Cell Tissue Organ Cult 137(3):587–597

Salvi P, Saxena SC, Petla BP, Kamble NU, Kaur H, Verma P, Rao V, Ghosh S, Majee M (2016) Differentially expressed galactinol synthase (s) in chickpea are implicated in seed vigor and longevity by limiting the age induced ROS accumulation. Sci Rep 6(1):1–15

Salvi P, Kamble NU, Majee M (2020) Ectopic over-expression of ABA-responsive Chickpea galactinol synthase (CaGolS) gene results in improved tolerance to dehydration stress by modulating ROS scavenging. Environ Exp Bot 171:103957

Saradhi PP, Mohanty P (1997) Involvement of proline in protecting thylakoid membranes against free radical-induced photodamage. J Photochem Photobiol B Biol 38(2–3):253–257

Sarafraz-Ardakani MR, Khavari-Nejad RA, Moradi F, Najafi F (2014) Abscisic acid and cytokinin-induced osmotic and antioxidant regulation in two drought-tolerant and drought-sensitive cultivars of wheat during grain filling under water deficit in field conditions. Not Sci Biol 6(3):354–362

Savchenko T, Tikhonov K (2021) Oxidative stress-induced alteration of plant central metabolism. Life 11(4):304

Seckin B, Sekmen AH, Türkan I (2009) An enhancing effect of exogenous mannitol on the antioxidant enzyme activities in roots of wheat under salt stress. J Plant Growth Regul 28(1): 12–20

Sengupta S, Patra B, Ray S, Majumder AL (2008) Inositol methyl tranferase from a halophytic wild rice, *Porteresia coarctata* Roxb.(Tateoka): regulation of pinitol synthesis under abiotic stress. Plant Cell Environ 31(10):1442–1459

Sengupta S, Mukherjee S, Basak P, Majumder AL (2015) Significance of galactinol and raffinose family oligosaccharide synthesis in plants. Front Plant Sci 6:656

Shafiq S, Akram NA, Ashraf M (2015) Does exogenously-applied trehalose alter oxidative defense system in the edible part of radish (*Raphanus sativus* L.) under water-deficit conditions? Sci Hortic 185:68–75

Sharma SS, Dietz KJ (2006) The significance of amino acids and amino acid-derived molecules in plant responses and adaptation to heavy metal stress. J Exp Bot 57(4):711–726

Sharma A, Shahzad B, Kumar V, Kohli SK, Sidhu GPS, Bali AS, Handa N, Kapoor D, Bhardwaj R, Zheng B (2019) Phytohormones regulate accumulation of osmolytes under abiotic stress. Biomolecules 9(7):285

Shemi R, Wang R, Gheith ES, Hussain HA, Hussain S, Irfan M, Cholidah M, Zhang K, Zhang S, Wang L (2021) Effects of salicylic acid, zinc and glycine betaine on morpho-physiological growth and yield of maize under drought stress. Sci Rep 11(1):1–14

Shen BO, Jensen RG, Bohnert HJ (1997) Increased resistance to oxidative stress in transgenic plants by targeting mannitol biosynthesis to chloroplasts. Plant Physiol 113(4):1177–1183

Sheteiwy MS, Shao H, Qi W, Hamoud YA, Shaghaleh H, Khan NU, Yang R, Tang B (2019) GABA-alleviated oxidative injury induced by salinity, osmotic stress and their combination by regulating cellular and molecular signals in rice. Int J Mol Sci 20(22):5709

Sheveleva EV, Marquez S, Chmara W, Zegeer A, Jensen RG, Bohnert HJ (1998) Sorbitol-6-phosphate dehydrogenase expression in transgenic tobacco: high amounts of sorbitol lead to necrotic lesions. Plant Physiol 117(3):831–839

Signorelli S, Arellano JB, Melø TB, Borsani O, Monza J (2013) Proline does not quench singlet oxygen: evidence to reconsider its protective role in plants. Plant Physiol Biochem 64:80–83

Signorelli S, Coitiño EL, Borsani O, Monza J (2014) Molecular mechanisms for the reaction between• OH radicals and proline: insights on the role as reactive oxygen species scavenger in plant stress. J Phys Chem B 118(1):37–47

Signorelli S, Imparatta C, Rodríguez-Ruiz M, Borsani O, Corpas FJ, Monza J (2016) In vivo and in vitro approaches demonstrate proline is not directly involved in the protection against superoxide, nitric oxide, nitrogen dioxide and peroxynitrite. Funct Plant Biol 43(9):870–879

Silva VM, Tavanti RFR, Gratão PL, Alcock TD, Dos Reis AR (2020) Selenate and selenite affect photosynthetic pigments and ROS scavenging through distinct mechanisms in cowpea (*Vigna unguiculata* (L.) walp) plants. Ecotoxicol Environ Saf 201:110777

Singh M, Singh VP, Dubey G, Prasad SM (2015) Exogenous proline application ameliorates toxic effects of arsenate in *Solanum melongena* L. seedlings. Ecotoxicol Environ Saf 117:164–173

Siracusa L, Napoli E, Ruberto G (2022) Novel chemical and biological insights of inositol derivatives in mediterranean plants. Molecules 27(5):1525

Sita K, Kumar V (2020) Role of Gamma Amino Butyric Acid (GABA) against abiotic stress tolerance in legumes: a review. Plant Physiol Rep 25(4):654–663

Smirnoff N, Cumbes QJ (1989) Hydroxyl radical scavenging activity of compatible solutes. Phytochemistry 28(4):1057–1060

Soares C, Carvalho ME, Azevedo RA, Fidalgo F (2019) Plants facing oxidative challenges—a little help from the antioxidant networks. Environ Exp Bot 161:4–25

Sorwong A, Sakhonwasee S (2015) Foliar application of glycine betaine mitigates the effect of heat stress in three marigold (*Tagetes erecta*) cultivars. Horticult J MI-038. https://doi.org/10.2503/hortj.MI-038

Steward FC (1949) γ-Aminobutyric acid: a constituent of the potato tuber? Science 110:439–440

Syeed S, Sehar Z, Masood A, Anjum NA, Khan NA (2021) Control of elevated ion accumulation, oxidative stress, and lipid peroxidation with salicylic acid-induced accumulation of glycine betaine in salinity-exposed *Vigna radiata* L. Appl Biochem Biotechnol 193(10):3301–3320

Takahama U (2004) Oxidation of vacuolar and apoplastic phenolic substrates by peroxidase: physiological significance of the oxidation reactions. Phytochem Rev 3(1):207–219

Tavanti RFR, David Queiroz G, Da Rocha C, Silva A, Moya Peres W, Pereira Paixão A, Galindo FS, Silva VM, Bossolani JW, Melero MM, Oliveira GDS, Junior EF, Dos Reis AR (2020) Changes in photosynthesis and antioxidant metabolism of cotton (*Gossypium hirsutum* L.) plants in response to manganese stress. Arch Agron Soil Sci 66(6):743–762

Tavanti TR, de Melo AAR, Moreira LDK, Sanchez DEJ, dos Santos Silva R, da Silva RM, Dos Reis AR (2021) Micronutrient fertilization enhances ROS scavenging system for alleviation of abiotic stresses in plants. Plant Physiol Biochem 160:386–396

Tisarum R, Theerawitaya C, Samphumphuang T, Singh HP, Cha-Um S (2020) Foliar application of glycinebetaine regulates soluble sugars and modulates physiological adaptations in sweet potato (Ipomoea batatas) under water deficit. Protoplasma 257(1):197–211

Toubiana D, Sade N, Liu L, Rubio Wilhelmi MDM, Brotman Y, Luzarowska U, Vogel JP, Blumwald E (2020) Correlation-based network analysis combined with machine learning techniques highlight the role of the GABA shunt in *Brachypodium sylvaticum* freezing tolerance. Sci Rep 10(1):1–16

Upadhyay R, Panda SK (2010) Zinc reduces copper toxicity induced oxidative stress by promoting antioxidant defense in freshly grown aquatic duckweed *Spirodela polyrhiza* L. J Hazard Mater 175(1-3):1081–1084

Valluru R, Van den Ende W (2011) Myo-inositol and beyond–emerging networks under stress. Plant Sci 181(4):387–400

Wang F, Zeng B, Sun Z, Zhu C (2009) Relationship between proline and Hg2+-induced oxidative stress in a tolerant rice mutant. Arch Environ Contam Toxicol 56(4):723–731

Wang Y, Luo Z, Huang X, Yang K, Gao S, Du R (2014) Effect of exogenous γ-aminobutyric acid (GABA) treatment on chilling injury and antioxidant capacity in banana peel. Sci Hortic 168: 132–137

Wang Y, Gu W, Meng Y, Xie T, Li L, Li J, Wei S (2017) γ-Aminobutyric acid imparts partial protection from salt stress injury to maize seedlings by improving photosynthesis and upregulating osmoprotectants and antioxidants. Sci Rep 7(1):1–13

Wang CX, Liu L, Li Y, Song X, Li J (2022) Exogenous sorbitol improves maize tolerance to drought through up-regulating antioxidant system and endogenous sorbitol biosynthesis. Available at SSRN 4079608

Wei D, Zhang W, Wang C, Meng Q, Li G, Chen TH, Yang X (2017) Genetic engineering of the biosynthesis of glycinebetaine leads to alleviate salt-induced potassium efflux and enhances salt tolerance in tomato plants. Plant Sci 257:74–83

Williamson JD, Jennings DB, Guo WW, Pharr DM, Ehrenshaft M (2002) Sugar alcohols, salt stress, and fungal resistance: polyols—multifunctional plant protection? J Am Soc Hortic Sci 127(4):467–473

Wu X, Jia Q, Ji S, Gong B, Li J, Lü G, Gao H (2020) Gamma-aminobutyric acid (GABA) alleviates salt damage in tomato by modulating Na+ uptake, the GAD gene, amino acid synthesis and reactive oxygen species metabolism. BMC Plant Biol 20(1):1–21

Wu Q, Su N, Huang X, Cui J, Shabala L, Zhou M, Yu M, Shabala S (2021) Hypoxia-induced increase in GABA content is essential for restoration of membrane potential and preventing ROS-induced disturbance to ion homeostasis. Plant Commun 2(3):100188

Xu J, Yin H, Li X (2009) Protective effects of proline against cadmium toxicity in micropropagated hyperaccumulator, *Solanum nigrum* L. Plant Cell Reports 28(2):325–333

Yan Z, Chen J, Li X (2013) Methyl jasmonate as modulator of Cd toxicity in *Capsicum frutescens* var. fasciculatum seedlings. Ecotoxicol Environ Saf 98:203–209

Yancey PH (2005) Organic osmolytes as compatible, metabolic and counteracting cytoprotectants in high osmolarity and other stresses. J Exp Biol 208(15):2819–2830

Yang X, Lu C (2005) Photosynthesis is improved by exogenous glycinebetaine in salt-stressed maize plants. Physiol Plant 124(3):343–352

Yang A, Cao S, Yang Z, Cai Y, Zheng Y (2011) γ-Aminobutyric acid treatment reduces chilling injury and activates the defence response of peach fruit. Food Chem 129(4):1619–1622

Yang Z, Yu J, Merewitz E, Huang B (2012) Differential effects of abscisic acid and glycine betaine on physiological responses to drought and salinity stress for two perennial grass species. J Am Soc Hortic Sci 137(2):96–106

Yildizli A, Çevik S, Ünyayar S (2018) Effects of exogenous myo-inositol on leaf water status and oxidative stress of *Capsicum annuum* under drought stress. Acta Physiol Plant 40(6):1–10

Zaid A, Ahmad B, Jaleel H, Wani SH, Hasanuzzaman M (2020) A critical review on iron toxicity and tolerance in plants: role of exogenous phytoprotectants. In: Plant micronutrients. Springer, Cham, pp 83–99

Zhang H, Sun X, Dai M (2021) Improving crop drought resistance with plant growth regulators and rhizobacteria: mechanisms, applications, and perspectives. Plant Communications 3:100228

Zhu X, Liao J, Xia X, Xiong F, Li Y, Shen J, Wen B, Ma Y, Wang Y, Fang W (2019) Physiological and iTRAQ-based proteomic analyses reveal the function of exogenous γ-aminobutyric acid (GABA) in improving tea plant (*Camellia sinensis* L.) tolerance at cold temperature. BMC Plant Biol 19(1):1–20

Zouari M, Hassena AB, Trabelsi L, Rouina BB, Decou R, Labrousse P (2019) Exogenous proline-mediated abiotic stress tolerance in plants: possible mechanisms. In: Osmoprotectant-mediated abiotic stress tolerance in plants. Springer, Cham, pp 99–121

Zulfiqar F, Ashraf M, Siddique KH (2022) Role of glycine betaine in the thermotolerance of plants. Agronomy 12(2):276

ROS as Signaling Molecule Under Unfavorable Conditions

11

11.1 Introduction

Plants exorbitantly generate reactive oxygen species (ROS) as an unavoidable consequence of their metabolic performance under both abiotic and biotic stress conditions. Generally, antioxidants and certain osmolytes maintain a threshold level of ROS in plant tissues. However, the prolonged exposure to stresses diminishes the intrinsic defense ability of plants to quench excess ROS. This results in an abundance of ROS in cells, leading to oxidative damage. Thus, ROS has always been regarded as the deleterious agent, inducing distress to vital biomolecules. Nevertheless, recent studies have evidently demonstrated the role of ROS as a signaling molecule, inducing "oxidative signaling," mediating stress tolerance, plant growth, autophagy, and programmed cell death (PCD) (Foyer and Noctor 2005; Kumar et al. 2017; Mittler 2017; Suzuki and Katano 2018; Turkan 2018). In fact, ROS are the redox molecules which display hormesis (Schieber and Chandel 2014). The transient accumulation of ROS in plant cells effectuates signal transduction pathway from within cell (organelle-to-nucleus) and cell-to-cell, eliciting adaptive response against adverse situations (Mittler et al. 2011). ROS-induced signaling cascade modulates components downstream or ROS may act downstream of other signaling molecules such as Ca^{2+}, RNS, and phytohormones to eventually produce desired response (Bhattacharjee 2005; Mignolet-Spruyt et al. 2016; Dvořák et al. 2021).

ROS including superoxide radical anions ($O_2^{\bullet -}$), hydroxyl ion ($^{\bullet}OH$), singlet oxygen ($^1O_2$), and hydrogen peroxide (H_2O_2) are generated in plant organelles via different routes. For instance, ROS are formed in chloroplast and mitochondria as a consequence of over-reduced electron transport chain (ETC), in peroxisomes as a by-product of photorespiration and activity of metabolic enzymes like xanthine oxidase (XO), and in apoplast by action of enzymes such as NADPH (nicotinamide adenine dinucleotide phosphate) oxidase (NOX) and class III peroxidases (POXs) (Mignolet-Spruyt et al. 2016; Martins et al. 2018; Sachdev et al. 2021, 2022). The role of different reactive species in signaling significantly deviates from each other

S. Sachdev et al., *Reactive Oxygen Species in Plants*,
https://doi.org/10.1007/978-981-19-9884-3_11

and shows high specificity. This specific nature of ROS signaling depends on each oxidant species' unique chemical properties like reactivity, stability, ability to diffuse through lipid membranes, and target molecules (Laloi et al. 2004; Choudhury et al. 2013; Reczek and Chandel 2015). ROS including $^{\bullet}OH$, $O_2^{\bullet-}$, and $^1O_2$ have short half-life and are unable to travel longer distances; however, these molecules partake in signaling pathways indirectly via their breakdown products (Mignolet-Spruyt et al. 2016; Turkan 2018). On the contrary, H_2O_2 is more stable due to longer half-life and can accumulate in high levels that allows it to travel longer distances and induce signaling (Sewelam et al. 2016; Sachdev et al. 2022). Thus, H_2O_2 is the crucial ROS, which acts as an effective signaling molecule in plants. Besides chemical properties, the production site and production rate of ROS under stress vary, which favors the signaling process, thus making ROS signaling a complex pathway (Jalmi and Sinha 2015; Reczek and Chandel 2015; Mignolet-Spruyt et al. 2016; Turkan 2018). Comprehensively, ROS are important signaling messengers, which are crucial for induction of response required for plants to sustain their growth. The information pertaining to ROS-mediated signaling under altered environmental regimes and complex ROS signaling cascades initiated from different organelles has been discussed.

11.2 ROS Signaling in Plants

The transient accumulation of ROS in plants under the impression of unfavorable circumstances triggers systemic and retrograde signaling. ROS generated at a particular site triggers genesis of ROS at other sites. This phenomenon is known as ROS-induced ROS release (RIRR) (Zandalinas and Mittler 2018). RIRR leads to cell-to-cell communication via systemic signaling, which initiates from stress-exposed regions known as "local" tissues and heads toward unaffected zones representing "systemic" tissues, to alarm and activate defense mechanisms (Kumar et al. 2017; Fichman and Mittler 2020; Sachdev et al. 2022) (Fig. 11.1). The ROS accumulation in the apoplast of stress-influenced cells is triggered by plasma membrane localized NOX (also known as respiratory burst oxidase homologs, RBOH), which initiates ROS auto-propagating wave that is transmitted to systemic tissues

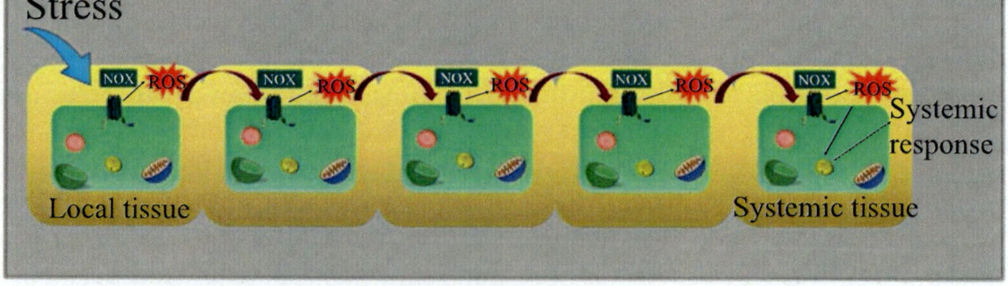

Fig. 11.1 Systemic signaling induced by ROS auto-propagating wave on perception of stress by local tissues, resulting in systemic response at systemic tissues

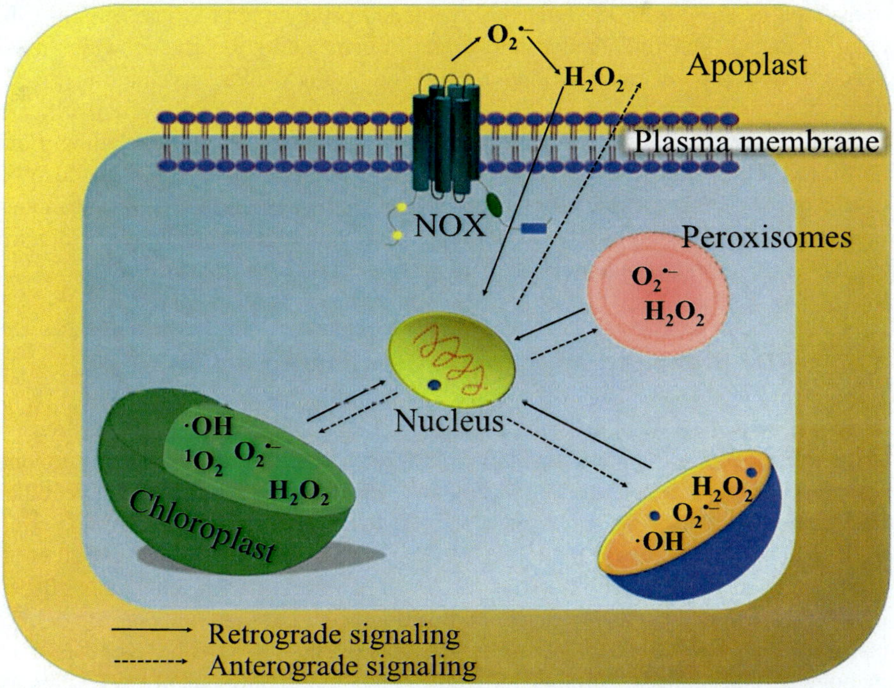

Fig. 11.2 Production of ROS in various cellular organelles triggers retrograde signaling toward nucleus that in turn induces anterograde signaling to produce acclimatization response in particular organelle

with the help of adjoining cells (Mignolet-Spruyt et al. 2016; Fichman and Mittler 2020; Sachdev et al. 2021). The auto-propagating ROS waves travel at very fast speed, corresponding to 8.4 cm min^{-1} (Podgórska et al. 2017; Sachdev et al. 2022). As plants recruit antioxidants to quench ROS, for long distance self-perpetuation of ROS signals, continuous generation of ROS by each cell along the path of ROS wave is required (Mittler et al. 2011) (Fig. 11.1). Therefore, each individual cell involved in ROS systemic signaling recruits its own ROS genesis mechanism for ROS production (Mittler et al. 2011; Mignolet-Spruyt et al. 2016). The systemic response elicited by local pathogen infection is known as systemic acquired resistance (SAR), whereas the systemic response developed in response to abiotic stress is termed systemic acquired acclimation (SAA) (Fichman and Mittler 2020).

Under adverse conditions, cell organelles such as chloroplast, peroxisomes, apoplast, and mitochondria communicate with the nucleus via ROS and ROS-modified targeted molecules initiated retrograde signaling (Choudhury et al. 2013; Cui et al. 2019; Dvořák et al. 2021; Li and Kim 2021; Sachdev et al. 2022) (Fig. 11.2). These retrograde signals modulate gene expression, triggering anterograde signaling back from nuclei to produce stress acclimatization response (Bhattacharjee 2012; Cui et al. 2019; Dvořák et al. 2021; Sachdev et al. 2021, 2022) (Fig. 11.2). The retrograde signaling involves perception, transduction, and translation of discrete ROS signals to eventually produce appropriate response. In chloroplast, $^1O_2$ generated under light stress is quenched by β-carotene, which itself gets oxidized and generates oxidative cleavage products like β-cyclocitral (β-CC) that act as a retrograde signal (Phua et al. 2021). Further, ROS participating in signaling pathways show cross-talk with other signaling pathways including plant hormones. The plant hormones not only act downstream of ROS, but can also be upstream of ROS, particularly in signaling pathways induced by hormones (Bhattacharjee 2005). Thus, ROS participate in both feedback and feedforward responses (Karuppanapandian et al. 2011; Kumar et al. 2017; Sachdev et al. 2022).

11.3 ROS Signaling on Plant–Pathogen Interaction

The pathogens capable of crossing cuticle and cell-wall barriers are recognized by two-level perception system including first-level cell surface localized immune receptor like pattern recognition receptors (PRRs) and second-level intracellular nucleotide binding-leucine-rich repeat (NB-LLR) receptor (NLRs) (Thomma et al. 2011; Bigeard et al. 2015; Qi et al. 2017; Leibman-Markus et al. 2018; Iqbal et al. 2021; Nguyen et al. 2021; Yuan et al. 2021). These receptors either directly or indirectly sense pathogens to induce innate immune and/or defense response (Zipfel et al. 2006; Qi et al. 2017; Leibman-Markus et al. 2018). The perception of pathogens takes place through two different recognition pathways, namely pathogen-associated molecular patterns (PAMP), also known as microbe-associated molecular patterns (MAMP), and pathogen effectors by PRRs and NLRs leading to pattern-triggered immunity (PTI) and effector-triggered immunity (ETI), respectively (Thomma et al. 2011; Bigeard et al. 2015; Leibman-Markus et al. 2018; Nguyen et al. 2021). After recognition of pathogens, plants induce several defense mechanisms including transient burst of ROS that deliver toxic effects on pathogens, stomatal closure and/or trigger PCD to limit pathogen progression at the site of infection (Mur et al. 2008; O'Brien et al. 2012; Bigeard et al. 2015). The locally induced PTI and ETI can trigger SAR to distant locations (Liu et al. 2021). The pathogens interacting with the plants are initially perceived by the plant via PRR localized at the plasma membrane by recognizing PAMP that induces appropriate response. However, certain pathogens produce virulence factors or effectors, which inhibit PAMP-induced immune response. To counter such pathogens, plants have evolved resistance (R) genes encoding NLRs that can detect pathogen virulence factors or effectors and mediate ETI by eliciting biphasic ROS burst, which results in hypersensitive response (HR), leading to PCD, terminating pathogen proliferation to non-infected sites (Zipfel et al. 2006; Tsuda and Katagiri 2010) (Fig. 11.3).

Apoplast is the primary site where ROS burst has been observed during plant–pathogen interaction in PTI (Qi et al. 2017). The plant cell plasma membrane is equipped with various receptors like receptor kinases (RKs) and receptor-like proteins (RLPs), which get activated on recognition of characteristic molecular structure of microbes, i.e., PAMPs, and in turn, activate cellular responses (Qi et al. 2017). The pathogen recognition by receptor modulates activity of ion

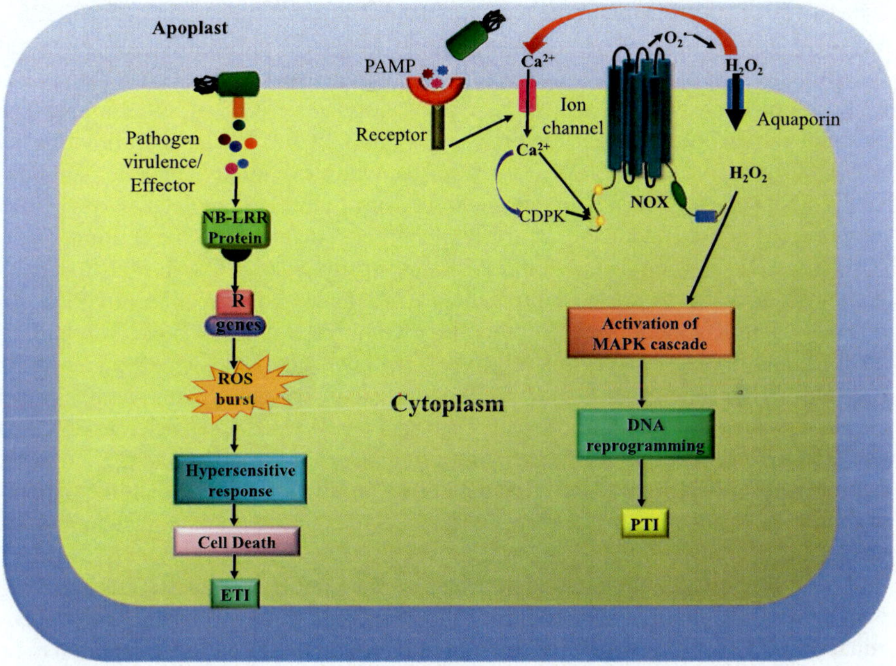

Fig. 11.3 ROS-mediated signaling in plant cell under biotic stress. In plants biotic stress signaling is mediated on recognition of either pathogen-associated molecular patterns (PAMP) and/or pathogen virulence/effectors, triggering pathogen-triggered immunity (PTI) and/or effector-triggered immunity (ETI). On perception of PAMP by receptor induces activation of ion channel resulting in increase in cytosolic calcium (Ca^{2+}) level, which activates calcium-dependent protein kinase (CDPK). Both Ca^{2+} and CDPK bind with NADPH oxidase (NOX) generating ROS burst in apoplast. H_2O_2 from apoplast moves within cytoplasm via aquaporin and activates mitogen-activated protein kinase (MAPK) cascade, which induces DNA reprogramming and PTI. Pathogens producing effectors are perceived by the nucleotide-binding site-leucine-rich repeat (NB-LRR) protein, subsequently followed by R genes, which cause ROS burst, triggering cell hypersensitive response, eventually resulting in ETI

channels, elevates cytosolic calcium (Ca^{2+}) level, activates Ca^{2+}-dependent protein kinases (CDPKs or CPKs), increases extracellular pH leading to alkalization, induces cell-wall reinforcement through callose deposition, promotes transient ROS burst via NOX or RBOH, activates mitogen-activated protein kinase (MAPK) cascade, and induces transcriptional reprogramming (Zipfel et al. 2006; Qi et al. 2017; Liu and He 2017; Ngou et al. 2021; Yuan et al. 2021) (Fig. 11.3). The receptor kinases (RK) such as Flagellin-Sensing 2 (FLG2) and Elongation Factor TU Receptor (EFR) are leucine-rich repeat-receptor kinases (LRR-RKs), which can recognize 22 amino-acid epitope of bacterial flagellin (flg22) (Cheng et al. 2021) and bacterial elongation factor (EF) Tu epitope elf 18 (Zipfel et al. 2006), respectively (Qi et al. 2017). The PRR FLG2 is associated with two cytosolic protein kinases, Botrytis-Induced Kinase1 (BIKI) and PBS1-Like1 (PBL1) during the pre-activation condition. The perception of PAMP flg22 by FLG2 is mediated via co-receptor BRI1-Associated Receptor Kinase 1 (BAK1), which triggers trans-phosphorylation of FLG2, leading to activation of cytosolic BIK1 and PBL1

(Qi et al. 2017). Activation of the receptor causes transient increase in cytosolic Ca^{2+} via activation of ion channels and simultaneously phosphorylates NOX at N-terminal (Qi et al. 2017). Moreover, transient increase in cytosolic Ca^{2+} activates CPK, particularly, CPK5 which also phosphorylates NOX at N-terminal. The elevated intracellular Ca^{2+} itself binds with NOX at the EF hand, promoting its activity. The activation of NOX via various pathways eventually generates superoxide in apoplast via transfer of an electron from cytosol to molecular oxygen in apoplast. The generated $O_2^{\bullet-}$ further undergoes dismutation generating H_2O_2 (Qi et al. 2017). Though Ca^{2+} and CPK are upstream of ROS in PTI, the ROS burst also mediates increase in intracellular Ca^{2+} level and activation of CPK5. As both Ca^{2+} and ROS amplify each other, their conjugated relation is likely to participate in systemic signaling from cell-to-cell via Ca^{2+} and ROS waves under stress (Qi et al. 2017). The generation of ROS on interaction of plants with pathogens can also be mediated by plant hormone abscisic acid (ABA). ABA can activate RBOHD and RBOHF to produce ROS, inducing immune response (Qi et al. 2017). ABA induces phosphorylation of RBOHF mediating production of ROS. Unlike pattern-triggered ROS generation, ABA-mediated ROS acts upstream of intracellular Ca^{2+} increase (Qi et al. 2017).

The transient ROS burst in apoplast mediates signaling to induce PTI by moving into cytosol via plasma membrane localized specific aquaporins. In *Arabidopsis*, aquaporin PIP1;4 (Plasma membrane Intrinsic Proteins, PIP) has been reported to be activated during PTI response and mediates movement of H_2O_2 into cytosol, which triggers induction of defense response (Tian et al. 2016). The accumulation of ROS in cytoplasm is sensed by a redox-based mechanism. The H_2O_2 reacts with the target protein that acts as an ROS sensor. H_2O_2 induces structural and functional alteration of target proteins, which eventually translocate to the nucleus, where they act as transcription factors (TFs) (Marcec et al. 2019). Further, ROS can also activate an MAPK cascade that mediates signals downstream to induce immunity response (Marcec et al. 2019).

11.4 Abiotic Stress-Mediated ROS Signaling

Abiotic constraints such as water scarcity, extreme temperature, high light intensity, heavy metals, and salinity can overwhelm production of ROS in plant cells that mediate signals for stress acclimatization (Tripathy and Oelmüller 2012; Choudhury et al. 2013; Kohli et al. 2017; Ansari et al. 2021). The ROS signals can elicit specific cellular response including apoptosis, stomatal closure, and enhanced antioxidant defense to address particular stress (Yan et al. 2007; Kohli et al. 2017). The ROS signal in plants under various abiotic stresses has been discussed in detail.

11.4.1 ROS Signaling and Alleviation of Salinity Stress

Salinity is one of the most deleterious abiotic stresses, affecting crop production worldwide. It induces ionic imbalance, osmotic stress, and ROS accumulation in plants, which reduce photosynthesis, and induces oxidative damage, premature senescence, nutritional imbalance, and genotoxicity (Kaleem et al. 2018; Yang and Guo 2018; Jalil and Ansari 2020; Tanveer and Ahmed 2020). Nevertheless, ROS have also been found linked with the salinity stress tolerance. ROS production in cells under salinity triggers redox signaling, which regulates expression of genes modulating tolerance against stress (Tanveer and Ahmed 2020). The production of ROS in the apoplast under salinity stress was shown to induce positive response on leaf elongation in maize whereas abridged ROS production reduces leaf elongation (Rodríguez et al. 2004). Plants exposed to salinity stress are often characterized by ionic imbalance and osmotic stress due to excess Na^+ influx in the cytosol and reduced water potential, preferably in roots (Kurusu et al. 2015). Salinity perception activates Ca^{2+} permeable mechanical stimuli (MS) channels localized on root cell plasma membrane, triggering Ca^{2+} influx. Increase in cytosolic Ca^{2+} content results in activation of multiple Ca^{2+}-binding proteins like calcium-dependent protein kinase (CDPKs or CPKs) (Kurusu et al. 2015). The increased cytosolic Ca^{2+} and activation of Ca^{2+} binding proteins together trigger activation of NOX/RBOH at N-terminal via direct binding to EF-hand motif and phosphorylation, respectively (Ogasawara et al. 2008; Kurusu et al. 2015). This causes NOX-mediated production of ROS in apoplast, which in turn activates plasma membrane Ca^{2+} channels, further enhancing cytosolic Ca^{2+} level (Tanveer and Ahmed 2020). The elevated cytosolic Ca^{2+} content may induce downstream events including activation of a salt overly sensitive (SOS) pathway component SOS1, a plasma membrane NA^+/H^+ exchanger which facilitates efflux of Na^+ (Chung et al. 2008) and NHX1 (Kurusu et al. 2015). H_2O_2 generated in apoplast is transported to cytosol via plasma membrane aquaporins, where it induces signaling by modifying target proteins, which in turn regulate the gene expression facilitating required response, promoting tolerance to salt stress (Kurusu et al. 2015). Moreover, H_2O_2 in cytosol may react with superoxide produced by mitochondria under stress and generate $^{\bullet}OH$ via Fenton reaction (Kurusu et al. 2015). Accumulation of $^{\bullet}OH$ in cytosol leads to activation of non-selective cation channels (NSCCs) permeable to Na^+, K^+, and Ca^{2+}, promoting Na^+ uptake and K^+ loss (Kurusu et al. 2015). The loss of K^+ mediates activation of caspase-like proteases and leads to PCD (Kurusu et al. 2015).

11.4.2 ROS Signaling and Alleviation of Heavy Metal Stress

Plants growing in media contaminated with heavy metals show uptake of metals from roots, which are further translocated throughout the plants tissues, get accumulated, and induce toxic effects (Berni et al. 2019). Plants alleviate heavy metal toxicity via multiple mechanisms including adsorption on the chelating agents and/or sequestering them into vacuole, facilitated by the cellular signaling (Jalmi et al. 2018). ROS generated in plant cells due to heavy metal induce NOX activity and/or over-reduction of ETC and activate signal transduction via MAPK cascade (Jalmi and Sinha 2015; Jalmi et al. 2018). In rice roots, excess iron accumulation reportedly induces cell death via ROS-mediated signaling involving activation of MAPK cascade (Tsai and Huang 2006). In *Arabidopsis*, ROS-mediated signaling facilitates transport of phytohormone auxin via AUX1, which induces tolerance against arsenite stress (Krishnamurthy and Rathinasabapathi 2013). Analogously, Liu et al. (2018) have reported a rapid production of H_2O_2 in leaves of maize plant on acute exposure to copper, which triggers signaling leading to upregulation of antioxidant enzymes (SOD, CAT, and APX) activity via MAPK cascade (ZmMPK3).

11.4.3 ROS-Mediated Tolerance to Heat Stress

Heat stress is becoming a significant problem for plant growth and development due to progressing global warming (Ohama et al. 2017). When the ambient heat or temperature rises above the threshold level, plants considerably adjust themselves to survive by modulating regulatory mechanism at the molecular level (Hemantaranjan et al. 2014). Plants rescue themselves from heat stress injuries with the help of molecular chaperones known as heat shock proteins (HSPs) (Feder and Hofmann 1999; Medina et al. 2021). Short-term exposure to heat stress causes H_2O_2 burst in plant cells via NOX activity, which the heat shock factors (HSFs) sense (Kotak et al. 2007). These HSFs mediate received signals further to the nucleus, activating HSP gene expression that eventually triggers immune response in plants (Scarpeci et al. 2008; Mittler et al. 2012; Medina et al. 2021). Thus, ROS generated under heat stress act as a signaling molecule and prompt a required response via a complex mechanism involving HSP (Medina et al. 2021). H_2O_2 generation under the early stage of heat stress induces signaling by activating HSF that alters expression of heat shock genes in *Arabidopsis* (Volkov et al. 2006). Heat stress accompanied with the anoxia stress triggers ROS production in plants, which induces transcription of HSFs, in particular *HSFA2*, and HSPs, promoting plant tolerance to stress (Pucciariello et al. 2012). The tomato plant overexpressing Brassinazole resistant 1 (BZR1) enhances RBOH1 activity, generating H_2O_2 in apoplast, which eventually induces signaling leading to heat stress tolerance (Yin et al. 2018).

11.5 Compartment-Specific ROS Signaling in Plants

ROS evidently mediate signaling via discrete sets of compartments (organelle)-specific signaling molecules (Mignolet-spruyt et al. 2016). The signaling initiated from particular organelle or compartment includes compartment-specific sensory system, which links ROS with other signaling pathways and relays message downstream, facilitating compartment-specific responses (Mignolet-spruyt et al. 2016).

Thus, the ROS-mediated signaling in plant cells, in addition to specific chemical nature of particular reactive species, also depends on spatio-temporal control of sensory mechanisms, the developmental stage, and distribution as well as relative position of a particular cellular organelle or compartment in a cell (Mhamdi et al. 2010; Shapiguzov et al. 2019; Suzuki et al. 2012; Sewelam et al. 2014; Mignolet-Spruyt et al. 2016).

11.5.1 Apoplast ROS and Signaling

Apoplast, the intercellular space present beyond the plasma membrane, is an interface for communication between environment and plant cells. Apoplast facilitates rapid systemic signaling due to fast diffusion of ROS from apoplast compared to cytosol (Mignolet-Spruyt et al. 2016). It possesses unique characteristics like lower pH than cytoplasm and less abundance of low-molecular- weight antioxidants (Mignolet-Spruyt et al. 2016). Though lower pH reduces redox sensitivity of protein cysteine (Cys) residue, low antioxidant buffering capacity facilitates accumulation of ROS in apoplast, which upregulates ROS-mediated signaling and counteracts low protein redox sensitivity (Mignolet-Spruyt et al. 2016). The apoplastic ROS signaling possibly engages two pathways. The first pathway includes diffusion of H_2O_2 molecules through specific aquaporins localized in plasma membranes, enhancing the intracellular ROS pool that enables activation of symplastic ROS signaling (Podgórska et al. 2017). The second pathway involves the H_2O_2 auto-propagating wave for long-distance systemic signaling (Mignolet-Spruyt et al. 2016; Podgórska et al. 2017).

The systemic signaling in apoplast under stress is generally mediated via two isoforms of RBOH, i.e., RBOHD and RHOHF (Waszczak et al. 2018). The signals are transduced to adjacent cells, where RBOH/NOX utilizes similar control mechanisms as employed by previous cell RBOH, like CDPKs, particularly, CPK5 (Mignolet-Spruyt et al. 2016). The phosphorylation of RBOH by CPK5 causes the activation of RBOH, leading to ROS perception in the adjacent cell, where Ca^{2+} signaling and CPK5 again get activated, spreading the RBOH activation throughout the signaling pathway (Mignolet-Spruyt et al. 2016). The process leads to systemic signaling from local tissues to systemic tissues, where it alters gene expression (Miller et al. 2009). The ROS wave is crucial to induce systemic acquired resistance in plants against biotic stress and systemic acquired acclimation to abiotic stress (Mignolet-Spruyt et al. 2016).

The signaling cascade triggered by apoplastic ROS from extracellular space to nucleus involves different components or elements like Ca^{2+}, receptor-like proteins, oxylipins, modification in ascorbate (AsA) pool, hormones, and others. Plasma membrane associated-NOX/RBOH is crucial for transducing apoplastic ROS signals at a fast pace to long distance as well as to the nucleus. ROS burst in apoplast increases cytosolic Ca^{2+} content, which promotes activity of NOX, yielding more ROS in apoplast (Podgórska et al. 2017). Further, elevated cytosolic Ca^{2+} induces further responses through action of Ca^{2+}-binding proteins that act as calcium sensor and assist in Ca^{2+}-induced signaling (Podgórska et al. 2017). During the Ca^{2+} influx, movement of other ions into the cytoplasm takes place via particular channels like anion channel, H^+ pump, stelar potassium (K^+) outward rectifier (SKOR), or permeable membrane. Thus, elevation in H^+ ions in cytosol and efflux of K^+ or Cl^- is considered as among the earliest responses in ROS-mediated signaling (Podgórska et al. 2017). In addition to variation in ion influx/efflux, pH of apoplast also fluctuates, as buffering capacity in this compartment is comparatively less efficient than symplast (Podgórska et al. 2017). The ROS present in extracellular environment cause lipid peroxidation and alkalization of the apoplast, facilitating propagation of signals via alkali-responsive peroxidases, or alteration in the redox status and/or content of apoplastic antioxidant pool (Zurbriggen et al. 2010). A shift in pH of apoplast affects the activity of the enzymes like cell-wall POX and modulates gene expression (Lager et al. 2010; Podgórska et al. 2017). ROS in apoplast can cause oxidation of AsA, enhancing the content of dehydroascorbate (DHA) that acts as a signaling molecule (Podgórska et al. 2017). Moreover, other enzymes containing sulfur residues like glutathione (GSH), peroxiredoxins (Prxs), and thioredoxins (Trxs) are susceptible to reversible redox reactions and can directly partake in modulation of cellular metabolism or accomplish their function through downstream signaling components (Podgórska et al. 2017).

The excessive accumulation of ROS in the apoplast under stress leads to oxidative damage to lipid membrane and proteins, and their breakdown products formed may act as receptor ligands (Mignolet-Spruyt et al. 2016). Oxidation of lipid membranes generates a peroxidation product oxylipins, which acts as a signaling molecule (Podgórska et al. 2017). Moreover, ROS can cause direct redox modification of peptides or receptor proteins domain such as Cys residue, located in extracellular space inducing conformational alterations which mediates activation of receptors and signal transduction (Mignolet-Spruyt et al. 2016). The specific receptor proteins located at the outer membrane of the plasma membrane, in particular, innate immune receptor and hormone receptor are found sensitive to ROS (Podgórska et al. 2017). The heterotrimer located in the plasma membrane on perceiving stimuli due to apoplastic ROS accumulation triggers a signal to the nucleus via downstream signaling events (Podgórska et al. 2017). The plasma membrane-associated RLKs, which are Cys-rich transmembrane proteins with an extracellular ligand binding domain and a conserved intracellular protein kinase domain, are activated by the apoplastic ROS (Podgórska et al. 2017). In plants, intracellular signaling is activated directly via protein phosphatases and kinases on perception of extracellular ROS (Neill et al. 2002).

11.5.2 Chloroplastic ROS-Mediated Signaling

Chloroplast is the major site for ROS generation compared to mitochondria and peroxisomes and a hub for ROS signaling (Mignolet-Spruyt et al. 2016). The perturbed photosynthetic activity including over-reduction of electron transport chain (ETC) results in genesis of ROS-mediated retrograde signaling. Chloroplast also responds to ROS signals generated by other organelles. The excess ROS generated in apoplast relay information to chloroplast, inducing production of chloroplastic ROS (Mignolet-Spruyt et al. 2016). For example, pathogen-associated molecular pattern (PAMP) recognition-related immunity responses trigger ROS burst in apoplast, which relay signals to the chloroplast, eventually inducing specific response like basal resistance and/or HR cell death (Nomura et al. 2012).

The retrograde signaling of chloroplastic ROS, particularly H_2O_2 and $^1O_2$ has been proposed to occur via three different pathways (Fig. 11.4). The first pathway includes induction of stromule projections (stroma-filled tubular extension of plastid) facilitating direct transfer of H_2O_2 and proteins from plastids to nuclei, bypassing the cytosol (Brunkard et al. 2015; Brunkard and Burch-Smith 2018; Waszczak et al. 2018; Mielecki et al. 2020; Mullineaux et al. 2020) (Fig. 11.4a). The second pathway involves rapid transfer of H_2O_2 to nuclei from chloroplasts present in the vicinity of the nuclei by crossing double membranes of both chloroplast and nucleus (Breeze and Mullineaux 2022). The transfer of H_2O_2 is facilitated via diffusion through membrane aquaporins (Fig. 11.4b) (Mielecki et al. 2020). This kind of signaling pathway participates in the regulation of high-light-responsive gene expression (Waszczak et al. 2018). The third pathway includes indirect signaling mediated by chloroplastic metabolites and/or their oxidative derivatives, generated via ROS (H_2O_2 and $^1O_2$)-induced oxidative PTM of chloroplastic lipids, peptides, and others (Waszczak et al. 2018; Kim 2020; Unal et al. 2020). For instance, H_2O_2 generated in chloroplast of *Arabidopsis* induces inactivation of the phosphatase enzyme SAL1, resulting in accumulation of 3′-phosphoadenosine 5′--phosphate (PAP), which serves as a messenger conveying message to nucleus (Chan et al. 2016) under drought and high light stress (Estavillo et al. 2011) (Fig. 11.4c).

The $^1O_2$ produced in chloroplast under stress acts as a potent retrograde signaling molecule owing to high reactivity and short half-life (Laloi and Havaux 2015; Mignolet-Spruyt et al. 2016; Unal et al. 2020). *flu* mutant of *Arabidopsis* has elucidated the $^1O_2$-mediated signaling as the $^1O_2$ burst in chloroplast during dark/light transition induces chlorosis and cell death along with intense reprogramming of nuclear gene expression (op den Camp et al. 2003). Moreover, $^1O_2$ burst in *flu* mutant under both biotic and abiotic stress evokes a wide range of responses including upregulation of enhanced disease susceptibility1 (EDS1) protein required for accumulation of SA, which facilitates synthesis of PR protein and resistance to biotrophic pathogens (Ochsenbein et al. 2006). $^1O_2$-mediated retrograde signaling response depends on two nuclear encoded plastid proteins, EXECUTER1 (EX1) and EX2, inducing PCD (Wagner et al. 2004; Kim and Apel 2013; Brunkard and Burch-Smith 2018). $^1O_2$ being a reactive molecule causes oxidation and cleavage of β-carotene, generating carbonyl products such as β-CC, which act as a secondary messenger and mediate retrograde signaling (Ramel et al. 2012a, b). β-CC is a volatile compound that acts as a rapid diffusible signal, resulting in modulation of nuclear transcriptome, eventually conferring tolerance against abiotic stress (Ramel et al. 2012b; Li and Kim 2021).

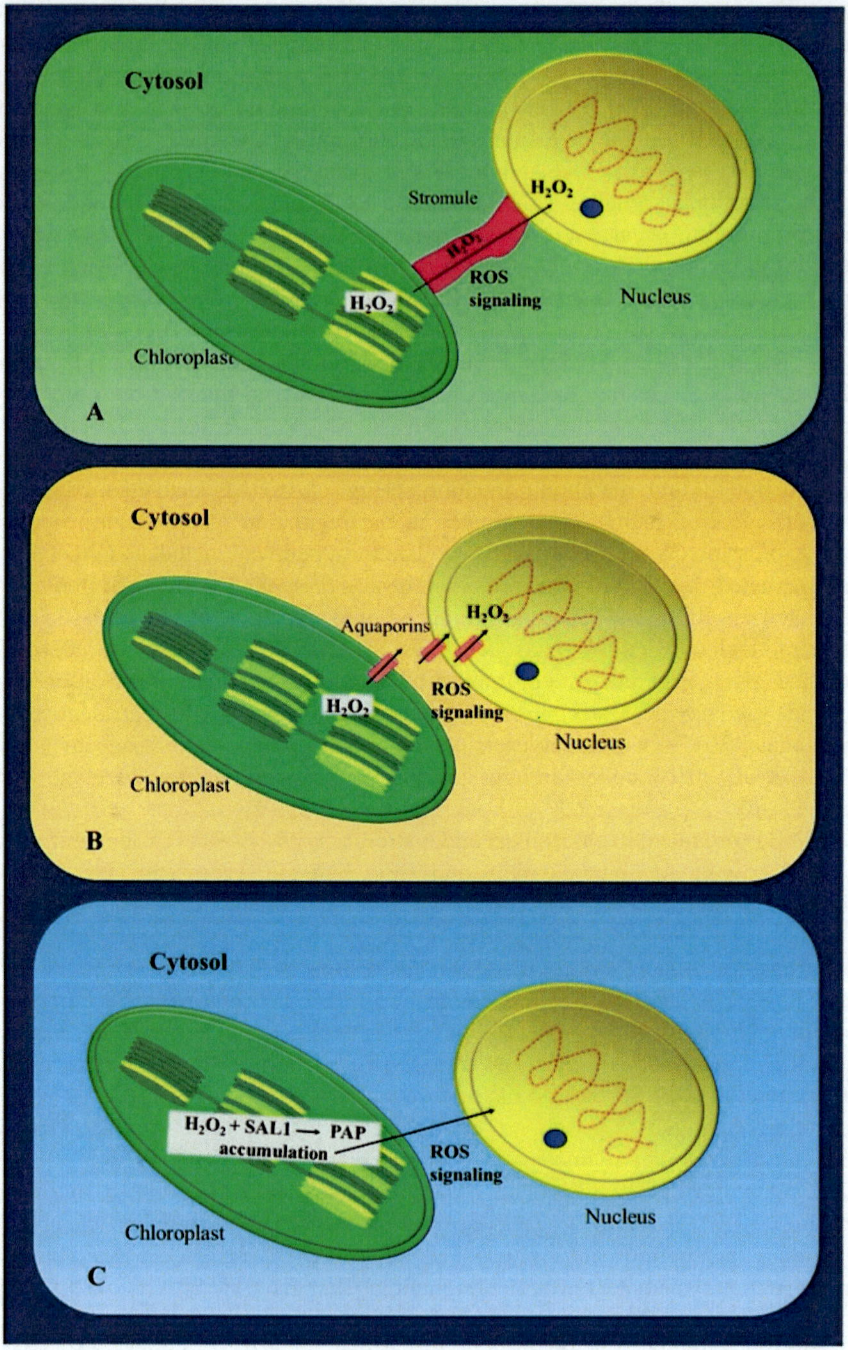

Fig. 11.4 Chloroplast ROS retrograde signaling involves three different pathways. (**a**) Retrograde signaling via formation of stromule projection; (**b**) Direct transfer of ROS generated in chloroplast in vicinity of nucleus via aquaporins; (**c**) Signaling mediated by intermediates formed in chloroplast on reaction with ROS

11.5.3 Mitochondrial ROS Signal Transduction

Accumulation of ROS in mitochondria also induces retrograde signaling. In *Arabidopsis*, ROS-mediated retrograde signaling upregulates the expression of genes belonging to mitochondrial dysfunction regulon (De Clercq et al. 2013), such as *Alternative oxidase1* (*AOX1*) (Vanlerberghe 2013). *AOX1* is considered as a hallmark for mitochondrial retrograde signaling events, which often get activated under stress conditions, particularly, those altering respiratory metabolism (Li et al. 2013; Mignolet-Spruyt et al. 2016). Under stressful circumstances, *AOX1* prevents over-reduction of mitochondrial ETC and oxidative stress (Vanlerberghe 2013; Mignolet-Spruyt et al. 2016). Enhanced activity of alternative oxidase (AOX) directs electrons toward molecular oxygen before complex III, which abridges ROS production and regulates altered redox balance. Two endoplasmic reticulum (ER)-localized TFs *Arabidopsis NAC domain containing protein13* (*ANAC013*) and *ANAC017* mediating the process are translocated to the nucleus in response to organelle-dependent stress signals, conferring tolerance against oxidative stress in plants (De Clercq et al. 2013; Ng et al. 2013).

11.5.4 Peroxisomal ROS-Induced Signaling

Peroxisome is an organelle that witnesses the production of ROS via photorespiratory pathway, which is considered as a key source of ROS signals in plants exposed to various environmental stresses (Foyer and Noctor 2009; Voss et al. 2013; Mignolet-Spruyt et al. 2016). H_2O_2 as a ROS generates/accumulates in peroxisomes and mediates retrograde signaling, inducing translational responses in plants (Vanderauwera et al. 2005; Queval et al. 2012; Mignolet-Spruyt et al. 2016). The antioxidant enzyme catalase (CAT) regulates the ROS homeostasis in peroxisomes. Thus, the ability of peroxisomal H_2O_2 to mediate signaling has been studied with the aid of *CAT* loss-of-function mutant (Mhamdi et al. 2012; Sandalio et al. 2013; Mignolet-Spruyt et al. 2016; Su et al. 2019). Transcriptomic study in *Arabidopsis CAT2* mutant reveals that peroxisomal H_2O_2-mediated signaling responses depend on photoperiod, upregulate the defense genes, and induce oxidative stress-responsive transcripts (Queval et al. 2007). The CAT-deficient mutants of *Arabidopsis* and *Nicotiana tabacum* elevate photorespiratory H_2O_2 level and alter gene expression profile, resulting in induction of pathogenesis-related (PR) protein genes, accumulation of salicylic acid (SA) and cell death (Takahashi et al. 2011; Chaouch et al. 2010; Queval et al. 2012). The gene expression profile induced by peroxisomal H_2O_2 signaling shows specificity with transcriptional responses, which differ from that produced by chloroplastic H_2O_2 (Sewelam et al. 2014). The double mutant of *Arabidopsis*, lacking *CAT2* and cytosolic *ascorbate peroxidase1* (*APX1*) displays responses including cell cycle regulation, activation of DNA repair, and inhibition of programmed cell death (Vanderauwera et al. 2011). A triple mutant of *Arabidopsis* lacking *CAT1, CAT2,* and *CAT3* produces intense transcriptional changes, resulting in differential expression of TF genes involved in regulation of plant growth and response against biotic and abiotic stress (Su et al. 2018). Further, peroxisomal H_2O_2 retrograde signaling upregulates several protein kinases like OXI1, cysteine-rich receptor-like kinases (CRKs), and some components of MAPK cascade (Su et al. 2018).

11.5.5 Inter-organellar Signaling

Although every cellular organelle regulates its own redox status and induces retrograde signaling, the ROS or other signaling components show inter-organellar communication. Chloroplast, peroxisomes, apoplast, and mitochondria are connected with each other via metabolic activities, energy production, and redox state (Dang et al. 2014; Mignolet-Spruyt et al. 2016; Shai et al. 2016; Su et al. 2019) and cooperatively mediate signals to the nucleus (Woodson and Chory 2008; Kopczewski and Kuźniak 2013). For instance, the accumulation of PAP in chloroplast under drought and light stress due to dephosphorylation of SAL1 induces retrograde signaling, which causes activation of TFs eventually leading to regulation of stress-responsive genes (Estavillo et al. 2011). Intriguingly, the SAL1 targets not only chloroplast but also mitochondria, suggesting PAP-mediated chloroplastic signaling having linked to the mitochondrial pathway (Van Aken and Whelan 2012). The peroxisomal redox state shows an intricate relation with the redox state of other organelles and can affect their metabolic regulation via cross-talk (Sousa et al. 2019). Peroxisomes show unique features of producing peroxules, a tubular extension, in response to alteration in ROS content under stress, facilitating contact of peroxisomes with other organelles including chloroplast and mitochondria (Sandalio et al. 2021). The low level of peroxisomal H_2O_2 can alter antioxidant enzymes in chloroplast as well as in cytosol, as observed in transgenic rice plants being impaired in peroxisomal *APX4* and *CAT* activity (Sousa et al. 2019). The transgenic plant exhibits better photosynthetic resilience than wild-type plants due to metabolic modifications that significantly alter various networks involved in photosynthetic action under oxidative stress (Sousa et al. 2019).

11.6 Conclusion

The transient accumulation of ROS in plant cells under altered environmental conditions is perceived and transduced to elicit acclimation response. The ROS-mediated signaling induces response at the local as well as distant sites. The signaling cascade downstream modifies the structure and function of different transcription factors, unique to particular stress, which modulate gene expression in the nucleus, inducing tolerance response. The signal transduction mediated by ROS depends on the chemical nature of particular reactive species that determines discrete signaling cascades, involving interaction with other signaling molecules. Moreover, the cellular compartment-specific production of ROS also influences their signaling prospect, eliciting compartment-specific responses. The transient accumulation of ROS in different cellular organelles/compartments displays a complex inter-organellar communication, addressing a wide range of adverse environmental cues, which modulate redox status and facilitate plants to adapt under persisting adverse conditions. The amalgamation of persisting knowledge on ROS signaling in plant cells with advanced molecular tool-based research could help to decipher the complexity in ROS signaling via identification of unknown molecular targets or unidentified mechanisms that could facilitate development of plants with ability to cope with severe circumstances.

References

Ansari MI, Jalil SU, Ansari SA, Hasanuzzaman M (2021) GABA shunt: a key-player in mitigation of ROS during stress. Plant Growth Regul 94(2):131–149

Berni R, Luyckx M, Xu X, Legay S, Sergeant K, Hausman JF, Lutts S, Cai G, Guerriero G (2019) Reactive oxygen species and heavy metal stress in plants: impact on the cell wall and secondary metabolism. Environ Exp Bot 161:98–106

Bhattacharjee S (2005) Reactive oxygen species and oxidative burst: roles in stress, senescence and signal transducation in plants. Curr Sci 89:1113–1121

Bhattacharjee S (2012) The language of reactive oxygen species signaling in plants. J Bot 2012: 985298

Bigeard J, Colcombet J, Hirt H (2015) Signaling mechanisms in pattern-triggered immunity (PTI). Mol Plant 8(4):521–539

Breeze E, Mullineaux PM (2022) The passage of H2O2 from chloroplasts to their associated nucleus during retrograde signalling: reflections on the role of the nuclear envelope. Plants 11(4):552

Brunkard JO, Burch-Smith TM (2018) Ties that bind: the integration of plastid signalling pathways in plant cell metabolism. Essays Biochem 62(1):95–107

Brunkard JO, Runkel AM, Zambryski PC (2015) Chloroplasts extend stromules independently and in response to internal redox signals. Proc Natl Acad Sci 112(32):10044–10049

Chan KX, Mabbitt PD, Phua SY, Mueller JW, Nisar N, Gigolashvili T, Stroeher E, Grassl J, Arlt W, Estavillo GM, Jackson CJ, Pogson BJ (2016) Sensing and signaling of oxidative stress in chloroplasts by inactivation of the SAL1 phosphoadenosine phosphatase. Proc Natl Acad Sci 113(31):E4567–E4576

Chaouch S, Queval G, Vanderauwera S, Mhamdi A, Vandorpe M, Langlois-Meurinne M, Van Breusegem F, Saindrenan P, Noctor G (2010) Peroxisomal hydrogen peroxide is coupled to biotic defense responses by ISOCHORISMATE SYNTHASE1 in a daylength-related manner. Plant Physiol 153(4):1692–1705

Cheng JH, Bredow M, Monaghan J, DiCenzo GC (2021) Proteobacteria contain diverse flg22 epitopes that elicit varying immune responses in *Arabidopsis thaliana*. Mol Plant-Microbe Interact 34(5):504–510

Choudhury S, Panda P, Sahoo L, Panda SK (2013) Reactive oxygen species signaling in plants under abiotic stress. Plant Signal Behav 8(4):e23681

Chung JS, Zhu JK, Bressan RA, Hasegawa PM, Shi H (2008) Reactive oxygen species mediate Na +-induced SOS1 mRNA stability in *Arabidopsis*. Plant J 53(3):554–565

Cui F, Brosché M, Shapiguzov A, He XQ, Vainonen JP, Leppälä J, Troota A, Kangasjarvi S, Salojarvi J, Kangasjarvi J, Overmyer K (2019) Interaction of methyl viologen-induced chloroplast and mitochondrial signalling in *Arabidopsis*. Free Radic Biol Med 134:555–566

Dang KV, Plet J, Tolleter D, Jokel M, Cuiné S, Carrier P, Auroy P, Richaud P, Johnson X, Alric J, Allahverdiyeva Y, Peltier G (2014) Combined increases in mitochondrial cooperation and oxygen photoreduction compensate for deficiency in cyclic electron flow in *Chlamydomonas reinhardtii*. Plant Cell 26(7):3036–3050

De Clercq I, Vermeirssen V, Van Aken O, Vandepoele K, Murcha MW, Law SR, Inze A, Ng S, Ivanova A, Rombaut D, van de Cotte B, Jaspers P, Van de Peer Y, Kangasjarvi J, Whelan J, Van Breusegem F (2013) The membrane-bound NAC transcription factor ANAC013 functions in mitochondrial retrograde regulation of the oxidative stress response in *Arabidopsis*. Plant Cell 25(9):3472–3490

Dvořák P, Krasylenko Y, Zeiner A, Šamaj J, Takáč T (2021) Signaling toward reactive oxygen species-scavenging enzymes in plants. Front Plant Sci 11:2178

Estavillo GM, Crisp PA, Pornsiriwong W, Wirtz M, Collinge D, Carrie C, Giraud E, Whelan J, David P, Javot H, Brearley C, Hell R, Marin E, Pogson BJ (2011) Evidence for a SAL1-PAP chloroplast retrograde pathway that functions in drought and high light signaling in *Arabidopsis*. Plant Cell 23(11):3992–4012

Feder ME, Hofmann GE (1999) Heat-shock proteins, molecular chaperones, and the stress response: evolutionary and ecological physiology. Annu Rev Physiol 61(1):243–282

Fichman Y, Mittler R (2020) Rapid systemic signaling during abiotic and biotic stresses: is the ROS wave master of all trades? Plant J 102(5):887–896

Foyer CH, Noctor G (2005) Oxidant and antioxidant signalling in plants: a re-evaluation of the concept of oxidative stress in a physiological context. Plant Cell Environ 28(8):1056–1071

Foyer CH, Noctor G (2009) Redox regulation in photosynthetic organisms: signaling, acclimation, and practical implications. Antioxid Redox Signal 11(4):861–905

Hemantaranjan A, Bhanu AN, Singh MN, Yadav DK, Patel PK, Singh R, Katiyar D (2014) Heat stress responses and thermotolerance. Adv Plants Agric Res 1(3):1–10

Iqbal Z, Iqbal MS, Hashem A, Abd Allah EF, Ansari MI (2021) Plant defense responses to biotic stress and its interplay with fluctuating dark/light conditions. Front Plant Sci 12:631810

Jalil SU, Ansari MI (2020) Physiological role of Gamma-aminobutyric acid in salt stress tolerance. In: Salt and drought stress tolerance in plants. Springer, Cham, pp 337–350

Jalmi SK, Sinha AK (2015) ROS mediated MAPK signaling in abiotic and biotic stress-striking similarities and differences. Front Plant Sci 6:769

Jalmi SK, Bhagat PK, Verma D, Noryang S, Tayyeba S, Singh K, Sharma D, Sinha AK (2018) Traversing the links between heavy metal stress and plant signaling. Front Plant Sci 9:12

Kaleem F, Shabir G, Aslam K, Rasul S, Manzoor H, Shah SM, Khan AR (2018) An overview of the genetics of plant response to salt stress: present status and the way forward. Appl Biochem Biotechnol 186(2):306–334

Karuppanapandian T, Moon JC, Kim C, Manoharan K, Kim W (2011) Reactive oxygen species in plants: their generation, signal transduction, and scavenging mechanisms. Aust J Crop Sci 5(6): 709–725

Kim C (2020) ROS-driven oxidative modification: its impact on chloroplasts-nucleus communication. Front Plant Sci 10:1729

Kim C, Apel K (2013) Singlet oxygen-mediated signaling in plants: moving from flu to wild type reveals an increasing complexity. Photosynth Res 116(2):455–464

Kohli SK, Handa N, Gautam V, Bali S, Sharma A, Khanna K, Arora S, Thukral AK, Ohri P, Karpets YV, Kolupaev YE, Bhardwaj R (2017) ROS signaling in plants under heavy metal stress. In: Reactive oxygen species and antioxidant systems in plants: role and regulation under abiotic stress. Springer, Singapore, pp 185–214

Kopczewski T, Kuźniak E (2013) Redox signals as a language of interorganellar communication in plant cells. Open Life Sci 8(12):1153–1163

Kotak S, Larkindale J, Lee U, von Koskull-Döring P, Vierling E, Scharf KD (2007) Complexity of the heat stress response in plants. Curr Opin Plant Biol 10(3):310–316

Krishnamurthy A, Rathinasabapathi B (2013) Auxin and its transport play a role in plant tolerance to arsenite-induced oxidative stress in *Arabidopsis thaliana*. Plant Cell Environ 36(10): 1838–1849

Kumar V, Khare T, Sharma M, Wani SH (2017) ROS-induced signaling and gene expression in crops under salinity stress. In: Reactive oxygen species and antioxidant systems in plants: role and regulation under abiotic stress. Springer, Singapore, pp 159–184

Kurusu T, Kuchitsu K, Tada Y (2015) Plant signaling networks involving Ca2+ and Rboh/Nox-mediated ROS production under salinity stress. Front Plant Sci 6:427

Lager IDA, Andréasson OLA, Dunbar TL, Andreasson E, Escobar MA, Rasmusson AG (2010) Changes in external pH rapidly alter plant gene expression and modulate auxin and elicitor responses. Plant Cell Environ 33(9):1513–1528

Laloi C, Havaux M (2015) Key players of singlet oxygen-induced cell death in plants. Front Plant Sci 6:39

Laloi C, Apel K, Danon A (2004) Reactive oxygen signalling: the latest news. Curr Opin Plant Biol 7(3):323–328

Leibman-Markus M, Pizarro L, Schuster S, Lin ZD, Gershony O, Bar M et al (2018) The intracellular nucleotide-binding leucine-rich repeat receptor (SlNRC4a) enhances immune signalling elicited by extracellular perception. Plant Cell Environ 41(10):2313–2327

Li M, Kim C (2021) Chloroplast ROS and stress signaling. Plant Commun 3.100264

Li S, Mhamdi A, Clement C, Jolivet Y, Noctor G (2013) Analysis of knockout mutants suggests that *Arabidopsis* NADP-MALIC ENZYME2 does not play an essential role in responses to oxidative stress of intracellular or extracellular origin. J Exp Bot 64(12):3605–3614

Liu Y, He C (2017) A review of redox signaling and the control of MAP kinase pathway in plants. Redox Biol *11*:192–204

Liu J, Wang J, Lee S, Wen R (2018) Copper-caused oxidative stress triggers the activation of antioxidant enzymes via ZmMPK3 in maize leaves. PLoS One 13(9):e0203612

Liu D, Luo D, He P (2021) ROS around RIPK. Mol Plant 14(10):1607–1609

Marcec MJ, Gilroy S, Poovaiah BW, Tanaka K (2019) Mutual interplay of Ca2+ and ROS signaling in plant immune response. Plant Sci 283:343–354

Martins L, Trujillo-Hernandez JA, Reichheld JP (2018) Thiol based redox signaling in plant nucleus. Front Plant Sci 9:705

Medina E, Kim SH, Yun M, Choi WG (2021) Recapitulation of the function and role of ROS generated in response to heat stress in plants. Plants 10(2):371

Mhamdi A, Queval G, Chaouch S, Vanderauwera S, Van Breusegem F, Noctor G (2010) Catalase function in plants: a focus on *Arabidopsis* mutants as stress-mimic models. J Exp Bot 61(15): 4197–4220

Mhamdi A, Noctor G, Baker A (2012) Plant catalases: peroxisomal redox guardians. Arch Biochem Biophys 525(2):181–194

Mielecki J, Gawroński P, Karpiński S (2020) Retrograde signaling: understanding the communication between organelles. Int J Mol Sci 21(17):6173

Mignolet-Spruyt L, Xu E, Idänheimo N, Hoeberichts FA, Mühlenbock P, Brosché M, Van Breusegem F, Kangasjärvi J (2016) Spreading the news: subcellular and organellar reactive oxygen species production and signalling. J Exp Bot 67(13):3831–3844

Miller G, Schlauch K, Tam R, Cortes D, Torres MA, Shulaev V, Dangl J, L., & Mittler, R. (2009) The plant NADPH oxidase RBOHD mediates rapid systemic signaling in response to diverse stimuli. Sci Signal 2(84):ra45

Mittler R (2017) ROS are good. Trends Plant Sci 22(1):11–19

Mittler R, Vanderauwera S, Suzuki N, Miller GAD, Tognetti VB, Vandepoele K, Gollery M, Shulaev V, Van Breusegem F (2011) ROS signaling: the new wave? Trends Plant Sci 16(6): 300–309

Mittler R, Finka A, Goloubinoff P (2012) How do plants feel the heat? Trends Biochem Sci 37(3): 118–125

Mullineaux PM, Exposito-Rodriguez M, Laissue PP, Smirnoff N, Park E (2020) Spatial chloroplast-to-nucleus signalling involving plastid–nuclear complexes and stromules. Philos Trans R Soc B 375(1801):20190405

Mur LA, Kenton P, Lloyd AJ, Ougham H, Prats E (2008) The hypersensitive response; the centenary is upon us but how much do we know? J Exp Bot 59(3):501–520
Neill SJ, Desikan R, Clarke A, Hurst RD, Hancock JT (2002) Hydrogen peroxide and nitric oxide as signalling molecules in plants. J Exp Bot 53(372):1237–1247
Ng S, Ivanova A, Duncan O, Law SR, Van Aken O, De Clercq I, Wang Y, Carrie C, Xu L, Kmiec B, Walker H, Van Breusegem F, Whelam J, Giraud E (2013) A membrane-bound NAC transcription factor, ANAC017, mediates mitochondrial retrograde signaling in *Arabidopsis*. Plant Cell 25(9):3450–3471
Ngou BPM, Ahn HK, Ding P, Jones JD (2021) Mutual potentiation of plant immunity by cell-surface and intracellular receptors. Nature 592(7852):110–115
Nguyen QM, Iswanto ABB, Son GH, Kim SH (2021) Recent advances in effector-triggered immunity in plants: new pieces in the puzzle create a different paradigm. Int J Mol Sci 22(9): 4709
Nomura H, Komori T, Uemura S, Kanda Y, Shimotani K, Nakai K, Furuichi T, Takebayashi K, Sugimoto T, Sano S, Suwastika NI, Fukusaki E, Yoshioka H, Nakahira Y, Shiina T (2012) Chloroplast-mediated activation of plant immune signalling in *Arabidopsis*. Nat Commun 3(1): 1–11
O'Brien JA, Daudi A, Butt VS, Paul Bolwell G (2012) Reactive oxygen species and their role in plant defence and cell wall metabolism. Planta 236(3):765–779
Ochsenbein C, Przybyla D, Danon A, Landgraf F, Göbel C, Imboden A, Feussner I, Apel K (2006) The role of EDS1 (enhanced disease susceptibility) during singlet oxygen-mediated stress responses of *Arabidopsis*. Plant J 47(3):445–456
Ogasawara Y, Kaya H, Hiraoka G, Yumoto F, Kimura S, Kadota Y, Hishinuma H, Senzaki E, Yamagoe S, Nagata K, Nara M, Suzuki K, Tanokura M, Kuchitsu K (2008) Synergistic activation of the *Arabidopsis* NADPH oxidase AtrbohD by Ca2+ and phosphorylation. J Biol Chem 283(14):8885–8892
Ohama N, Sato H, Shinozaki K, Yamaguchi-Shinozaki K (2017) Transcriptional regulatory network of plant heat stress response. Trends Plant Sci 22(1):53–65
op den Camp RG, Przybyla D, Ochsenbein C, Laloi C, Kim C, Danon A, Wagner D, Hideg E, Gobel C, Feussner I, Nater M, Apel K (2003) Rapid induction of distinct stress responses after the release of singlet oxygen in *Arabidopsis*. Plant Cell 15(10):2320–2332
Phua SY, De Smet B, Remacle C, Chan KX, Van Breusegem F (2021) Reactive oxygen species and organellar signaling. J Exp Bot 72(16):5807–5824
Podgórska A, Burian M, Szal B (2017) Extra-cellular but extra-ordinarily important for cells: apoplastic reactive oxygen species metabolism. Front Plant Sci 8:1353
Pucciariello C, Banti V, Perata P (2012) ROS signaling as common element in low oxygen and heat stresses. Plant Physiol Biochem 59:3–10
Qi J, Wang J, Gong Z, Zhou JM (2017) Apoplastic ROS signaling in plant immunity. Curr Opin Plant Biol 38:92–100
Queval G, Issakidis-Bourguet E, Hoeberichts FA, Vandorpe M, Gakière B, Vanacker H, Miginiac-Maslow M, Van Breusegem F, Noctor G (2007) Conditional oxidative stress responses in the *Arabidopsis* photorespiratory mutant cat2 demonstrate that redox state is a key modulator of daylength-dependent gene expression, and define photoperiod as a crucial factor in the regulation of H2O2-induced cell death. Plant J 52(4):640–657
Queval G, Neukermans J, Vanderauwera S, Van Breusegem F, Noctor G (2012) Day length is a key regulator of transcriptomic responses to both CO2 and H2O2 in *Arabidopsis*. Plant Cell Environ 35(2):374–387
Ramel F, Birtic S, Cuiné S, Triantaphylidès C, Ravanat JL, Havaux M (2012a) Chemical quenching of singlet oxygen by carotenoids in plants. Plant Physiol 158(3):1267–1278
Ramel F, Birtic S, Ginies C, Soubigou-Taconnat L, Triantaphylidès C, Havaux M (2012b) Carotenoid oxidation products are stress signals that mediate gene responses to singlet oxygen in plants. Proc Natl Acad Sci 109(14):5535–5540
Reczek CR, Chandel NS (2015) ROS-dependent signal transduction. Curr Opin Cell Biol 33:8–13

Rodríguez AA, Córdoba AR, Ortega L, Taleisnik E (2004) Decreased reactive oxygen species concentration in the elongation zone contributes to the reduction in maize leaf growth under salinity. J Exp Bot 55(401):1383–1390
Sachdev S, Ansari SA, Ansari MI, Fujita M, Hasanuzzaman M (2021) Abiotic stress and reactive oxygen species: generation, signaling, and defense mechanisms. Antioxidants 10(2):277
Sachdev S, Jaiswal P, Ansari MI (2022) Coordinated of reactive oxygen functions species metabolism and defense systems in abiotic stress tolerance advancements in developing abiotic stress-resilient plants: basic mechanisms to trait improvements, p 23
Sandalio LM, Rodríguez-Serrano M, Romero-Puertas MC, Río LAD (2013) Role of peroxisomes as a source of reactive oxygen species (ROS) signaling molecules. In: Peroxisomes and their key role in cellular signaling and metabolism. Springer, Dordrecht, pp 231–255
Sandalio LM, Peláez-Vico MA, Molina-Moya E, Romero-Puertas MC (2021) Peroxisomes as redox-signaling nodes in intracellular communication and stress responses. Plant Physiol 186(1):22–35
Scarpeci TE, Zanor MI, Valle EM (2008) Investigating the role of plant heat shock proteins during oxidative stress. Plant Signal Behav 3(10):856–857
Schieber M, Chandel NS (2014) ROS function in redox signaling and oxidative stress. Curr Biol 24(10):R453–R462
Sewelam N, Jaspert N, Van Der Kelen K, Tognetti VB, Schmitz J, Frerigmann H, Elia S, Jurgen Z, Van Breusegem F, Maurino VG (2014) Spatial H2O2 signaling specificity: H2O2 from chloroplasts and peroxisomes modulates the plant transcriptome differentially. Mol Plant 7(7): 1191–1210
Sewelam N, Kazan K, Schenk PM (2016) Global plant stress signaling: reactive oxygen species at the cross-road. Front Plant Sci 7:187
Shai N, Schuldiner M, Zalckvar E (2016) No peroxisome is an island—peroxisome contact sites. Biochim Biophys Acta, Mol Cell Res 1863(5):1061–1069
Shapiguzov A, Vainonen JP, Hunter K, Tossavainen H, Tiwari A, Järvi S, Hellman M, Aarabi F, Alseekh S, Wybouw B, Van Der Kelen K, Nikkanen L, Krasensky-Wrzaczek J, Sipari N, Keinänen M, Tyystjärvi E, Rintamäki E, De Rybel B, Salojärvi J, Van Breusegem F, Fernie AR, Brosché M, Permi P, Aro EM, Wrzaczek M, Kangasjärvi J (2019) *Arabidopsis* RCD1 coordinates chloroplast and mitochondrial functions through interaction with ANAC transcription factors. elife 8:e43284
Sousa RH, Carvalho FE, Lima-Melo Y, Alencar VT, Daloso DM, Margis-Pinheiro M, Komatsu S, Silveira JA (2019) Impairment of peroxisomal APX and CAT activities increases protection of photosynthesis under oxidative stress. J Exp Bot 70(2):627–639
Su T, Wang P, Li H, Zhao Y, Lu Y, Dai P, Ren T, Wang X, Li X, Shao Q, Zhao D, Zhao Y, Ma C (2018) The *Arabidopsis* catalase triple mutant reveals important roles of catalases and peroxisome-derived signaling in plant development. J Integr Plant Biol 60(7):591–607
Su T, Li W, Wang P, Ma C (2019) Dynamics of peroxisome homeostasis and its role in stress response and signaling in plants. Front Plant Sci 10:705
Suzuki N, Katano K (2018) Coordination between ROS regulatory systems and other pathways under heat stress and pathogen attack. Front Plant Sci 9:490
Suzuki N, Koussevitzky SHAI, Mittler RON, Miller GAD (2012) ROS and redox signalling in the response of plants to abiotic stress. Plant Cell Environ 35(2):259–270
Takahashi F, Mizoguchi T, Yoshida R, Ichimura K, Shinozaki K (2011) Calmodulin-dependent activation of MAP kinase for ROS homeostasis in *Arabidopsis*. Mol Cell 41(6):649–660
Tanveer M, Ahmed HAI (2020) ROS signalling in modulating salinity stress tolerance in plants. In: Salt and drought stress tolerance in plants. Springer, Cham, pp 299–314
Thomma BP, Nürnberger T, Joosten MH (2011) Of PAMPs and effectors: the blurred PTI-ETI dichotomy. Plant Cell 23(1):4–15
Tian S, Wang X, Li P, Wang H, Ji H, Xie J, Qiu Q, Shen D, Dong H (2016) Plant aquaporin AtPIP1; 4 links apoplastic H2O2 induction to disease immunity pathways. Plant Physiol 171(3): 1635–1650

Tripathy BC, Oelmüller R (2012) Reactive oxygen species generation and signaling in plants. Plant Signal Behav 7(12):1621–1633
Tsai TM, Huang HJ (2006) Effects of iron excess on cell viability and mitogen-activated protein kinase activation in rice roots. Physiol Plant 127(4):583–592
Tsuda K, Katagiri F (2010) Comparing signaling mechanisms engaged in pattern-triggered and effector-triggered immunity. Curr Opin Plant Biol 13(4):459–465
Turkan I (2018) ROS and RNS: key signalling molecules in plants. J Exp Bot 69(14):3313
Unal D, García-Caparrós P, Kumar V, Dietz KJ (2020) Chloroplast-associated molecular patterns as concept for fine-tuned operational retrograde signalling. Philos Trans R Soc B 375(1801): 20190443
Van Aken O, Whelan J (2012) Comparison of transcriptional changes to chloroplast and mitochondrial perturbations reveals common and specific responses in *Arabidopsis*. Front Plant Sci 3:281
Vanderauwera S, Zimmermann P, Rombauts S, Vandenabeele S, Langebartels C, Gruissem W, Inze D, Van Breusegem F (2005) Genome-wide analysis of hydrogen peroxide-regulated gene expression in *Arabidopsis* reveals a high light-induced transcriptional cluster involved in anthocyanin biosynthesis. Plant Physiol 139(2):806–821
Vanderauwera S, Suzuki N, Miller G, Van De Cotte B, Morsa S, Ravanat JL, Hegie A, Triantaphylides C, Shulaev V, Van Montagu MCE, Van Breusegem F, Mittler R (2011) Extranuclear protection of chromosomal DNA from oxidative stress. Proc Natl Acad Sci 108(4):1711–1716
Vanlerberghe GC (2013) Alternative oxidase: a mitochondrial respiratory pathway to maintain metabolic and signaling homeostasis during abiotic and biotic stress in plants. Int J Mol Sci 14(4):6805–6847
Volkov RA, Panchuk II, Mullineaux PM, Schöffl F (2006) Heat stress-induced H2O2 is required for effective expression of heat shock genes in *Arabidopsis*. Plant Mol Biol 61(4):733–746
Voss I, Sunil B, Scheibe R, Raghavendra AS (2013) Emerging concept for the role of photorespiration as an important part of abiotic stress response. Plant Biol 15(4):713–722
Wagner D, Przybyla D, op den Camp R, Kim C, Landgraf F, Lee KP, Wursch M, Laloi C, Nater M, Hideg E, Apel K (2004) The genetic basis of singlet oxygen induced stress responses of *Arabidopsis thaliana*. Science 306(5699):1183–1185
Waszczak C, Carmody M, Kangasjärvi J (2018) Reactive oxygen species in plant signaling. Annu Rev Plant Biol 69:209–236
Woodson JD, Chory J (2008) Coordination of gene expression between organellar and nuclear genomes. Nat Rev Genet 9(5):383–395
Yan J, Tsuichihara N, Etoh T, Iwai S (2007) Reactive oxygen species and nitric oxide are involved in ABA inhibition of stomatal opening. Plant Cell Environ 30(10):1320–1325
Yang Y, Guo Y (2018) Unraveling salt stress signaling in plants. J Integr Plant Biol 60(9):796–804
Yin Y, Qin K, Song X, Zhang Q, Zhou Y, Xia X, Yu J (2018) BZR1 transcription factor regulates heat stress tolerance through FERONIA receptor-like kinase-mediated reactive oxygen species signaling in tomato. Plant Cell Physiol 59(11):2239–2254
Yuan M, Jiang Z, Bi G, Nomura K, Liu M, Wang Y, Cai B, Zhao JM, He SY, Xin XF (2021) Pattern-recognition receptors are required for NLR-mediated plant immunity. Nature 592(7852):105–109
Zandalinas SI, Mittler R (2018) ROS-induced ROS release in plant and animal cells. Free Radic Biol Med 122:21–27
Zipfel C, Kunze G, Chinchilla D, Caniard A, Jones JD, Boller T, Felix G (2006) Perception of the bacterial PAMP EF-Tu by the receptor EFR restricts *Agrobacterium*-mediated transformation. Cell 125(4):749–760
Zurbriggen MD, Carrillo N, Hajirezaei MR (2010) ROS signaling in the hypersensitive response: when, where and what for? Plant Signal Behav 5(4):393–396

Molecular Mechanisms Underpinning Signaling Function of ROS

12

12.1 Introduction

Plants surviving in natural conditions, to attune their metabolic and physiological performance under both normal and altered environmental cues have evolved a complex signaling pathway, which regulates and induces required responses via signaling components (Kreslavski et al. 2012; Saed-Moucheshi et al. 2014; Mignolet-Spruyt et al. 2016). The specific signaling components are integrated together in a circuit to trigger adaptive responses in plants, which depends on the type, severity, and combination of prevailing stress (Mignolet-Spruyt et al. 2016). Intriguingly, a particular set of signaling components may impart tolerance against more than one stress, referred to as cross-tolerance (Suzuki et al. 2014; Mignolet-Spruyt et al. 2016). In plants, the signaling process is mediated by a number of signaling molecules such as calcium (Ca^{2+}) ions, phytohormones, reactive nitrogen species (RNS), and reactive oxygen species (ROS) (Mignolet-Spruyt et al. 2016; Iqbal et al. 2021; Sachdev et al. 2021), which ultimately influence the gene expression through modification of transcription factors (TFs) (Choudhury et al. 2013).

ROS are the key signaling molecules in plants that play a central role in plant growth, development, and defense in response to abiotic and biotic cues (Zandalinas and Mittler 2018). The content of ROS in different cell organelles varies, at any given time, due to difference in ROS production and quenching capacity of particular organelle (Zandalinas et al. 2018). Moreover, ROS production in particular cell organelle may induce different signaling pathways (Van Breusegem et al. 2008), which integrate ROS to a complex signaling network. The transient increase in ROS under stress induces oxidation of downstream signaling molecules such as apoplastic signaling components, phosphatases, kinases, redox-sensitive proteins, and TFs mediating signal transduction (Mignolet-Spruyt et al. 2016; Dvořák et al. 2021). Further, ROS signal transduction pathway shows a cross-talk with other signaling pathways mediated by RNS, Ca^{2+}, and phytohormones (Bahin et al. 2011; Mittler et al. 2011; Molassiotis and Fotopoulos 2011; Considine et al. 2015;

S. Sachdev et al., *Reactive Oxygen Species in Plants*,
https://doi.org/10.1007/978-981-19-9884-3_12

Mignolet-Spruyt et al. 2016; Hempel and Trebak 2017; Kumar et al. 2017; Czarnocka and Karpiński 2018; Mullineaux et al. 2018; Suzuki and Katano 2018; Sachdev et al. 2022). ROS signals interacting with other signaling pathways have been demonstrated to induce complementary, synergistic, or antagonistic effects on regulatory mechanisms (Bhattacharjee 2019). A clear understanding of cellular language linked with ROS-mediated complex signaling networks is of immense significance, which aids in shaping future research for sustainable food production. The present chapter explores the molecular mechanism of ROS signaling pathways, ROS-mediated modulation of gene expression, and its interaction with other signaling components/molecules during signal transduction.

12.2 Mechanism of ROS Signaling in Plants

The genesis and accumulation of ROS in plants in small titers induce tolerance response by modifying a nuclear transcriptome. Three pathways have been proposed for ROS-induced alteration in gene expression (Apel and Hirt 2004) (Fig. 12.1). Firstly, ROS activate sensors or receptors, triggering signaling cascades that eventually change gene expression. Secondly, ROS involve in direct oxidation of signaling components. Thirdly, ROS modulate activity of TFs, ultimately changing gene expression (Apel and Hirt 2004; Miller and Mittler 2006; Tripathy and Oelmüller 2012). The events occurring in sensor-based ROS signaling involve perception of ROS signals, which are then processed, transmitted to the target site, and eventually translated to induce optimal response (Podgórska et al. 2017). The ROS signals are sensed by the sensors/receptors, which are generally proteins, either linked to the membranes or present intracellularly in cytoplasm (Podgórska et al. 2017). ROS or their modified products bind with receptors to induce conformational changes that initiate enzymatic activity and/or switch on or switch off ion channels like Ca^{2+} and K^+ channel (Czarnocka and Karpiński 2018). The receptors that can sense ROS include histidine kinases, redox-regulated ion channels, and ROS-sensitive phosphatases (Steffens 2014). The perception of stress by receptors leads to generation of Ca^{2+}, which binds with their respective target like mitogen-activated protein kinases (MAPK) signaling components inducing phosphorylation cascade (Huang et al. 2012; Yuan et al. 2018). Phosphorylation of signaling components, in turn, targets stress-responsive genes and TFs, which ultimately modify gene expression, eventually triggering stress response (Huang et al. 2012; Podgórska et al. 2017; Czarnocka and Karpiński 2018). Major TFs activated in response to ROS and participate in intracellular signaling comprise different families such as WRKY, NAC, zinc finger TFs (ZAT), heat shock proteins (HSP), and others (Singh et al. 2002; Van Breusegem et al. 2008). The activation of TFs by ROS is also thought to be induced via direct oxidation (Podgórska et al. 2017). In plant cells, various ROS mediate transcription of definite sets of genes (Gill and Tuteja 2010; Podgórska et al. 2017). A common TF-binding site for all oxidative stress-responsive genes so far eludes its recognition, but several ROS-induced genes specific redox-sensitive promoters have been identified (Desikan et al. 2001). In *Arabidopsis* genome,

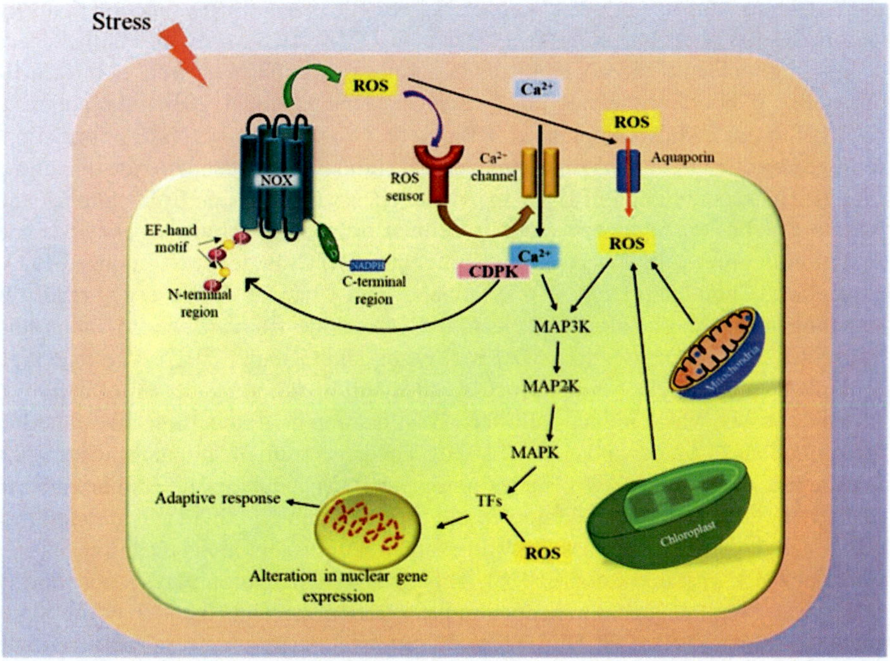

Fig. 12.1 Mechanisms of ROS signaling in plants under stress mediating adaptive response via activation of mitogen-activated protein kinases (MAPKs) cascade and modulating calcium flux (Ca^{2+}). Onset of stress on perception results in reactive oxygen species (ROS) production, which is sensed by the sensor and in turn activates Ca^{2+} channel resulting in Ca^{2+} influx. Increase in Ca^{2+} concentration in cytosol is sensed by calcium-dependent protein kinase (CDPK), resulting in activation of NADPH oxidase (NOX) at EF-hand motif of cytosolic N-terminal via phosphorylation, which in turn generates ROS. Elevated ROS diffuses from apoplast and cellular organelles (mitochondria, chloroplast) into cytosol via aquaporin where it mediates signaling by phosphorylating MAPKs cascade (MAP3K, MAP2K, and MAPK), then transcription factors (TFs), eventually causing alteration in gene expression and induce response

more specific cis-regulatory elements (CRE) in gene promoter regions, which function as binding site for TFs have been recognized (Geisler et al. 2006).

The mechanism of ROS sensing and signal transduction is a characteristic of their oxidizing properties (Waszczak et al. 2018; Dvořák et al. 2021), resulting in modification of signaling targets like redox-sensitive protein molecules, TFs, and receptor-proteins, inhibition of phosphatases, and alteration of metabolic pathways (Harir and Mittler 2009; Dvořák et al. 2021). ROS signaling events take place via oxidative post-translational modification (PTMs) of proteins (Waszczak et al. 2018). The common examples of PTMs include protein phosphorylation, glycosylation, sulfonylation, *S*-nitrosylation, and ubiquitylation/ubiquitination (Waszczak et al. 2018; Tanveer and Ahmed 2020). ROS can alter the redox status of protein by oxidizing methionine residues and thiol group of cysteines (Cys) and/or triggering protein carboxylation (Kumar et al. 2017; Waszczak et al. 2018; Yalcinkaya et al.

2019; Dvořák et al. 2021). The PTMs alter protein structure and function, causing their activation or deactivation (Waszczak et al. 2015, 2018). H_2O_2 primarily acts as a signaling molecule and accumulates in apoplast, cytoplasm as well as in cellular organelles (Choudhury et al. 2013; Schieber and Chandel 2014). In apoplast, NADPH oxidase (also known as RBOH or NOX) catalyzes H_2O_2 formation, followed by dismutation of superoxide radical anion ($O_2^{\bullet-}$) by superoxide dismutase (SOD). The accumulation of H_2O_2 in cytosol occurs on diffusion from apoplast and other organelles via aquaporins, and in cellular organelles, activity of enzymes and alteration in process like electron transport chain (ETC) under stress-induced H_2O_2 generation (Choudhury et al. 2013; Schieber and Chandel 2014). H_2O_2 mediates signaling through oxidation of selective Cys residue (thiolate anion) in protein molecules or phosphatases like MAPK (Reczek and Chandel 2015). The oxidation of thiolate anion by H_2O_2 via a process called sulfonylation results in formation of sulfenic species, which induces allosteric modification in protein molecules, altering their functions (Tanveer and Ahmed 2020). The generation of sulfenic species is the first degree of protein (Cys residue) oxidative PTM, which is reversible and can easily reduce back to thiolate anion by the action of thioredoxin (Trx), glutaredoxin (Grx), and disulfide reductases, restoring function of protein molecules (Reczek and Chandel 2015; Waszczak et al. 2018). This reversible oxidation plays a noteworthy role in the reversible signal transduction process (Reczek and Chandel 2015). Since small fraction (nM) of H_2O_2 undergoes thiolate oxidation to sulfenic species, presence of H_2O_2 in excess amount leads to oxidation of thiolate anions to sulfenic and sulfonic species, which show irreversible alteration, eventually causing permanent oxidative damage to proteins (Reczek and Chandel 2015; Waszczak et al. 2018). Thus, small titers of H_2O_2 facilitate signaling while its built-up concentration induces oxidative stress. Moreover, the susceptibility of Cys residues of all ROS-sensitive protein is not the same and highly dependent on local electrostatic environment. Consequently, an ambient pH majorly determines the susceptibility of Cys residues to oxidation (Roos et al. 2013; Waszczak et al. 2018). The plant transduction mechanism also involves negative regulation as one of the specific characteristics, in which response is triggered by inactivating repressor proteins. The protein dephosphorylation or ubiquitination is involved in negative regulation, which is then proceeded by transcription, resulting in faster transduction process (Rosenfeld et al. 2002). As ROS have potential to induce oxidation of protein molecules, abundance of ROS results in metabolic response via oxidized proteins. Some of the oxidized proteins are degraded via a ubiquitination mechanism while others directly act as signaling molecules. Thus, presence of different kinds of oxidized protein in excess facilitates regulation of plant responses (Podgórska et al. 2017).

Recently, a mechanism for extracellular H_2O_2-induced signaling has been documented in *Arabidopsis*, which is mediated through a plasma membrane localized H_2O_2 receptor capable of sensing and transducing signals into an intracellular signaling cascade (Dvořák et al. 2021). The HYDROGEN-PEROXIDE-INDUCED Ca^{2+} INCREASES 1 (HPCA1), a membrane bound receptor kinase belonging to a protein family of leucine-rich repeat (LRR), can perceive apoplastic

H_2O_2 through covalent modification of Cys residues (oxidation of two pairs) in its extracellular domain, which causes its autophosphorylation. This accelerates Ca^{2+} influx via activation of Ca^{2+} channels in guard cells, ultimately inducing stomatal closure (Wu et al. 2020). Under stress, ROS-mediated transcription of stress-responsive genes such as heat shock gene (HSG) mediates tolerance to stress by inducing specific responses (Pucciariello et al. 2012). The HSG transcripted by ROS signaling can activate heat shock proteins (HSPs), which function as molecular chaperons and regulate oxidative damage under heat, anoxia, light, and cold stress (Heidarvand and Maali Amiri 2010; Pucciariello et al. 2012). The HSP prevents aggregation, misfolding, denaturation, and degradation of protein molecules and facilitates protein refolding (Pucciariello et al. 2012). Volkov et al. (2006) reported hydrogen peroxide (H_2O_2) induced modulation of expression of HSG, triggering production of ascorbate peroxidase 2 (APX2), HSP17.6, and HSP 18.2 on exposure of *Arabidopsis* cell culture to heat stress. Analogously, the unrestricted accumulation of H_2O_2 in *Arabidopsis* mediates upregulation of HSPs and production of ascorbate peroxidase 1 (APX1), which facilitate effective quenching of excess H_2O_2, triggering tolerance against high light stress (Pnueli et al. 2003) and combined heat and drought stress (Koussevitzky et al. 2008). H_2O_2-induced upregulation of gene expression encoding HSPs, other transcription factors such as *ZAT10* and *ZAT12*, fermentation genes like *ALCOHOL DEHYDROGENASE* (ADH), and proteins under anoxia/hypoxia stress confers acclimatization to stress (Banti et al. 2010; Pucciariello et al. 2012; Yang and Hong 2015; Sasidharan et al. 2018). *Arabidopsis thaliana* roots exposed to low temperature demonstrate expression of *AtSRC2* gene (*A. thaliana* homolog of soybean gene regulated by cold 2 (*SRC2*)), which results in AtRBOHF (NOX enzyme) activation by interacting at the cytosolic N-terminal region, eventually generating ROS that mediate appropriate acclimatization response (Kawarazaki et al. 2013).

12.3 Mitogen-Activated Protein Kinase Cascade Activation by ROS

Protein kinases as well as phosphatases are crucial components of signaling cascade that involves phosphorylation and/or dephosphorylation of the protein molecules (Taj et al. 2010). The mitogen-activated protein kinases (MAPK) pathway is the crucial and well-studied signaling mechanism in cells and includes a class of serine/threonine (ser/thr) protein kinases (Taj et al. 2010). MAPK signaling components either directly perceive external cues or present downstream of second messengers, and on perception of signals, MAPK cascade mediates stress response by relaying messages to the nucleus (Taj et al. 2010; Raja et al. 2017). The MAPK signaling cascade comprises at least three interlinked kinases, i.e., mitogen-activated protein kinase kinase kinase (MAPKKK, MEKK/MKKK, MAP3K), mitogen-activated protein kinase kinase (MAPKK, MEK/MKK, MAP2K), and mitogen-activated protein kinase (MAPK, MPK), which get phosphorylated consecutively (Sinha et al. 2011; Liu and He 2017; Raja et al. 2017). The activation of MAPK cascade initially involves the phosphorylation of two ser/thr residue present within the MAPKK activation loop by the activated MAPKKK, resulting in activation of MAPKK, which further activates MAPK by dual phosphorylation of threonine/ tyrosine (thr/tyr) residue along the TXY motif in highly conserved activation loop (Jonak et al. 2002; Hamel et al. 2012; Liu and He 2017). A fourth type of protein kinase, mitogen-activated protein kinase kinase kinase kinase (MAPKKKK, MAP4K or MAP3K kinase) reportedly acts as an adaptor to link upstream signals with the main MAPK cascade (Raja et al. 2017). The study of *Arabidopsis* genome also reveals a complex signaling network of kinases comprising 23 MAPK, 10 MAPKK, 80 MAPKKK, and 10 MAPKKKK (Jonak et al. 2002). Due to the presence of a large number of MAPK, MAPKK, MAPKKK, and MAPKKKK, thousands of combinations can occur, which potentially participate in the signaling process (Cvetkovska et al. 2005; Taj et al. 2010).

ROS possess the capacity to mediate phosphorylation of MAPKs under stress and/or developmental cues (Komis et al. 2018; Dvořák et al. 2021). In ROS-mediated signaling, certain kinases, in particular, MAPKKK, MEKK1, MPK3, MPK4, and MPK6 have been identified to play a specific role (Colcombet and Hirt 2008). In *Arabidopsis* seedlings, exogenous application of H_2O_2 activated MAPK cascade involving MEKK1-MPK4 kinase (Nakagami et al. 2006) and AtMPK3 and AtMPK6 via *Arabidopsis homologs of nucleus and phragmoplast localized kinases* 1 (ANP1-a type of MAPKKK) (Kovtun et al. 2000). The activated (phosphorylated) MAPK interacts and modulates phosphorylation status of target proteins like TFs, enzymes, and other proteins, eventually altering the gene expression, inducing required response against stress (Hamel et al. 2012; Liu and He 2017; Dvořák et al. 2021). The proteins belonging to the WRKY family have been reported to work downstream of MAPK in several circumstances and upregulate gene transcription (Eulgem and Somssich 2007; Chen et al. 2012). ROS-activated MAPK cascade comprising MEKK1, MKK1/2, and MPK4 induces initial plant defense against biotic stress (Zhang et al. 2012). A pathogen-induced H_2O_2 activates the MPK7 downstream of MKK3 (MAP2K), inducing expression of pathogenesis-related (PR) genes (Dóczi et al. 2007). Exposure of *Arabidopsis* to O_2 deprivation condition or treatment with Antimycin A (AA) and/or potassium cyanide (inhibitors of cell respiration) induces ROS genesis and activation of AtMPK3 and AtMPK6 (Chang et al. 2012). Presumably, O_2 deprivation leads to production of ROS in mitochondria, which get transferred to the cytosol, where it activates MAPK cascade, eventually generating retrograde signaling from mitochondria to nucleus. However, deprivation of O_2 or inhibition of cell respiration can reduce CO_2 in green cells, affecting photosynthesis. Thus, alteration in photosynthetic activity could trigger production of ROS in chloroplast, which may partake in activation of retrograde signaling via activation of AtMPK3 and AtMPK6 in cytoplasm. In another study, exposure of *Arabidopsis* to high light intensity activates AtMPK6, triggering chloroplast-induced retrograde signaling via TFs like *ERF6* and *ERF104* (*ethylene responsive element binding factors*) (Vogel et al. 2014).

ROS also modulate MAPK signaling via direct interaction that triggers oxidation of Cys residue of MAPKs, altering their stability and protein–protein interaction, promoting aggregation (Dvořák et al. 2021). This leads to either activation or deactivation of MAPKs. For example, H_2O_2 activates *oxidative signal inducible1* (OXI1) *kinase*, which is crucial for full activation of MPK3 and MPK6 (Rentel et al. 2004). H_2O_2 promotes sulfenylation of MPK2, MPK4, and MPK7 (Waszczak et al. 2014) and activates BnMAPK of *Brassica napus* via oxidation of Cys232 residue followed by aggregation (Zhang et al. 2015).

MAPKs are effective regulators of antioxidants. MAPKs show capacity to modify the expression of several antioxidant enzymes under stress conditions (Dvořák et al. 2021). Xing et al. (2013) have reported MKK5 (MAP2K) triggered expression of SODs in *Arabidopsis* under high light stress by activating downstream MPK3 and MPK6. Further, Xing et al. (2007) have also observed that after application of abscisic acid (ABA) and exposure to drought stress, MKK1 induces activation of only CAT1 (catalase), not CAT2 and CAT3 in plants during salinity stress. The expression of CAT1 is induced by MPK6, which is downstream of MKK1 and is crucial for ABA-induced H_2O_2 production (Xing et al. 2008). Moreover, MAPK signaling could alter the expression of NOX. Adachi et al. (2015) have reported activation of RBOHD leading to ROS burst in *Nicotiana benthamiana* exposed to bacterial effector R3a/AVR3a and protein elicitor INF1 on phosphorylation of WRKY TFs by MEK2 cascade.

12.4 Cross-Talk Between ROS and Other Signaling Molecules

12.4.1 Dialogue Between ROS and Ca^{2+} Signals

The cytosolic Ca^{2+} is a crucial secondary messenger involved in various signaling networks occurring either intracellularly or extracellularly under stress conditions (Kohli et al. 2017; Tian et al. 2020). Ca^{2+} signature in a plant cell (in cytoplasm and/or cellular organelles) occurring under particular stimuli is specific and depends upon frequency, amplitude as well as duration of occurrence (Mazars et al. 2010). High accumulation of Ca^{2+} in a cell can be detrimental, as excess Ca^{2+} can bind with protein molecules and organic acids, forming insoluble complexes (Kohli et al. 2017). Therefore, the proficient signaling requires an adequate level of calcium only, but not its excess (Kudla et al. 2010; Kohli et al. 2017). An adequate content of Ca^{2+} is regulated by storing Ca^{2+} in vacuole, apoplast, and endoplasmic reticulum (Kohli et al. 2017). A response to stress conditions contemplates an increase in content of cytosolic Ca^{2+} (Stael et al. 2012; Kohli et al. 2017). Elevation in cytosolic Ca^{2+} is regulated via environmental stimulus, Ca^{2+} channels, ion pumps, Ca^{2+} transporter embedded in plasma or organelle membrane, and Ca^{2+} binding proteins that function as Ca^{2+} sensors including calmodulins (CaMs), CaM-like proteins (CMLs), calcineurin B-like (CBL) protein, and calcium-dependent protein kinase (CDPKs or CPKs) (Bouché et al. 2005; Das and Pandey 2010; Sewelam et al. 2016; Podgórska et al. 2017; Yuan et al. 2018; Zhang et al. 2018; Dvořák et al. 2021). The ser/thr-CDPKs are important components of a stress-induced signaling network. These CDPKs include a large number of members in the multigene family in various plant species (Dvořák et al. 2021). The Ca^{2+} sensors along with other signaling components facilitate downstream signaling resulting in phosphorylation, activation of MAPKs, transient accumulation of ROS or nitric oxide (NO), and/or transcriptional reprogramming (Yuan et al. 2018).

The cytosolic Ca^{2+} and ROS show interlinking. For instance, ROS regulate Ca^{2+} content in cytosol while Ca^{2+} mediates ROS production (Gaupels et al. 2017). The change in Ca^{2+} flux in a cell induced by H_2O_2 is crucial for triggering several biological responses, in particular, stress adaptation, stomatal closure, and different types of programmed cell death (PCD) (Rentel and Knight 2004; Mazars et al. 2010). Production of ROS by NOX in extracellular space opens channels and transporters localized at the plasma membrane, enabling influx of Ca^{2+} into cytoplasm (Gilroy et al. 2016; Podgórska et al. 2017; Waszczak et al. 2018). This elevates cytosolic Ca^{2+} concentration and triggers membrane depolarization (Mittler et al. 2011). In turn, Ca^{2+} upregulates the activity of NOX by binding directly at the EF-hand motif of cytosolic N-terminal region, which acts as a Ca^{2+} sensor, and induces a feedback loop, eventually propagating the signal wave to neighboring cells (Mazars et al. 2010; Gilroy et al. 2016; Podgórska et al. 2017). Apart from Ca^{2+}, CDPK also activates the NOX enzyme by inducing phosphorylation of two N-terminal serine (ser) residues (Ogasawara et al. 2008; Mazars et al. 2010). CaM senses an increase in Ca^{2+} flux in transgenic tobacco cell under biotic and abiotic stress, resulting in Ca^{2+}/CaM complex. The complex activates NAD kinase, eventually enhancing the content of NADPH that promoted ROS production due to activation of NOX enzyme (Harding et al. 1997). ROS-induced increase in Ca^{2+} flux under stress conditions mediates stress-specific response in plants. Low temperature reportedly increases the ROS accumulation, which subsequently triggers transient Ca^{2+}elevation (Suzuki and Mittler 2006; Guo et al. 2018; Yuan et al. 2018). In potato, StCDPK5 activates StRBOH via direct phosphorylation of N-terminal resulting in H_2O_2 burst, which eventually leads to hypersensitive response (HR) like cell death and elevated resistance toward biotrophic pathogen *Phytophthora infestans*, a causal agent of late blight disease (Yoshioka et al. 2009). In leaves of *Zea mays*, production of antioxidants by ABA facilitates cross-talk between ROS and Ca^{2+} via activation of NOX/RBOH. In the ROS signaling process, Ca^{2+} performs both upstream and downstream functions (Jiang and Zhang 2003). The modulation of NOX by Ca^{2+} is also triggered via CDPKs such as CPK/CBL-CBL interacting protein kinase (CBL-CIPKs) (Dubiella et al. 2013; Monaghan et al. 2014; Gilroy et al. 2016; Dvořák et al. 2021). Ca^{2+} not only induces oxidative burst in the cell, but also functions downstream of the oxidative burst, triggering responses like cellular defense against stress (Yoshioka et al. 2009). The Ca^{2+} signaling under stress modulates antioxidant enzymes like ascorbate peroxidase (APX) and catalase (CAT), enabling tolerance against stress in plants. The elevated cytosolic Ca^{2+} content activates calmodulin, which mediates signals to promote activity of CAT, enhancing detoxification of H_2O_2 (Yang and Poovaiah 2002; Kohli et al. 2017). Thus, an increased cytosolic Ca^{2+} facilitates H_2O_2 balance in the cell. For instance, increased activity of peroxisomes to scavenge H_2O_2 on exposure to heavy metal stress increases cytosolic Ca^{2+} accumulation, which promotes CAT3 activity (Costa et al. 2010). Moreover, an increased activity of *OsCPK12* in rice plants elevates tolerance against salt stress by an enhanced expression of *OsAPX2*, *OsAPX8,* and a decreased expression NOX gene, reducing ROS titer (Asano et al. 2012). Thus, the cross-talk between ROS and Ca^{2+} signaling is a crucial strategy adopted by plants to mediate stress tolerance.

12.4.2 ROS and Phytohormones Cross-Talk

ROS signaling shows extensive cross-talk with phytohormones signaling (Fig. 12.2) to induce required molecular and physiological responses in plants under stress (Choudhury et al. 2017; Kanojia and Dijkwel 2018; Bhattacharjee 2019; Devireddy et al. 2021). The ability of ROS, particularly, H_2O_2 to interact with plant hormones like salicylic acid (SA), jasmonic acid (JA), abscisic acid (ABA), brassinosteroid (BR), and ethylene (ET) is well known (Xia et al. 2015). H_2O_2 participates in the upregulation of stress hormone synthesis like SA (León et al. 1995). The production of ROS triggered by adverse conditions is further altered by hormones via activation of NOX (Devireddy et al. 2021). The plant hormone ABA activates RBOHD and RBOHF via cross-talk between ABA and ROS (Mittler and Blumwald 2015; Devireddy et al. 2021). The hormone signaling is mediated by the hormone receptor present on a plasma membrane, which further results in MAPK signaling and

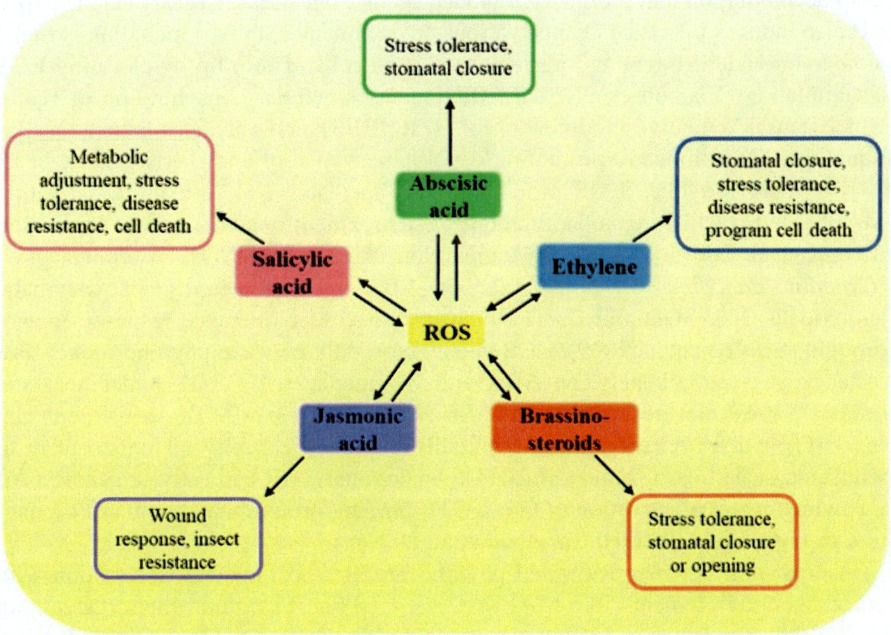

Fig. 12.2 Cross-talk between ROS and phytohormone signaling mediating stress tolerance in plants

activates gene expression (Santner and Estelle 2009). The signaling pathways regulated by phytohormones involve ROS downstream to trigger stress acclimation responses like stomatal closure and hydraulic conductivity (Bhattacharjee 2019).

The oxidative signal transduction pathway triggered by ROS involves phytohormones like JA, SA, and ET, downstream. Conversely, phytohormones mediated signaling, relays message downstream by involving ROS molecules (Bhattacharjee 2019). Thus, signaling events induced by phytohormones result in production of ROS, and ROS-mediated signaling triggers production of phytohormones, forming a self-amplifying loop (Han et al. 2013; Xia et al. 2015) that ultimately regulates gene expression and mediates appropriate response (Devireddy et al. 2021). Auxin and SA exhibit the ability to induce production of $O_2^{\bullet -}$ in extracellular space (Mori et al. 2001, 2009; Schopfer et al. 2002). The genesis of ROS by auxin facilitates cell cycle and maintains ROS homeostasis. The generation of ROS in turn inhibits auxin signaling via activation of MAPK cascade and triggers oxidative stress signaling (Tognetti et al. 2012). Thus, ROS displays both feedback and feedforward interaction with phytohormones (Bhattacharjee 2019). In the SA signaling pathway, ROS lie both upstream and downstream and induce CDPKs signaling to induce closure of stomata (Prodhan et al. 2018). ET induces ROS production in plants under salt stress, mediating stress regulation. For instance, loss-of-function of *Ethylene Overproducer1* (ETO1) mutant of *Arabidopsis* with ability to generate ET under salt stress promotes formation of ROS via RBOHF (Jiang et al. 2013). Similarly, an increase in ET content enhances ROS accumulation that promotes tolerance against salt stress in rice (Li et al. 2014). BR also induces tolerance against various environmental stresses including extreme temperature and drought and also controls stem cells of root tip by elevating ROS accumulation (Yaqoob et al. 2022). BR increases cellular concentration of H_2O_2, which causes oxidative modification of BZR1/BES1 and enhances their transcriptional activity, ultimately promoting development of root and meristem (Lv et al. 2018). ABA provokes ROS production that mediates NO formation, activating MAPK and facilitating upregulation of antioxidant enzyme gene expression (Zhang et al. 2007). Increased accumulation of ROS and NO in Bermuda grass (*Cynodon dactylon*) on application of ABA induces activity of enzymatic antioxidants like SOD and CAT, which in turn confer tolerance response toward drought stress (Lu et al. 2009). ROS initiate cross-talk between phytohormones. For instance, a cross-talk between ABA and BR mediated by ROS under heat and oxidative stress elevates expression of *RBOH1* and increases NOX activity, enhancing H_2O_2 and ABA content that eventually improved capacity of tomato plant to withstand heat stress (Zhou et al. 2014). Analogously, BR can activate synthesis of ET, which triggers activation of Gα, a GTP-binding protein that stimulates production of H_2O_2 via AtRBOHF (Yaqoob et al. 2022).

A cross-talk between ROS and phytohormones also facilitates interaction with other signaling elements like MAPK, Ca^{2+}, CDPKs, TFs, and others (Devireddy et al. 2021). For instance, oxidative stress and phytohormones like JA and ABA can activate MPK1 and MPK2 (Ortiz-Masia et al. 2007). MPK9 and MPK12 activated by ROS and ABA act upstream of anion channels in guard cells, leading to stomatal closure (Jammes et al. 2009). The accumulation of ROS prompts salt stress tolerance by regulating Na^+ influx in root, reducing Na^+ loading into xylem and maintaining root K^+ retention (Jiang et al. 2013). Production of H_2O_2 in the presence of a pathogen triggers oxidation and accumulation of glutathione (GSH), which subsequently activates SA signaling pathway, inducing defense response in *Arabidopsis* (Han et al. 2013). Application of 24-epibrassibolide (EBR) to tomato plant subjected to heat stress confers tolerance to oxidative stress by elevating H_2O_2 content in apoplast and transcript level of *RBOH1*, MPK1, and MPK2 as well as activation of MPK1/2 (Nie et al. 2013). Likewise, ABA induces signaling cascade via phosphorylation of AtMKK1 and AtMPK6, which leads to generation of ROS and CAT1 expression, eliciting response to drought stress (Xing et al. 2008). A well-tuned interaction of ROS and hormones with these elements is crucial to induce appropriate response in plants at molecular, physiological, and metabolic level, facilitating acclimatization of plants under unfavorable conditions (Zhu 2016; Kollist et al. 2019; Zandalinas et al. 2019, 2020; Devireddy et al. 2021).

12.4.3 Interplay Between ROS and RNS

The reactive nitrogen species (RNS) are nitric oxide free radical (NO) and NO derived molecules including both free radicals and non-radicals such as nitrogen dioxide ($^{\bullet}NO_2$), S-nitrosoglutathione (GSNO), nitroxyl anion (NO^-), peroxynitrite ($ONOO^-$), dinitrogen trioxide (N_2O_3), *S*-nitrosothiols (RSNOs), and others (Halliwell 2006; Molassiotis and Fotopoulos 2011; Airaki et al. 2012; del Río 2015; Phaniendra et al. 2015; Saddhe et al. 2019). Similar to ROS, RNS plays a dual role in plants, inducing nitrosative stress as well as mediating signaling cascade (del Río et al. 2006; Saddhe et al. 2019). It has been observed that the oxidative burst triggered by pathogen infection in plant cells is simultaneously accompanied by the nitrosative burst (Molassiotis and Fotopoulos 2011; Wang et al. 2013). The cross-talk between ROS-induced oxidative and RNS-triggered nitrosative signaling mediates acclimatory responses in plants against abiotic and biotic stress (Molassiotis and Fotopoulos 2011). The interplay between ROS and RNS has been well demonstrated under biotic stress leading to HR response, a form of PCD, resulting in rapid death of cells in vicinity of infection sites (Molassiotis and Fotopoulos 2011; Wang et al. 2013). After recognition of a pathogen, cells induce HR by concomitant generation of ROS/H_2O_2 and RNS/NO, which react chemically to form $^1O_2$ or hydroxyl radical ($^{\bullet}OH$), mediating cell death (Wang et al. 2013). Moreover, ROS or NO may trigger release of cytochrome c from mitochondria, altering caspase-like signaling cascade, resulting in HR (Wang et al. 2013). The interaction of ROS and RNS such as $O_2^{\bullet-}$ and NO results in formation of non-radical $ONOO^-$. $ONOO^-$ has a short half-life but can cross the biological membrane and is more reactive than parent molecules leading to critical alteration of biomolecules (Molassiotis and Fotopoulos 2011). These properties allow $ONOO^-$ to mediate signaling, eventually inducing response against abiotic stress and nitrosylation of Prxs (Molassiotis and Fotopoulos 2011). Further, gene expression induced by ROS and RNS signaling under stress shows overlapping. The proteomic study conducted by Tanou et al. (2009) to investigate the impact of pretreatment of citrus plant with H_2O_2 or NO donating sodium nitroprusside (SNP) in alleviating salinity stress reveals overlapping between both H_2O_2- and NO-mediated signaling pathway, leading to acclimatization against salinity by reducing protein carbonylation and *S*-nitrosylation.

12.5 Conclusion

The ability of ROS to oxidize cellular biomolecules not only results in oxidative stress but also aids in signal transduction. ROS are important redox molecules that facilitate plants to tolerate unfavorable cues by inducing signaling cascades that activate protein kinases and transcription factors (TFs) and upregulate gene expressions. The signaling process of ROS is not simple but shows complexity as spatio-temporal production of ROS under varied stresses shows a discrete pathway resulting in upregulation of specific gene expression. Further, ROS signaling pathways integrate with other signaling cascades either upstream or downstream to relay messages to the nucleus, facilitating appropriate response. Exploring the precise mechanism of specific ROS signaling under different environmental conditions, variation in spatio-temporal production, and cross-talk with other signaling pathways will aid in improving crop productivity. Moreover, deciphering the mechanism of how plants build memories of previously encountered adverse stimuli at molecular level could pave a new direction for designing resilient crops.

References

Adachi H, Nakano T, Miyagawa N, Ishihama N, Yoshioka M, Katou Y, Yaeno T, Shirasu K, Yoshioka H (2015) WRKY transcription factors phosphorylated by MAPK regulate a plant immune NADPH oxidase in *Nicotiana benthamiana*. Plant Cell 27(9):2645–2663

Airaki M, Leterrier M, Mateos RM, Valderrama R, Chaki M, Barroso JB, del Rio LA, Palma JM, Corpas FJ (2012) Metabolism of reactive oxygen species and reactive nitrogen species in pepper (*Capsicum annuum* L.) plants under low temperature stress. Plant Cell Environ 35(2):281–295

Apel K, Hirt H (2004) Reactive oxygen species: metabolism, oxidative stress, and signaling transduction. Annu Rev Plant Biol 55:373

Asano T, Hayashi N, Kobayashi M, Aoki N, Miyao A, Mitsuhara I, Ichikawa H, Komatsu S, Hirochika H, Kikuchi S, Ohsugi R (2012) A rice calcium-dependent protein kinase OsCPK12 oppositely modulates salt-stress tolerance and blast disease resistance. Plant J 69(1):26–36

Bahin E, Bailly C, Sotta B, Kranner I, Corbineau F, Leymarie J (2011) Crosstalk between reactive oxygen species and hormonal signalling pathways regulates grain dormancy in barley. Plant Cell Environ 34(6):980–993

Banti V, Mafessoni F, Loreti E, Alpi A, Perata P (2010) The heat-inducible transcription factor HsfA2 enhances anoxia tolerance in *Arabidopsis*. Plant Physiol 152(3):1471–1483

Bhattacharjee S (2019) ROS: central component of signaling network in plant cell. In: Reactive oxygen species in plant biology. Springer, New Delhi, pp 127–153

Bouché N, Yellin A, Snedden WA, Fromm H (2005) Plant-specific calmodulin-binding proteins. Annu Rev Plant Biol 56(1):435–466

Chang R, Jang CJ, Branco-Price C, Nghiem P, Bailey-Serres J (2012) Transient MPK6 activation in response to oxygen deprivation and reoxygenation is mediated by mitochondria and aids seedling survival in *Arabidopsis*. Plant Mol Biol 78(1):109–122

Chen L, Song Y, Li S, Zhang L, Zou C, Yu D (2012) The role of WRKY transcription factors in plant abiotic stresses. Biochim Biophys Acta Gene Regul Mech 1819(2):120–128

Choudhury S, Panda P, Sahoo L, Panda SK (2013) Reactive oxygen species signaling in plants under abiotic stress. Plant Signal Behav 8(4):e23681

Choudhury FK, Rivero RM, Blumwald E, Mittler R (2017) Reactive oxygen species, abiotic stress and stress combination. Plant J 90(5):856–867

Colcombet J, Hirt H (2008) Arabidopsis MAPKs: a complex signalling network involved in multiple biological processes. Biochem J 413(2):217–226

Considine MJ, María Sandalio L, Helen Foyer C (2015) Unravelling how plants benefit from ROS and NO reactions, while resisting oxidative stress. Ann Bot 116(4):469–473

Costa A, Drago I, Behera S, Zottini M, Pizzo P, Schroeder JI, Pozzan T, Schiavo FL (2010) H2O2 in plant peroxisomes: an in vivo analysis uncovers a Ca2+-dependent scavenging system. Plant J 62(5):760–772

Cvetkovska M, Rampitsch C, Bykova N, Xing T (2005) Genomic analysis of MAP kinase cascades in *Arabidopsis* defense responses. Plant Mol Biol Report 23(4):331–343

Czarnocka W, Karpiński S (2018) Friend or foe? Reactive oxygen species production, scavenging and signaling in plant response to environmental stresses. Free Radic Biol Med 122:4–20

Das R, Pandey GK (2010) Expressional analysis and role of calcium regulated kinases in abiotic stress signaling. Curr Genomics 11(1):2–13

del Río LA (2015) ROS and RNS in plant physiology: an overview. J Exp Bot 66(10):2827–2837

del Río LA, Sandalio LM, Corpas FJ, Palma JM, Barroso JB (2006) Reactive oxygen species and reactive nitrogen species in peroxisomes. Production, scavenging, and role in cell signaling. Plant Physiol 141(2):330–335

Desikan RAH, Mackerness S, Hancock JT, Neill SJ (2001) Regulation of the *Arabidopsis* transcriptome by oxidative stress. Plant Physiol 127(1):159–172

Devireddy AR, Tschaplinski TJ, Tuskan GA, Muchero W, Chen JG (2021) Role of reactive oxygen species and hormones in plant responses to temperature changes. Int J Mol Sci 22(16):8843

Dóczi R, Brader G, Pettkó-Szandtner A, Rajh I, Djamei A, Pitzschke A, Teige M, Hirt H (2007) The Arabidopsis mitogen-activated protein kinase kinase MKK3 is upstream of group C mitogen-activated protein kinases and participates in pathogen signaling. Plant Cell 19(10):3266–3279

Dubiella U, Seybold H, Durian G, Komander E, Lassig R, Witte CP, Schulze WX, Romeis T (2013) Calcium-dependent protein kinase/NADPH oxidase activation circuit is required for rapid defense signal propagation. Proc Natl Acad Sci 110(21):8744–8749

Dvořák P, Krasylenko Y, Zeiner A, Šamaj J, Takáč T (2021) Signaling toward reactive oxygen species-scavenging enzymes in plants. Front Plant Sci 11:2178

Eulgem T, Somssich IE (2007) Networks of WRKY transcription factors in defense signaling. Curr Opin Plant Biol 10(4):366–371

Gaupels F, Durner J, Kogel KH (2017) Production, amplification and systemic propagation of redox messengers in plants? The phloem can do it all! New Phytol 214(2):554–560

Geisler M, Kleczkowski LA, Karpinski S (2006) A universal algorithm for genome-wide in silicio identification of biologically significant gene promoter putative cis-regulatory-elements; identification of new elements for reactive oxygen species and sucrose signaling in *Arabidopsis*. Plant J 45(3):384–398

Gill SS, Tuteja N (2010) Reactive oxygen species and antioxidant machinery in abiotic stress tolerance in crop plants. Plant Physiol Biochem 48(12):909–930

Gilroy S, Białasek M, Suzuki N, Górecka M, Devireddy AR, Karpiński S, Mittler R (2016) ROS, calcium, and electric signals: key mediators of rapid systemic signaling in plants. Plant Physiol 171(3):1606–1615

Guo X, Liu D, Chong K (2018) Cold signaling in plants: insights into mechanisms and regulation. J Integr Plant Biol 60(9):745–756

Halliwell B (2006) Phagocyte-derived reactive species: salvation or suicide? Trends Biochem Sci 31(9):509–515

Hamel LP, Nicole MC, Duplessis S, Ellis BE (2012) Mitogen-activated protein kinase signaling in plant-interacting fungi: distinct messages from conserved messengers. Plant Cell 24(4): 1327–1351

Han Y, Chaouch S, Mhamdi A, Queval G, Zechmann B, Noctor G (2013) Functional analysis of *Arabidopsis* mutants points to novel roles for glutathione in coupling H2O2 to activation of salicylic acid accumulation and signaling. Antioxid Redox Signal 18(16):2106–2121

Harding SA, Oh SH, Roberts DM (1997) Transgenic tobacco expressing a foreign calmodulin gene shows an enhanced production of active oxygen species. EMBO J 16(6):1137–1144

Harir Y, Mittler R (2009) The ROS signaling network of cells. In: Reactive oxygen species in plant signaling. Springer, Berlin, Heidelberg, pp 165–174

Heidarvand L, Maali Amiri R (2010) What happens in plant molecular responses to cold stress? Acta Physiol Plant 32(3):419–431

Hempel N, Trebak M (2017) Crosstalk between calcium and reactive oxygen species signaling in cancer. Cell Calcium 63:70–96

Huang A, Yang YM, Yan C, Kaley G, Hintze TH, Sun D (2012) Altered MAPK signaling in progressive deterioration of endothelial function in diabetic mice. Diabetes 61(12):3181–3188

Iqbal Z, Iqbal MS, Hashem A, Abd Allah EF, Ansari MI (2021) Plant defense responses to biotic stress and its interplay with fluctuating dark/light conditions. Front Plant Sci 12:631810

Jammes F, Song C, Shin D, Munemasa S, Takeda K, Gu D, Cho D, Lee S, Giordo R, Sritubtim S, Leonhardt N, Ellis BE, Murata Y, Kwak JM (2009) MAP kinases MPK9 and MPK12 are preferentially expressed in guard cells and positively regulate ROS-mediated ABA signaling. Proc Natl Acad Sci 106(48):20520–20525

Jiang M, Zhang J (2003) Cross-talk between calcium and reactive oxygen species originated from NADPH oxidase in abscisic acid-induced antioxidant defence in leaves of maize seedlings. Plant Cell Environ 26(6):929–939

Jiang C, Belfield EJ, Cao Y, Smith JAC, Harberd NP (2013) An *Arabidopsis* soil-salinity–tolerance mutation confers ethylene-mediated enhancement of sodium/potassium homeostasis. Plant Cell 25(9):3535–3552

Jonak C, Ökrész L, Bögre L, Hirt H (2002) Complexity, cross talk and integration of plant MAP kinase signalling. Curr Opin Plant Biol 5(5):415–424

Kanojia A, Dijkwel PP (2018) Abiotic stress responses are governed by reactive oxygen species and age. Annu Plant Rev Online:295–326. https://doi.org/10.1002/9781119312994.apr0611

Kawarazaki T, Kimura S, Iizuka A, Hanamata S, Nibori H, Michikawa M, Imai A, Abe M, Kaya H, Kuchitsu K (2013) A low temperature-inducible protein AtSRC2 enhances the ROS-producing activity of NADPH oxidase AtRbohF. Biochim Biophys Acta Mol Cell Res 1833(12): 2775–2780

Kohli SK, Handa N, Gautam V, Bali S, Sharma A, Khanna K, Arora S, Thukral AK, Ohri P, Karpets YV, Kolupaev YE, Bhardwaj R (2017) ROS signaling in plants under heavy metal stress. In: Reactive oxygen species and antioxidant systems in plants: role and regulation under abiotic stress. Springer, Singapore, pp 185–214

Kollist H, Zandalinas SI, Sengupta S, Nuhkat M, Kangasjärvi J, Mittler R (2019) Rapid responses to abiotic stress: priming the landscape for the signal transduction network. Trends Plant Sci 24(1): 25–37

Komis G, Šamajová O, Ovečka M, Šamaj J (2018) Cell and developmental biology of plant mitogen-activated protein kinases. Annu Rev Plant Biol 69:237–265

Koussevitzky S, Suzuki N, Huntington S, Armijo L, Sha W, Cortes D, Shulaev V, Mittler R (2008) Ascorbate peroxidase 1 plays a key role in the response of *Arabidopsis thaliana* to stress combination. J Biol Chem 283(49):34197–34203

Kovtun Y, Chiu WL, Tena G, Sheen J (2000) Functional analysis of oxidative stress-activated mitogen-activated protein kinase cascade in plants. Proc Natl Acad Sci 97(6):2940–2945

Kreslavski VD, Los DA, Allakhverdiev SI, Kuznetsov VV (2012) Signaling role of reactive oxygen species in plants under stress. Russ J Plant Physiol 59(2):141–154

Kudla J, Batistič O, Hashimoto K (2010) Calcium signals: the lead currency of plant information processing. Plant Cell 22(3):541–563

Kumar V, Khare T, Sharma M, Wani SH (2017) ROS-induced signaling and gene expression in crops under salinity stress. In: Reactive oxygen species and antioxidant systems in plants: role and regulation under abiotic stress. Springer, Singapore, pp 159–184

León J, Lawton MA, Raskin I (1995) Hydrogen peroxide stimulates salicylic acid biosynthesis in tobacco. Plant Physiol 108(4):1673–1678

Li CH, Wang G, Zhao JL, Zhang LQ, Ai LF, Han YF, Sun DY, Zhang SW, Sun Y (2014) The receptor-like kinase SIT1 mediates salt sensitivity by activating MAPK3/6 and regulating ethylene homeostasis in rice. Plant Cell 26(6):2538–2553

Liu Y, He C (2017) A review of redox signaling and the control of MAP kinase pathway in plants. Redox Biol 11:192–204

Lu S, Su W, Li H, Guo Z (2009) Abscisic acid improves drought tolerance of triploid bermudagrass and involves H2O2-and NO-induced antioxidant enzyme activities. Plant Physiol Biochem 47(2):132–138

Lv W, Sui L, Yan X, Xie H, Jiang L, Geng C, Li Q, Yao X, Kong Y, Cao J (2018) ROS-dependent Atg4 upregulation mediated autophagy plays an important role in Cd-induced proliferation and invasion in A549 cells. Chem Biol Interact 279:136–144

Mazars C, Thuleau P, Lamotte O, Bourque S (2010) Cross-talk between ROS and calcium in regulation of nuclear activities. Mol Plant 3(4):706–718

Mignolet-Spruyt L, Xu E, Idänheimo N, Hoeberichts FA, Mühlenbock P, Brosché M, Van Breusegem F, Kangasjärvi J (2016) Spreading the news: subcellular and organellar reactive oxygen species production and signalling. J Exp Bot 67(13):3831–3844

Miller GAD, Mittler RON (2006) Could heat shock transcription factors function as hydrogen peroxide sensors in plants? Ann Bot 98(2):279–288

Mittler R, Blumwald E (2015) The roles of ROS and ABA in systemic acquired acclimation. Plant Cell 27(1):64–70

Mittler R, Vanderauwera S, Suzuki N, Miller GAD, Tognetti VB, Vandepoele K, Gollery M, Shulaev V, Van Breusegem F (2011) ROS signaling: the new wave? Trends Plant Sci 16(6): 300–309

Molassiotis A, Fotopoulos V (2011) Oxidative and nitrosative signaling in plants: two branches in the same tree? Plant Signal Behav 6(2):210–214

Monaghan J, Matschi S, Shorinola O, Rovenich H, Matei A, Segonzac C, Gro Malinovsky F, Rathjen JP, MacLean D, Romeis T, Zipfel C (2014) The calcium-dependent protein kinase CPK28 buffers plant immunity and regulates BIK1 turnover. Cell Host Microbe 16(5):605–615

Mori IC, Pinontoan R, Kawano T, Muto S (2001) Involvement of superoxide generation in salicylic acid-induced stomatal closure in *Vicia faba*. Plant Cell Physiol 42(12):1383–1388

Mori IC, Murata Y, Uraji M (2009) Integration of ROS and hormone signaling. In: Reactive oxygen species in plant signaling. Springer, Berlin, pp 25–42

Mullineaux PM, Exposito-Rodriguez M, Laissue PP, Smirnoff N (2018) ROS-dependent signalling pathways in plants and algae exposed to high light: comparisons with other eukaryotes. Free Radic Biol Med 122:52–64

Nakagami H, Soukupová H, Schikora A, Zárský V, Hirt H (2006) A mitogen-activated protein kinase kinase kinase mediates reactive oxygen species homeostasis in *Arabidopsis*. J Biol Chem 281(50):38697–38704

Nie WF, Wang MM, Xia XJ, Zhou YH, Shi K, Chen Z, Yu JQ (2013) Silencing of tomato RBOH1 and MPK2 abolishes brassinosteroid-induced H2O2 generation and stress tolerance. Plant Cell Environ 36(4):789–803

Ogasawara Y, Kaya H, Hiraoka G, Yumoto F, Kimura S, Kadota Y, Hishinuma H, Senzaki E, Yamagoe S, Nagata K, Nara M, Suzuki K, Tanokura M, Kuchitsu K (2008) Synergistic activation of the *Arabidopsis* NADPH oxidase AtrbohD by Ca2+ and phosphorylation. J Biol Chem 283(14):8885–8892

Ortiz-Masia D, Perez-Amador MA, Carbonell J, Marcote MJ (2007) Diverse stress signals activate the C1 subgroup MAP kinases of *Arabidopsis*. FEBS Lett 581(9):1834–1840

Phaniendra A, Jestadi DB, Periyasamy L (2015) Free radicals: properties, sources, targets, and their implication in various diseases. Indian J Clin Biochem 30(1):11–26

Pnueli L, Liang H, Rozenberg M, Mittler R (2003) Growth suppression, altered stomatal responses, and augmented induction of heat shock proteins in cytosolic ascorbate peroxidase (Apx1)-deficient *Arabidopsis* plants. Plant J 34(2):187–203

Podgórska A, Burian M, Szal B (2017) Extra-cellular but extra-ordinarily important for cells: apoplastic reactive oxygen species metabolism. Front Plant Sci 8:1353

Prodhan MY, Munemasa S, Nahar MNEN, Nakamura Y, Murata Y (2018) Guard cell salicylic acid signaling is integrated into abscisic acid signaling via the Ca2+/CPK-dependent pathway. Plant Physiol 178(1):441–450

Pucciariello C, Banti V, Perata P (2012) ROS signaling as common element in low oxygen and heat stresses. Plant Physiol Biochem 59:3–10

Raja V, Majeed U, Kang H, Andrabi KI, John R (2017) Abiotic stress: interplay between ROS, hormones and MAPKs. Environ Exp Bot 137:142–157

Reczek CR, Chandel NS (2015) ROS-dependent signal transduction. Curr Opin Cell Biol 33:8–13

Rentel MC, Knight MR (2004) Oxidative stress-induced calcium signaling in *Arabidopsis*. Plant Physiol 135(3):1471–1479

Rentel MC, Lecourieux D, Ouaked F, Usher SL, Petersen L, Okamoto H, Knight H, Peck HC, Grierson CS, Hirt H, Knight MR (2004) OXI1 kinase is necessary for oxidative burst-mediated signalling in *Arabidopsis*. Nature 427(6977):858–861

Roos G, Foloppe N, Messens J (2013) Understanding the p K a of redox cysteines: the key role of hydrogen bonding. Antioxid Redox Signal 18(1):94–127

Rosenfeld N, Elowitz MB, Alon U (2002) Negative autoregulation speeds the response times of transcription networks. J Mol Biol 323(5):785–793

Sachdev S, Ansari SA, Ansari MI, Fujita M, Hasanuzzaman M (2021) Abiotic stress and reactive oxygen species: generation, signaling, and defense mechanisms. Antioxidants 10(2):277

Sachdev S, Jaiswal P, Ansari MI (2022) Coordinated functions of reactive oxygen species metabolism and defense systems in abiotic stress tolerance. In: Advancements in developing abiotic stress-resilient plants. CRC Press, Boca Raton, pp 23–44

Saddhe AA, Malvankar MR, Karle SB, Kumar K (2019) Reactive nitrogen species: paradigms of cellular signaling and regulation of salt stress in plants. Environ Exp Bot 161:86–97

Saed-Moucheshi A, Pakniyat H, Pirasteh-Anosheh H, Azooz MM (2014) Role of ROS as signaling molecules in plants. In: Oxidative damage to plants. Academic Press, Amsterdam, pp 585–620

Santner A, Estelle M (2009) Recent advances and emerging trends in plant hormone signalling. Nature 459(7250):1071–1078

Sasidharan R, Hartman S, Liu Z, Martopawiro S, Sajeev N, van Veen H, Yeung E, Voesenek LA (2018) Signal dynamics and interactions during flooding stress. Plant Physiol 176(2): 1106–1117

Schieber M, Chandel NS (2014) ROS function in redox signaling and oxidative stress. Curr Biol 24(10):R453–R462

Schopfer P, Liszkay A, Bechtold M, Frahry G, Wagner A (2002) Evidence that hydroxyl radicals mediate auxin-induced extension growth. Planta 214(6):821–828

Sewelam N, Kazan K, Schenk PM (2016) Global plant stress signaling: reactive oxygen species at the cross-road. Front Plant Sci 7:187

Singh KB, Foley RC, Oñate-Sánchez L (2002) Transcription factors in plant defense and stress responses. Curr Opin Plant Biol 5(5):430–436

Sinha AK, Jaggi M, Raghuram B, Tuteja N (2011) Mitogen-activated protein kinase signaling in plants under abiotic stress. Plant Signal Behav 6(2):196–203

Stael S, Wurzinger B, Mair A, Mehlmer N, Vothknecht UC, Teige M (2012) Plant organellar calcium signalling: an emerging field. J Exp Bot 63(4):1525–1542

Steffens B (2014) The role of ethylene and ROS in salinity, heavy metal, and flooding responses in rice. Front Plant Sci 5:685

Suzuki N, Katano K (2018) Coordination between ROS regulatory systems and other pathways under heat stress and pathogen attack. Front Plant Sci 9:490

Suzuki N, Mittler R (2006) Reactive oxygen species and temperature stresses: a delicate balance between signaling and destruction. Physiol Plant 126(1):45–51

Suzuki N, Rivero RM, Shulaev V, Blumwald E, Mittler R (2014) Abiotic and biotic stress combinations. New Phytol 203(1):32–43

Taj G, Agarwal P, Grant M, Kumar A (2010) MAPK machinery in plants: recognition and response to different stresses through multiple signal transduction pathways. Plant Signal Behav 5(11): 1370–1378

Tanou G, Job C, Rajjou L, Arc E, Belghazi M, Diamantidis G, Molassiotis A, Job D (2009) Proteomics reveals the overlapping roles of hydrogen peroxide and nitric oxide in the acclimation of citrus plants to salinity. Plant J 60(5):795–804

Tanveer M, Ahmed HAI (2020) ROS signalling in modulating salinity stress tolerance in plants. In: Salt and drought stress tolerance in plants. Springer, Cham, pp 299–314

Tian W, Wang C, Gao Q, Li L, Luan S (2020) Calcium spikes, waves and oscillations in plant development and biotic interactions. Nat Plants 6(7):750–759

Tognetti VB, Mühlenbock PER, Van Breusegem F (2012) Stress homeostasis–the redox and auxin perspective. Plant Cell Environ 35(2):321–333

Tripathy BC, Oelmüller R (2012) Reactive oxygen species generation and signaling in plants. Plant Signal Behav 7(12):1621–1633

Van Breusegem F, Bailey-Serres J, Mittler R (2008) Unraveling the tapestry of networks involving reactive oxygen species in plants. Plant Physiol 147(3):978–984

Vogel MO, Moore M, König K, Pecher P, Alsharafa K, Lee J, Dietz KJ (2014) Fast retrograde signaling in response to high light involves metabolite export, MITOGEN-ACTIVATED PROTEIN KINASE6, and AP2/ERF transcription factors in *Arabidopsis*. Plant Cell 26(3): 1151–1165

Volkov RA, Panchuk II, Mullineaux PM, Schöffl F (2006) Heat stress-induced H2O2 is required for effective expression of heat shock genes in *Arabidopsis*. Plant Mol Biol 61(4):733–746

Wang Y, Loake GJ, Chu C (2013) Cross-talk of nitric oxide and reactive oxygen species in plant programed cell death. Front Plant Sci 4:314

Waszczak C, Akter S, Eeckhout D, Persiau G, Wahni K, Bodra N, Van Molle I, De Smet B, Vertommen D, Gevaert K, De Jaeger G, Van Montagu M, Messens J, Van Breusegem F (2014) Sulfenome mining in *Arabidopsis thaliana*. Proc Natl Acad Sci USA 111(31):11545–11550

Waszczak C, Akter S, Jacques S, Huang J, Messens J, Van Breusegem F (2015) Oxidative post-translational modifications of cysteine residues in plant signal transduction. J Exp Bot 66(10): 2923–2934

Waszczak C, Carmody M, Kangasjärvi J (2018) Reactive oxygen species in plant signaling. Annu Rev Plant Biol 69:209–236

Wu F, Chi Y, Jiang Z, Xu Y, Xie L, Huang F, Wan D, Ni J, Yuan F, Wu X, Zhang Y, Wang L, Ye R, Byeon B, Wang W, Zhang S, Sima M, Chen S, Zhu M, Pei J, Johnson DM, Zhu S, Cao X, Pei C, Zai Z, Liu Y, Liu T, Swift GB, Zhang W, Yu M, Hu Z, Siedow JN, Chen X, Pei ZM (2020) Hydrogen peroxide sensor HPCA1 is an LRR receptor kinase in *Arabidopsis*. Nature 578(7796):577–581

Xia XJ, Zhou YH, Shi K, Zhou J, Foyer CH, Yu JQ (2015) Interplay between reactive oxygen species and hormones in the control of plant development and stress tolerance. J Exp Bot 66(10):2839–2856

Xing Y, Jia W, Zhang J (2007) AtMEK1 mediates stress-induced gene expression of CAT1 catalase by triggering H2O2 production in *Arabidopsis*. J Exp Bot 58(11):2969–2981

Xing Y, Jia W, Zhang J (2008) AtMKK1 mediates ABA-induced CAT1 expression and H2O2 production via AtMPK6-coupled signaling in *Arabidopsis*. Plant J 54(3):440–451

Xing Y, Cao Q, Zhang Q, Qin L, Jia W, Zhang J (2013) MKK5 regulates high light-induced gene expression of Cu/Zn superoxide dismutase 1 and 2 in *Arabidopsis*. Plant Cell Physiol 54(7): 1217–1227

Yalcinkaya T, Uzilday B, Ozgur R, Turkan I, Mano JI (2019) Lipid peroxidation-derived reactive carbonyl species (RCS): their interaction with ROS and cellular redox during environmental stresses. Environ Exp Bot 165:139–149

Yang CY, Hong CP (2015) The NADPH oxidase Rboh D is involved in primary hypoxia signalling and modulates expression of hypoxia-inducible genes under hypoxic stress. Environ Exp Bot 115:63–72

Yang TPBW, Poovaiah BW (2002) Hydrogen peroxide homeostasis: activation of plant catalase by calcium/calmodulin. Proc Natl Acad Sci 99(6):4097–4102

Yaqoob U, Jan N, Raman PV, Siddique KH, John R (2022) Crosstalk between brassinosteroid signaling, ROS signaling and phenylpropanoid pathway during abiotic stress in plants: does it exist? Plant Stress 4:100075

Yoshioka H, Asai S, Yoshioka M, Kobayashi M (2009) Molecular mechanisms of generation for nitric oxide and reactive oxygen species, and role of the radical burst in plant immunity. Mol Cells 28(4):321–329

Yuan P, Yang T, Poovaiah BW (2018) Calcium signaling-mediated plant response to cold stress. Int J Mol Sci 19(12):3896

Zandalinas SI, Mittler R (2018) ROS-induced ROS release in plant and animal cells. Free Radic Biol Med 122:21–27

Zandalinas SI, Mittler R, Balfagón D, Arbona V, Gómez-Cadenas A (2018) Plant adaptations to the combination of drought and high temperatures. Physiol Plant 162(1):2–12

Zandalinas SI, Sengupta S, Burks D, Azad RK, Mittler R (2019) Identification and characterization of a core set of ROS wave-associated transcripts involved in the systemic acquired acclimation response of *Arabidopsis* to excess light. Plant J 98(1):126–141

Zandalinas SI, Fichman Y, Devireddy AR, Sengupta S, Azad RK, Mittler R (2020) Systemic signaling during abiotic stress combination in plants. Proc Natl Acad Sci 117(24):13810–13820

Zhang A, Jiang M, Zhang J, Ding H, Xu S, Hu X, Tan M (2007) Nitric oxide induced by hydrogen peroxide mediates abscisic acid-induced activation of the mitogen-activated protein kinase cascade involved in antioxidant defense in maize leaves. New Phytol 175(1):36–50

Zhang Z, Wu Y, Gao M, Zhang J, Kong Q, Liu Y, Ba H, Zhou J, Zhang Y (2012) Disruption of PAMP-induced MAP kinase cascade by a *Pseudomonas syringae* effector activates plant immunity mediated by the NB-LRR protein SUMM2. Cell Host Microbe 11(3):253–263

Zhang T, Zhu M, Song WY, Harmon AC, Chen S (2015) Oxidation and phosphorylation of MAP kinase 4 cause protein aggregation. Biochim Biophys Acta Proteins Proteom 1854(2):156–165

Zhang H, Zhang Y, Deng C, Deng S, Li N, Zhao C, Zhao R, Liang S, Chen S (2018) The *Arabidopsis* Ca2+-dependent protein kinase CPK12 is involved in plant response to salt stress. Int J Mol Sci 19(12):4062

Zhou J, Wang J, Li X, Xia XJ, Zhou YH, Shi K, Chen Z, Yu JQ (2014) H2O2 mediates the crosstalk of brassinosteroid and abscisic acid in tomato responses to heat and oxidative stresses. J Exp Bot 65(15):4371–4383

Zhu JK (2016) Abiotic stress signaling and responses in plants. Cell 167(2):313–324